Autopoiesis 19

THINKING ABOUT BIOLOGY

THINKING ABOUT BIOLOGY
An Invitation to Current Theoretical Biology

Editors

Wilfred D. Stein
Hebrew University
Jerusalem, Israel

Francisco J. Varela
CREA Ecole Polytecnique
Paris, France

Lecture Notes Volume III

Santa Fe Institute
Studies in the Sciences of Complexity

Addison-Wesley Publishing Company
The Advanced Book Program

Reading, Massachusetts   Menlo Park, California   New York
Don Mills, Ontario   Wokingham, England   Amsterdam   Bonn
Sydney   Singapore   Tokyo   Madrid   San Juan
Paris   Seoul   Milan   Mexico City   Taipei

Publisher: *David Goehring*
Editor-in-Chief: *Jack Repcheck*
Production Manager: *Michael Cirone*
Production Supervisor: *Lynne Reed*

Director of Publications, Santa Fe Institute: *Ronda K. Butler-Villa*
Publications Assistant, Santa Fe Institute: *Della L. Ulibarri*

Library of Congress Cataloging-in-Publication Data

Thinking about biology : an invitation to current theoretical biology
   / editors, Wilfred Stein, Francisco J. Varela.
       p.   cm. — (Santa Fe Institute studies in the sciences of
complexity. Lecture notes ; v. 3)
    Includes bibliographical references and index.
    ISBN 0-201-62453-2. — ISBN 0-201-62454-0 (pbk.)
    1. Biology. I. Stein, Wilfred D. II. Varela, Francisco J.,
1945–   . III. Series.
QH311.T53   1993
574'.01— dc20

                                           92-42853
                                             CIP

This volume was typeset using TEXtures on a Macintosh II computer. Camera-ready
output from a Hewlett-Packard LaserJet 4M Printer.

1 2 3 4 5 6 7 8 9 10 - MA - 96959493
First printing, July 1993

About the Santa Fe Institute

The *Santa Fe Institute* (SFI) is a multidisciplinary graduate research and teaching institution formed to nurture research on complex systems and their simpler elements. A private, independent institution, SFI was founded in 1984. Its primary concern is to focus the tools of traditional scientific disciplines and emerging new computer resources on the problems and opportunities that are involved in the multidisciplinary study of complex systems—those fundamental processes that shape almost every aspect of human life. Understanding complex systems is critical to realizing the full potential of science, and may be expected to yield enormous intellectual and practical benefits.

All titles from the *Santa Fe Institute Studies in the Sciences of Complexity* series will carry this imprint which is based on a Mimbres pottery design (circa A.D. 950–1150), drawn by Betsy Jones. The design was selected because the radiating feathers are evocative of the outreach of the Santa Fe Institute Program to many disciplines and institutions.

Santa Fe Institute
Studies in the Sciences of Complexity

Contributors to This Volume

Per Bak, Brookhaven National Laboratory

L. V. Beloussov, Moscow State University

Bertrand Clarke, Purdue University

G. Cocho, Universidad Nacional Autónoma de México

Antonio Coutinho, Institut Pasteur, France

Rob J. de Boer, Bioinformatica RUU, The Netherlands

Brian Goodwin, The Open University, Milton Keynes, England

Mae-Wan Ho, The Open University, Milton Keynes, England

Pauline Hogeweg, Bioinformatica RUU, The Netherlands

M. A. Jimenez-Montaño, Universidad de Las Américas, México

Stuart A. Kauffman, University of Pennsylvania and Santa Fe Institute

Mark Levinthal, Purdue University

Pier Luigi Luisi, Institut für Polymere, ETH, Switzerland

Jay E. Mittenthal, University of Illinois

F. Lara-Ochoa, Universidad Nacional Autónoma de México

Fritz-Albert Popp, International Institute of Biophysics, F.R.G.

J. L. Ruis, Universidad Nacional Autónoma de México

Peter T. Saunders, Kings College, London

Wilfred D. Stein, Hebrew University, Israel

John Stewart, Institut Pasteur, France

L. E. H. Trainor, University of Toronto, Canada

Jan D. van der Laan, Bioinformatica RUU, The Netherlands

Francisco J. Varela, CREA, Ecole Polytecnique, France

E. Vargas, Universidad de las Américas

Lewis Wolpert, University College, and Middlesex School of Medicine, London

Contents

THINKING ABOUT BIOLOGY

Wilfred D. Stein† and Francisco J. Varela‡
†Department of Biophysics, Hebrew University, Jerusalem, Israel and ‡CREA, Ecole Polytecnique, 1 rue Descartes, 75005 Paris, France

Thinking About Biology:
An Introductory Essay

ON THEORY-LADEN FACTS AND EMPIRICALLY LADEN CONCEPTS

No science has ever been done without a sustained and far-reaching imaginative effort. The images, myths, and patterns of ideas become, in due time, sedimented into coherent discourse, formulae, golden rules of work—in brief, what we identify as normal science. The naive picture of science as the steady, linear accumulation of true propositions by the sheer weight of empirical evidence is most certainly false and gone with the nineteenth century. Indeed, the main outcome of the modern field of the history of science has been to settle that issue once and for all, starting with the work of Alexandre Koyre, and later followed by the widely read *The Structure of Scientific Revolutions* by Thomas Kuhn. These historians of science, among many others, had the great merit of actually looking into the rich flows of the establishment of a concept in a particular discipline. Koyre dealt with the process of the origins of mechanics and Galileo's physics.[8] A more recent study has addressed, for instance, the emergence of immunology during Pasteur's time, in all its sociological complexity.[9] The message from these investigations has become unavoidable for all but the most recalcitrant empiricists: science is *never* done without an indissoluble link between theory and fact. Facts are made by the theoretical spectacles one puts

Thinking About Biology, Eds. W. D. Stein and F. J. Varela, SFI Studies in the
Sciences of Complexity, Lect. Note Vol. III, Addison-Wesley, 1993

1

on (the expression is Kuhn's) just as much as theory is shaped by the results of empirical observations.

Only rarely are these bubblings of imagination as sweeping and far reaching as Galileo's or Pasteur's. In practice, a scientist (in his or her own daily work) instead faces the reality of theory-laden facts and empirically laden thought at a smaller scale. It is not a matter of doing theory on Sunday and experiments during weekdays. Whether one explicitly calls it thinking theory or not, *every* scientific gesture presupposes a theory and an orientation: designing and doing an experiment, writing a paper, giving a seminar, and commenting on a student's project. This is not problematic, of course; it's just what science is all about. Problems arise when a scientist is too lopsided into one or the other extreme, when experimentation becomes blind repetition of implicit theoretical reasoning or when theory becomes mere talk expressing individual whims.

An excellent example of a balanced experimenter/thinker of biology is the late Conrad H. Waddington. He was an embryologist and evolutionary biologist, working at the University of Edinburgh until the 1970s, whose scientific production was considerable and wide-ranging. He devoted much effort to *thinking about* what he was doing, alone and in concert with others, linking his biological interest with emerging concepts in other sciences such as differential topology, to give birth to the now widely used notion of morphogenetic fields in developmental systems. Waddington was the organizer of a series of annual conferences between 1967–1970, which gathered an eclectic combination of biologists, mathematicians, and physicists from many fields, for informal discussion. These discussions were then published and widely read as a series of books under the general title *Towards a Theoretical Biology*.[22] The "towards" is particularly significant for us here, since the book the reader now holds in his or her hands is the result of discussions that have taken place over three meetings in Cuernavaca (Mexico), Solignac (France), and Santa Fe (USA) entitled *The Waddington Conferences in Theoretical Biology*, an explicit attempt to revive or rather re-enact Waddington's vision of 20 years before.[1]

One the of main purposes of this book is to offer a *sample* of thinking about biology in Waddington's balanced manner, one that avoids both the extremes of empiricist myopia and of theoretical aloofness. We offer this sample not as a survey of the "entire field" of theoretical biology. It is more like a good menu of examples, where a would-be young biologist can watch the process of thinking about biology in action. We would like to dispel as much as possible the impression that theoretical biology is merely a weekend entertainment for the lazy or the mathematically inclined biologist. Instead, in the various essays that follow, we see the process of thought as being an integral part of *asking* questions that make science both productive and more fun. In fact, this book was born to a large extent out of the desire

---

[1]The meetings were first convened under the initiative of Brian Goodwin (Open University, Milton Keynes) and Stuart Kauffman (Santa Fe Institute). A first volume of essays has already appeared.[5]

of many of those who are active biology thinkers to communicate their enthusiasm and pleasure in this dimension of their scientific activity.

MULTIPLE LEVELS IN THEORETICAL BIOLOGY

Thinking about biology is done in various ways. One of the most useful approaches to this variety, we have found, is expressed in the notion of *levels* of theory. We distinguish between *macro-, meso-*, and *micro-*theories.

Macrotheories are the large, imaginative canvases that constitute the conceptual scaffolding of a large portion of biology, and that appear only rarely in history. One ideal example is due to Darwin. Evolutionary biology as a field of study existed, of course, before 1859, the date of the publication of the *Origin of Species*. Yet Darwin's publication so radically altered the fundamental parameters of understanding evolution, in terms of modification by descent, that it provided the beginning of a new science altogether. Another more recent macrotheory is the founding of modern neuroscience by Ramón y Cajal early in this century. Here is a brilliant example of a scientist whose painstaking habits of histological observation coexisted happily with a wild imagination that mixed literature, dreams, and chess.[16]

Macrotheories are usually not that interesting to most practicing biologists. As students, they absorb the major dominant macrotheories, which become second nature in their minds, and they leave it at that. The large issues, such as what is life and its origins, how can a brain secrete a mind, or can evolution be sustained on a planetary scale, are left for occasional reading at best. Most of the time, when confronted with somebody who brings along a purported new macrotheory, most biologists are left cold and basically bored. "Philosophy," they utter as they leave the seminar room. A recent example: J. Lovelock's proposal that the entire biosphere is a regulatory device—the so-called Gaia hypothesis.[1,10] Like most macro thinkers, Lovelock had a very hard time finding listeners, let alone finding active supporters. For over ten years the debate was little more than a small fire on the periphery of the mainstream scientific community. Yet slowly, Lovelock's ideas do seem to have gotten hold of ecologists and geochemists, and they are now discussed in international meetings and featured in *Nature*. This is, then, one rare example of a macrotheory that does not die with its proponent. The mortality rate of macrotheories is high, and their proponents are often relegated to that limbo that practicing biologists call "just philosophy."

In this book, macrotheories are, for these reasons, not very strongly represented, except in section I on the origin of life. Here some rather large-scale thinking is proposed to address the issue, which falls just about at the right level to qualify as macro.[11,15] Other examples occur in section III with the notion of self-organized criticality.[1,7] These are some excellent examples of budding macrotheories that have yet to make their way into the larger world, but which are tempting enough to be included here for the benefit and pleasure of the reader.

Mesotheories, on the other hand, are more numerous. Their level is more modest, as they attempt to cover only a selected domain of biology. A good example of classical mesotheory is the introduction of the notion of a morphogenetic gradient to account for various developmental phenomena.[2] Clearly this was an idea that could be stated in sufficiently general terms to be applied to different developmental systems. Yet its range is not large enough to cover many other areas of biology. Another good example of mesotheory extensively dealt with in this book is immune networks.[4,21] Until recently, immunology was based on a macrotheory, the so-called clonal selection theory. The basis of clonal selection was that immune phenomena are, fundamentally, antibody responses brought about by the expansion of lymphocyte populations that have some specificity against an invading foreign antigen. In contrast, in the immune network approach, this notion is severely challenged: the main issue becomes not immune responses, but intra-organismic communication mediated by the dynamic link between the antibody species themselves.

Mesotheories are more popular among practicing biologists because they see, for them, a more immediate applicability. It *does* make a difference for an experiment to be done tomorrow, for instance, if I consider one or another mesotheory. Accordingly, mesotheories can be included in a seminar without the lecturer being considered impolite or vain. By contrast, presenting a macrotheory in a typical seminar would be rather impolite, unless the speaker has some impressive credentials to enable him to get away with it (a Nobel prize, for instance). There are even some mesotheoretical journals in several areas of biology (we will return to the issue of sources in a moment).

Microtheories are phenomenon-specific. Their purpose is to account in some conceptually clean or analytically astute manner for a given set of observations. For instance, a very famous microtheory was Hodgkin's and Huxley's equations for the action potential in nerve conduction. These equations represented a brilliant insight into a unique phenomenon, with wide implications to be sure, but nevertheless specific to the local phenomenology of a nerve action potential. In this book a few microtheories are included, such as the behavior of ant colonies[18] and low-level radiation in biological tissues.[6]

Microtheories are tolerated and respected among practicing biologists and are widely seen as very useful, although still not very many researchers spend enough time directly engaged in them. They represent a most common source of coffee conversation and seminar presentations, and most normal journals publish such contributions as a matter of fact. The dark side of microtheories is that they may be so low-flying as to be boring, except for a limited circle of initiates.

Do we *need* to say that the dividing lines between these different levels of theory are arbitrary and only heuristic? The point here is that such lines seem to be a useful way to address thinking about biology as a whole, and we have organized this book accordingly.

HEROES (AND ANTI-HEROES)

When it comes to thinking about biology, we all have our favorite paradigmatic heroes. This is perhaps a good way to describe theoretical biology, since we are not talking about some general rules, but about individuals and their style. As we mentioned before, Waddington has been a guiding figure for these meetings and a *de facto* example of a hero. Let us give two more examples.

Warren McCulloch started his days as a scientist in the 1920s in Chicago with a fascination with literature and mathematics. Quickly, however, his interest turned to a typically macro issue: what is mind, and how can a brain display one? This led him to medical school and to many years of using strychnine (a convulsive agent) to map interconnected regions of the brain. From this empirical basis, McCulloch took the bold step of conceiving such brain interconnectivity as logical circuits and of expressing them in the language of mathematical logic. With the collaboration of Walter Pitts, he wrote a classic paper entitled "The Logical Calculus Immanent in Nervous Activity,"[14] a paper that provided the basis for the concept of logical circuits as well as for brain theory. Like many other mesotheorists, he wrote considerably on other subjects[13] and was the center of the important Macy Conferences on Cybernetics, which laid the basis for modern cognitive science.

James Danielli was a contemporary of Waddington, trained originally as a physical chemist but, stimulated by the active school of biologists at University College, London, became himself a biologist. As a very young man, in his mid-twenties, he arrived at a bold and realistic concept of the structure of the cell membrane as a lipid bilayer. The insight is a perfect example of a mesotheoretical leap, a reshaping of current ideas and models by a synthesis of available data. Danielli once said that any physical chemist familiar with the work of the Dutch and French soap chemists could have reached the conclusions that he reached about the structure of the cell membrane. But, of course, none did; it was left to Jim Danielli to enunciate the now-so-familiar bilayer model. His interest in theoretical biology was maintained throughout his full and active scientific life. He founded the *Journal of Theoretical Biology* (discussed below) and was its active editor until his death. He always felt that theoretical biology was a discipline that should have as much status as theoretical chemistry, or perhaps even theoretical physics, but would probably have agreed with most of us that the great development of the discipline lies in the future. One of Danielli's great gifts was his futuristic vision, his ability to see new developments, new frontiers, long before the rest of us. Particularly striking are his last bold insights, his vision of the world, or even the Universe, as a single gene pool with organisms passing on genes from one another, across the boundaries of the species or phyla.

To speak of anti-heroes is also a good way of communicating the same point in a negative sense, but it is also more sensitive, and we shall invoke no names. But here is a typical profile. An excellent researcher obtains glory and academic power by steady production in one narrow field of biology, say, the characterization of a particular kind of enzyme. Once installed in this position of power, and usually in his early sixties, our scientist begins to do what he calls "thinking." Instead of there being an imaginative productive circulation between his laboratory experience and mature imagination, this "thinking" turns into a rather flat repetition of common sense or, worse still, into some very high level speculation about the nature of the soul or the about the future of mankind. This is an anti-hero who neither does a service to the field he is familiar with nor to biology at large. Instead "theory" becomes a mere sociological loophole that allows for a public tribune. We wish to distance theoretical biology from such secondary production. Thinking about biology is the very stuff of *doing* research, not an afterthought in someone's career or a social adornment.

GETTING PERSONAL

As editors of this volume, we are, and have been for some time, actively involved in theoretical biology. Since, as we said, there are only individual styles and no general hard and fast rule, it is perhaps useful for the reader to have some ideas as to what makes *us* personally interested in editing this volume and what we consider to be a significant example of our own scientific lines of thinking about biology.

Stein would like to make it clear that his bent is for work at the borderline between mesotheory and microtheory rather than something deeper. Indeed, his only real mesotheory, in which he proposed the fluid, amphilic model of the cell membrane, was so easy to get to that a whole host of theoreticians and even experimentalists were thinking in his way before the model was so elegantly, and convincingly, pictured by Singer and Nicholson. What Stein likes to do is to play with data, think hard about them, and try to build simple models which account for them. For instance, he has spent many years analyzing the data collected in the massive compendia *Cancer Incidence in Five Continents*.[17] The problem here is to test whether the classical model that cancer arises as a result of a number of successive mutations is adequate for all cancers, or whether some other model provides a better explanation in some cases. Also, one wants to know whether some useful ideas as to the mechanisms of carcinogenesis can be gleaned by analyzing the accepted or some other model. Comparing incidence rates of different cancers in different countries and life-styles has already given some important insights into this problem and should give more. Knudson, for example, suggested that the time course of appearance of the childhood cancer, retinoblastoma, could be accounted for by the mutation of both copies of a single gene, which has been identified and cloned since. What other cancer genes can be found by such, but more complex, analyses? But Stein, as a butterfly in theoretical biology, has also attempted to model the development of the chick's limb (with Lewis Wolpert and James Murray), the basis for calcium homeostasis (with Felix Bronner[3]), the basis for multi-drug resistance in cancer (with Hagai Ginsburg), and the linkage between proton pumping and ATP hydrolysis (with Peter Läuger), to list a few topics!

Varela wishes to select, as an example, his work on immune networks, which is partly covered in this volume. He started as a young biology student working in neuroscience, most particularly on visual perception and action, but very early on he was tempted into taking a theoretical stance by the then-budding field of neural networks and cybernetics. Over the years this inclination matured into a full-scale mesotheory of neural phenomena in which the autonomy of the organism figures pre-eminently.[12,19,20] In fact, one of the key concepts in this view is the fact that, more than as an input/output device, the brain needs to be conceived as an active unresting ensemble, which is modulated by sensory-motor couplings but not driven by them. At some point Varela met an immunologist who seemed to confront a similar dissatisfaction in his own discipline. As we have already said, traditional immunology is antigen-driven, and the immune system is seen only as a defense against an infectious challenge. Yet for an emerging new trend in immunology, this is unsatisfactory for a number of reasons discussed in this volume, and what one really needs is an organism-centered view of immunology. It is here that the key notion of immune networks enters, for it is the carrier of an internal dynamics. Varela found that the very same notions he (and others, of course) had developed for neural processes could, with the appropriate modifications, be applied to immune networks as well. What ensued were several years of collaborative work, where Varela played the theoretician amongst A. Coutinho's group at the Pasteur Institute. The results went beyond the expectations of either of them, for the apparently abstract ideas with which this project began rapidly became explicit models. These models had direct applications to experiments and yielded a rich crop of new results and even applications to the treatment of autoimmunity.[2] In a few years, these ideas had gone from vague intuitions to a full-blown research approach, complete with applications and new experimental results. The ease and speed with which this happened, and the quick impact that this new way of thinking about immune events had on the field at large, is one of the most rewarding scientific experiences Varela can record in his life as a scientist.

HOW TO USE THIS BOOK

The eleven chapters in this book have been chosen so as to give the student of theoretical biology some idea of the flavor of current research in the field. We have not attempted a systematic treatment of the subject. Indeed, the time might not yet be right for such an approach. Instead, we have divided up the chapters into three broad sections: the emergence of life, the development of the individual, and

---

[2] For a review see Varela and Coutinho.[21]

the study of the interaction between individuals and species. It would not do to work one's way through the book from beginning to end. The various chapters are too different in character and in difficulty (theoretical biology is not easy!) to encourage this. Rather, one should pick subjects that might be to one's immediate taste, sample these, and then come back for more. If the student is interested in, for example, ecology, then Bak's chapter on the Gaia hypothesis might be a good beginning or Trainor's chapter on the behavior of ant colonies. The properties of networks and the application of network theory to biology are discussed in these two chapters and in those chapters by Kauffman, by Varela and his colleagues, and by de Boer and his colleagues. This might also be a good sequence in which to read those chapters. The theoretical biologists' approach to embryological development might be enjoyed by reading Goodwin, Beloussov, and Wolpert, perhaps also in that order. One might want to dip into Saunders' chapter on the way since it covers both embryological and evolutionary development. Kauffman's chapter can be read for its importance to the theory of evolution as well as to network theory. One will read Luisi for the evolution of life itself; for the design of bacteria read Mittenthal's chapter. Finally, Cocho and his colleagues, and Ho and Popp represent quite different styles of working in theoretical biology, styles that might well attract a would-be theoretician. The idea is to dip and dip again!

Let us now see how the authors of the chapters in this book go about the task of Thinking about Biology. The overall topic of the first four chapters is "The Emergence of Life." Appropriately, we start our thinking by trying to define life. Luisi, in the first chapter, bases his definition of minimal life on the notion of autopoiesis, where an autopoietic unit is a system that makes itself, through a network of reactions that takes place within its own well-defined boundary. Consider, especially, his Figure 2 which is a particularly appealing schema of a minimal living system. Thus Luisi emphasizes the fundamental operation of the life process, rather than the chemical structures that make up the living cell, but he also considers various actual model systems that point to possible paths along the road to life's origins. In the next chapter, we encounter the organism and, in particular, its embryological development. Here, Saunders introduces us gently to some solutions of systems of differential equations and shows us that systems of nonlinear, but not of linear, differential equations have sufficient complexity to be good models for embryological, and even for evolutionary, development. We learn about the "attractor," that stable set of values of the parameters of a system to which the system evolves and to which it will return if perturbed. (Many chapters in this book deal with such attractors as defining the states of the cells and organisms that we see around us.) Saunders analyzes for us Waddington's seminal model of the epigenetic landscape and gives this model a firm mathematical underpinning. In our next chapter, Mittenthal, Clarke, and Levinthal look at the organization of a bacterium, approaching this question in a fundamental way, trying to find out whether the principles of design can help us understand why such an organism possesses its particular organizational form. They show that the life processes of, say, *Escherichia coli* can be considered as a system of two complementary nets, one in which key metabolites of the cell are synthesized and degraded, the other in which the cell's macromolecules are polymerized from these metabolites. For the student, this chapter shows how a deep, formal analysis that considers the basic design of such nets enables us to see that the most effective nets have a hub-and-branch structure, corresponding closely to that of the known metabolic paths. Thus, theory enables us to make sense of the vast catalogue of information that has been produced by the experimentalists. In the final chapter in this volume, Cocho and his colleagues continue this theme, while introducing us to a different way of thinking about biology. They take the raw data that the experimentalists have accumulated (here on DNA sequences in introns, exons, and intragenic regions, and on amino acid distributions at the binding sites of immunoglobulins) and analyze these statistically, asking questions as to what might be the fundamental forces on which evolution has worked, at the level of the sequences of the cell's macromolecules. The authors show that not all possible permutations of DNA sequences and amino acid distributions are indeed accessible. Certain physicochemical constraints (still to be found and, therefore, another task still awaiting its theoretician) must determine which of the many possible outcomes were, indeed, selected in the evolution of the cell's coding molecules.

We now come to the second and largest section of the book where our authors, in six chapters, consider "Development and the Individual." Goodwin argues that much of what we see around us as the forms of cells and organisms are essentially "natural"; that is, they are the structures that cells and collections of cells must produce as a result of mechanical forces arising between cellular components. Genes, the stuff on which evolution works its changes, determine merely the parameters of the generative dynamics of organisms, not the processes themselves. He considers, in turn, the regeneration of the single-celled organism, *Acetabularia*, the leaf patterns in higher plants, and gastrulation and the development of the eye in animals. In all cases, he argues, we can see the outlines of an understanding of the dynamics of the development of the organism or of its parts and can see evolution as an unfolding of the possible generic states of such complex dynamic systems. The student theoretical biologist must, surely, be inspired by the challenge to work out the consequences of this way of thinking. Beloussov, in the next chapter, helps us along this route with a full discussion of the concept of "self-organization." He shows us how the Curie principle—that negligibly small "causes" cannot generate noticeable events—is systematically broken in the biological world where, instead, spontaneous dissymmetrization is frequently found. Beloussov introduces us to a remarkably fruitful approach to embryological development in which the shapes of a variety of biological systems are well reproduced by a model based on the generation of internal pressures in sheets of cells. The student will note that the model leads to structures (a trefoil pattern, for example) which appear over and over again in the development of different animals. This suggests that a general principle of organization is here at work, one that is being revealed by the analyses. Wolpert's contribution, that follows, provides the necessary phenomenological background to the theoretical studies of Goodwin and Beloussov. He describes the phenomenon of gastrulation in which the single-layered ball of cells that is the blastula of the early embryo invaginates to give the three germ layers (ectoderm, mesoderm, and endoderm) from which all the organs of the body develop. He emphasises the remarkable conservation of gastrulation movements throughout the animal kingdom and considers the various views, classical and modern, that have been proposed for its evolutionary origin. Next, Ho and Popp introduce us to a variety of topics organized around the general concept of "coherence." To what extent can we consider a cell, an animal, or a flock of birds as a coherent object and what are the mechanisms by which this coherence is maintained? Thermodynamic analysis emphasizes the necessary rapidity of high-efficiency living processes and suggests that many of these may take place in a coherent fashion. It might be useful to consider the cell in terms of the concepts of solid-state physics and to assess the possible role of electromagnetic radiation in maintaining coherence. Finally, Ho and Popp discuss some experimental data suggesting that practically all animals emit light photons and suggest that these biophotons may be the determinants of the organism's coherence.

The next two chapters are concerned with the process whereby the antigenic identity of the individual is determined. It is known to us all that the immune system of the vertebrate is capable of mounting a response to "foreign" antigens. But how is "nonself" to be distinguished from "self"? An early view was that the immune system operates as a network in the sense that a particular protein displays a number of antigenic sites, each activating a different clone of antibody-producing cells, each of which then will be active against other proteins that share the same antigenic site. This schema takes on a deep theoretical significance when it is broadened to include antibodies that link to other antibodies as well as the interactions between the different classes of antibody-producing, and antigen—and antibody-directed, cells. An immune network will learn to recognize "self" antigens in the same way as a neural network will learn a particular task. In their chapter, Varela, Coutinho, and Stewart explore the provocative position that the central self-recognition properties of the immune system are more fundamental (and certainly more unexplored) than the peripheral responses in infections. Here the student will find an introduction to the ways in which immune networks can be modeled and how experiments can be designed based on the predictions of such modeling. The chapter by de Boer, van der Laan, and Hogeweg continues this theme. They provide a full introduction to the technique of modeling immune networks using the cellular automata approach, based on Conway's "Game of Life," discussed in detail in this chapter. They introduce us, too, to the concept of "shape space" by which the interaction between antigen and immune-cell receptor is categorized in terms of several molecular characteristics such as hydrophobicity, specific protuberances, and so on. Such a system, of interacting cellular automata, self-organizes into a stable cycle of patterns, elegantly pictured in this chapter and determined by variables such as the rate of recruitment of the immune-cell clones.

Our final chapter deals with "Species and Societies," that is, the interactions that can take place between individuals in a society and between species. A species does not evolve in a static environment, slowly improving its fitness. Rather, it interacts with its environment, moulding it, and hence moulding the environment of all other, necessarily evolving, species in the global community. Bak introduces us to the Gaia hypothesis of Lovelock (that life on Earth forms an integrated entity) and shows that in such a situation the global community will demonstrate the phenomenon of self-organized criticality, with avalanches of extinction being followed by periods of slow evolution. Bak links this to the pattern of appearance of earthquakes and points out how similar the Gutenberg-Richter law for the distribution of earthquake sizes is to the paleontologists' findings of small and large waves of extinctions of biological species. Does this formal similarity express an underlying similarity in behavior? The chapter provides a clear introduction to the different models of interacting evolving species, models giving predictions according with the Gutenberg-Richter formalism. Kauffman builds on this to explore the properties of random Boolean (on/off) networks as models of disordered dynamical systems. The question is, again, as in many of the chapters in this volume, whether such a random network will evolve, self-organize, to some kind of structure. Will one find stable, cycling attractors in such a system? The conclusions are of breathtaking broadness. Not only does one find stable cycles, but the properties of such cycles are good candidate models for organisms themselves. If one considers the genes as the Boolean elements, the stable cycles are the cell types and their number depends on the square root of the number of genes, much as does the number of cell types in an organism. The student who finds delight here in the way that theory can make sense of the data of biology is already infected with the virus of theoretical biology. Kauffman defines for us the properties of the "landscape" on which organisms evolve. This might be one in which many states of similar fitness exist (a rugged landscape) or one in which only isolated peaks exist (the smooth landscape). Finally, the properties of this evolutionary landscape themselves change with the evolution of competing and cooperating species, so that it is best to think of coupled, dancing landscapes! In the final chapter, Trainor demonstrates for us a different mode of theoretical biology, but one that must become increasingly common as the subject develops: the close integration of theory and data accumulation. He works together with Gordon, a naturalist who studies the behavior of ants in their colonies, accumulating data on the numbers of patrollers, foragers, maintenance workers, and midden workers. The problem is to understand the structure of the colony and to formulate laws that predict its behavior. The model chosen is a network, by analogy with neural or immune networks, in which the ants interact pairwise passing on information that can modify their behavior and role in the colony. A matrix of possible states is set up, with rules that govern transitions between matrix elements. The calculations are complex but the model already gives meaningful predictions and it is being elaborated upon as the work progresses.

WHERE TO GO FOR MORE

We wish to finish with some brief remarks on places where one can expect to get more ideas at all levels of thinking about biology in an ongoing manner. We refer here only to periodicals easily available in the North American and European context.

There are two excellent general sources for new ideas in theoretical biology. One is appropriately called the *Journal of Theoretical Biology* (JTB). The publication quality is acceptable, and the editorial policy admits contributions ranging from totally conceptual to detailed mathematical micromodeling. (For instance, most of the important contributions to immune networks over the last five years have appeared in JTB.) A similarly interesting medium is *Biological Cybernetics*, which was born from the tradition that gave its title but has evolved into a rather fine forum for most areas of theoretical biology. The rather narrow sounding *Physica D* often carries papers of general interest to biologists, particularly in the field of complex dynamics. A new journal, *Artificial Life*, promises to be a good medium to cover all theoretical issues in biology pertinent to the circulation between natural and artificial living systems. There are also those journals that concentrate on theory but are more restricted in scope. *Neural Computation*, a recent addition to the list, covers a large range of topics often in review form.

But, for the moment, we invite you to sample the chapters in this present volume as examples of current theoretical biology!

REFERENCES

1. Bak, P. "Self-Organized Criticality and Gaia." This volume.
2. Belousov, L. V. "Generation of Morphological Patterns: The Mechanical Ways to Create Regular Structure in Embryonic Development." This volume.
3. Bronner, F., and W. D. Stein. "Modulation of Bone Calcium-Binding Sites Regulates Plasma Calcium: An Hypothesis." *Calc. Tissue Intl.* **50** (1992): 483–489.
4. DeBoer, R. J., J. D. van der Laan, and P. Hogeweg. "Randomness and Pattern Scale in the Immune Network: A Cellular Automata Approach." This volume.
5. Goodwin, B., and P. Saunders, eds. *Theoretical Biology: Epigenetic and Evolutionary Order (A Waddington Memorial Conference).* Edinburgh University Press, 1989.
6. Ho, M.-W., and F.-A. Popp. "Biological Organization, Coherence, and Light Emission from Living Organisms." This volume.

7. Kauffman, S. A. "Requirements for Evolvability in Complex Systems: Orderly Dynamics and Frozen Components." This volume.

8. Koyré, A. *Galilean Studies*. Cambridge, MA: Harvard University Press, 1967.

9. Latour, B. *Pasteurization of French Science*. Cambridge, MA: Harvard University Press, 1988.

10. Lovelock, J. *Gaia: A New Look at Life on Earth*. Oxford: Oxford University Press, 1981.

11. Luisi, P. L. "Defining the Transition to Life: Self-Replicating Bounded Structures and Chemical Autopoiesis." This volume.

12. Maturana, H., and F. Varela. *The Tree of Knowledge*. Boston, MA: New Science Library, 1989.

13. McCulloch, W. *Embodiments of the Mind*. Cambridge, MA: MIT Press, 1982.

14. McCulloch W., and W. Pitts. *Bull. Math. Biophys.* **5** (1943): 25–42.

15. Mittenthal, J. E., B. Clark, and M. Levinthal. "Designing Bacteria." This volume.

16. Ramón y Cajal, S. *Recollections of My Life*, English trans. Cambridge, MA: MIT Press, 1990.

17. Stein, W. D. "Analysis of Cancer Incidence Data on the Basis of Multistage and Colonal Growth Models." *Adv. Canc. Res.* **56** (1991): 161–213.

18. Trainor, L. E. H. "Modeling the Behavior of Ant Colonies as an Emergent Property of a System of Ant-Ant Interactions." This volume.

19. Varela, F. *Principles of Biological Autonomy*. New York: North Holland, 1979.

20. Varela, F., and A. Coutinho. "Second-Generation Immune Networks." *Immunol. Today* **12** (1991): 159–167.

21. Varela, F. J., A. Coutinho, and J. Stewart. "What is the Immune Network For?" This volume.

22. Waddington, C. H., ed. *Towards a Theoretical Biology*. Vols. 1–4. Edinburgh: Edinburgh University Press, 1968 (1), 1969 (2), 1970 (3), 1972 (4).

I. The Emergence of Life

Pier Luigi Luisi
Institut für Polymere, ETH, Zürich, Switzerland

Defining the Transition to Life: Self-Replicating Bounded Structures and Chemical Autopoiesis

INTRODUCTION

Since the seminal work of Oparin,[42,43] the question of the origin of life has usually been discussed within the frame of molecular Darwinism. The main ingredient of molecular Darwinism is a mechanism which leads to selection in a large variety of structures: starting from small molecules, compounds with increasing molecular complexity and with emergent novel properties (binding, recognition, catalysis, information) would have evolved, until the most extraordinary of emergent properties—life itself—originated. A key step is the emergence of self-replication, since accidentally built structures would decay and disappear under prebiotic conditions. Based on our present knowledge, we ascribe self-replication to DNA and most of all the other functions (catalysis, recognition and transport, energy storage, and transduction) to proteins, although this last notion has changed somewhat with the discovery of the catalytic power of RNA.[1,11] Coevolution of nucleic acids and proteins in a system of mutually dependent macromolecules, able to replicate, is thus considered as the genesis of life.

This general view is represented in the literature by various approaches, which vary according to which component or process is thought to be essential.[12,14,15,20,21,22,23,24,25,26,30,34,35,37,44,45,46,55,64] One possible way to generalize all this is given in the following scheme, which includes two somewhat arbitrarily located couplings of variation/selection:

small molecules $\Longrightarrow$
variation of random macromolecules (DNA, RNA, proteins) $\Longrightarrow$
selection and coevolution of mutually interacting macromolecules $\Longrightarrow$
macromolecular complexes with lipidic membranes $\Longrightarrow$
selection of self-replicative unities $\Longrightarrow$
further selection and living cell

[Scheme 1]

Everybody agrees that the probability of the spontaneous coevolution of functional macromolecules is vanishingly small, but molecular Darwinism contains the elements to get around this problem: chance and time (within the frame of the coupling variation/selection). Given enough chance and time, anything may happen.

For some, this statement is difficult to accept, even in the elegant and sophisticated forms proposed by Eigen and coworkers.[15] Since the advent of the "RNA world,"[1,11] this hesitation is, for some researchers, somewhat reduced: ribozymes are able to replicate and to display catalytic function at the same time, so that the critical step of coevolution in the above scheme appears somewhat easier to conceive. But even now, when the action of ribozymes is no longer limited to the phosphodiester bond, the process of going from this function to the other minimal functions needed in the simplest cell is necessarily still confused in a sort of generous Darwinian fog.

In fact, it may be difficult to answer the question as to whether Scheme 1 brings us any closer to an understanding of the living. One may even assert that Scheme 1 is just a tautology of molecular Darwinism, and that, as such, it is unable to elicit concepts other than those from which it has been originally constructed. And one wonders, then, whether Scheme 1 is the only way to approach the problem of life in chemical terms. Do we really have to start from the edge of the spontaneous formation of functional DNA, RNA, or proteins, in order to understand or mimic the living?

An alternative approach is in principle possible. This consists of ignoring altogether the above macromolecules and the question of the terrestrial origin of life, and focusing attention instead on the *definition* of life as we now see it at work. In other words: instead of asking whether and how a molecular progression may bring about the emergence of novel properties complex enough for life, let us ask first what are the minimal mechanistic features of an already living entity. What is life in terms of basic structures and basic processes?

This shift in perspective from "how did life originate" to "how does life work" is an important one. We will see in this article that the choice of one or the other view brings about an important difference in the approach to the chemistry of life. One can even say that the difference in the two views corresponds to two main philosophical and experimental streams in the life sciences. The first approach hinges necessarily on the chemistry of DNA (or RNA) and can be epitomized as a "DNA-centered" view.[20,46] The second approach is necessarily based on the properties of the living cell and can be referred to as the "cell-centered" view. The latter approach is the essence of the notion of autopoiesis, a concept developed by Maturana and Varela in the 1970s,[31,32,57,58,59] which forms one main concept of this article. I will review the concept of autopoiesis in the next section, and then apply it to concrete chemical processes such as self-replicating micelles. I will then examine self-replication based on template mechanisms developed by von Kiedrowski[60,61] and independently by Rebek and coworkers.[39,50,51,56]

It is perhaps useful to end this introduction with a qualification of the word "life." With this word, when we are talking about the origin of life, or the transition to life, I mean *minimal* life, i.e., the simplest possible system that can be mechanistically defined as living. Obviously, from this definition of minimal life, we exclude all those subtle properties (intelligence, mind, conscience) that characterize higher organisms, and that make human life so particular. Although this statement may sound trivial to many, one should keep stressing it, appreciating the sensitivity of laymen to the word "life," because it is far from "trivial" for many scientists as well; it still represents the psychological stumbling block to accepting the field of the chemistry of life as an integral part of natural science.

THE NOTION OF AUTOPOIESIS

The approach of Maturana and Varela is a pragmatic one. Life, as we know it, is cellular and, although in principle other life forms are possible, let us ignore them. A cell is characterized, first of all, by a boundary which discriminates the "self" from the environment. Within this boundary, life is a metabolic network. Based on nutrients coming from the outside world, a cell sustains itself by a network of reactions that take place inside the boundary and that produce the cell components, including those which are assembled as the boundary itself. Thus, the cell is self-generating (i.e., cellular components are synthesized by the cell) and self-perpetuating (i.e., these components are in turn transformed into new compounds, which are part of the reaction network that continuously replaces them).

Maturana and Varela tried to translate these general observations into a scientific definition of "minimal life." The starting proposition is to define life as a unitary operation rather than as a structure. This unitary operation is what they have called an *autopoietic unit*. An autopoietic unit is a system that determines its own making, due to a network of reactions which take place within its own well-defined boundary. To use Varela's words: "an autopoietic machine continuously

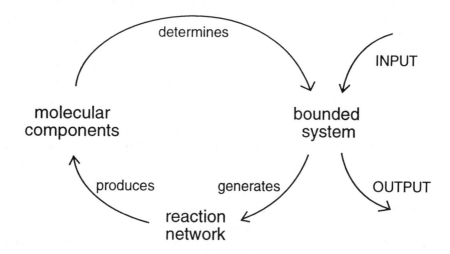

FIGURE 1 The circularity of processes in autopoietic self-production. This figure is redrawn with slight modifications from Fleischaker[18] and is reprinted by permission.

generates and specifies its own organization through its operation as a system of production of its own components."[57]

A schematic representation of an autopoietic unit is shown in Figure 1 (taken with slight modifications from Fleischaker[18]). This figure shows the circularity of the "self" in autopoietic self-production. In their operational relationships, molecular components determine a bounded organized system that generates reaction pathways, which in turn produce molecular components that determine system organization, which generates pathways, and so on. The "autopoietic product" is the generation of the metabolic network, and the "autopoietic self" is the dynamic and unitary operation of that network.[1]

It follows that autopoiesis cannot be ascribed to any single component, but rather to the operational unity of the whole system, echoing the intuition that life cannot be ascribed to any single molecular component (not even to DNA or RNA!) but only to the entire bounded metabolic network. It is apparent at this point that the notion of autopoiesis has its matrix in biology, but that it reaches into the more general issues raised by the theory of complex systems. Note also that it is the *organization* of the system which is stressed: structure (i.e., the actual material components) is almost secondary, in the sense that an autopoietic unity can be realized with several different specific structures.

---

[1]One remark about the self-generation of all components in the autopoietic unit. From this list are excluded, of course, primary components such as oxygen, water, CO_2, and metal ions; as well as the molecules entering through the boundary as energy sources.

Another important system-theoretic concept in this context is that of operational closure. As developed in the primary literature,[57] closure refers in general to the containment of system operations within a system boundary: in living systems, closure means that the autopoietic network of component operations follows the circular logic of Figure 1—the living system is materially and energetically open, but operationally closed. In contexts other than autopoiesis, the importance of a boundary for the definition and maintenance of life has been emphasized by several researchers—starting from the coacervate droplets of Oparin,[42,43] to the proteinoid microspheres of Fox,[21,22,23] and to the liposomes of Deamer,[14] just to note a few. Oro and Lazcano in a recent article argue "...that life in the absence of membranes is impossible."[46]

There is then a certain circularity in the definition of autopoiesis: we start from the experience of minimal life as cellular life, and we conclude by saying that we need a kind of minimal cell in order to define life. This is true and, far from being a weak point, emphasizes the down-to-earth and pragmatic view of autopoiesis.

Let us ask now: what are the criteria to establish whether a given entity is, or is not, autopoietic? This question has been addressed by the original proponents[58] in four summary notions and a six-point key of criteria. As proposed by Fleischaker,[18] this can be simplified to three basic questions:

1. Is the system self-bounded?
2. Is the system self-generating?
3. Is the system self-perpetuating?

If we can answer positively to all three of these questions, the system is autopoietic. We will see in one of the next sections that these three basic criteria can be further qualified.

SOME IMPLICATIONS OF THE THEORY OF AUTOPOIESIS

The simplicity of the notion of autopoiesis is misleading, as actually the acceptance of this definition of minimal life has profound biological and philosophical implications. The first implication is the relationship between autopoiesis and the living. Varela and Maturana state boldly that autopoiesis is the necessary *and* sufficient condition for the minimal units of life: all that is living must be based on autopoietic units, and if a system is discovered to be autopoietic, that system is defined as living[31,32]; i.e., it must correspond to the definition of minimal life.

To further understand the implications of the definition of autopoiesis for minimal life, one should bear in mind what the theory of autopoiesis does *not* contain.

1. It does not contain reference to functional macromolecules; these are, of course, not excluded—but are not strictly needed for the definition.

2. It does not make direct reference to the origin of life. The picture above (Figure 1) represents life as it is "here and now," to use a fashionable idiom.
3. It does not make reference to reproduction or to physical growth.
4. It does not make explicit reference to evolution.
5. It does not contain any reference to the kinetics or thermodynamics of the processes. That is, self-generation, self-perpetuation, boundary formation, etc., are not specified in terms of relative reaction rates. And there is no specification as to whether the underlying thermodynamics must be that which characterizes open systems, or not, nor from where the energy/fuel comes.

There are two points to make about these missing parameters and missing properties. The first is that, partly, the missing data (fuel, structural and kinetic properties, conditions, etc.) must be specified in each actual system, according to its particular structure. The second point is more subtle: some of such missing properties will emerge *as a consequence* of the original definition of autopoiesis. In other words, once the organization of the autopoietic unit is in operation, it entails the potentiality for self-replication and evolution. We will see this shortly for the case of self-replication, whereas evolution will not be considered in this article. This predetermined capability can actually be taken as a strong point in favor of autopoiesis as a theory of the living.

Of course, this definition of autopoiesis is a restricted one: it necessarily operates a selection and an elimination of terms and concepts. This is true for any definition, but nevertheless it is very important for all scientists who work in the field of the chemistry of life to have a definition of what is minimal life. This is important, first, as a basis of intellectual clarity, in order to ask consistent experimental questions, and, second, so as to eliminate a body of irrelevant semantic discussions.

There are several attempts in the literature to define life (for example, see Morowitz, Crick, and others[3,7,8,9,13,36,40,47,48,49,52]), and it might be instructive to compare them with each other. This is, however, outside the scope of this article. Here, I intend to show the simplicity and clarity of the definition of autopoiesis in one important case: the discrimination of the living from the nonliving. For example: according to the above definition, a virus is not living; a virus is a bounded structure, but, when it reproduces in a host cell, the new viruses are not built within the boundary of the virus itself but outside of it, in the host cell. Likewise, a robot which assembles other robot units utilizing parts that are built by some other machines, is not living. An amoeba is an autopoietic unit, however, as replication of its structural components takes place within its own organizational domain. Another implication of the definition of autopoiesis is its use in devising synthetic models of the living. The ideas on self-replicating micelles and liposomes, which will be discussed in this article, are a direct demonstration of this, as they have emerged from a collaboration between Varela and Luisi.[28] One should also mention that the notion of autopoiesis has been applied to biological[30] as well as to inorganic systems.[63]

What about higher organisms? The attempt to apply the notion of autopoiesis to a higher organism such as a cat or a man is an interesting one. It raises, first of all, the notion of the hierarchy of autopoietic structures: from a unicellular to a multicellular organism, and, in a series of steps of increasing complexity, we may even consider extending the notion of autopoiesis to the family, to the village, to the society. Clearly, such an extension is at the same time challenging and difficult. One sees immediately, for example, the difficulty of defining the boundary in these more complex, nonspatial cases. This discussion, again, lies outside the scope of the present article, which concentrates on the minimal units of life. Some of the questions dealing with the hierarchy of autopoiesis are discussed in the recent book by Maturana and Varela.[33]

CHEMICAL INCARNATIONS OF A MINIMAL AUTOPOIETIC UNIT

Let us now turn to consider an autopoietic unit in chemical terms, qualifying for example the notion of boundary in terms of chemical structure; and then qualifying the processes of self-generation and self-perpetuation in terms of relative rate constants. As already mentioned, the notion of operational closure is central: it refers to the containment of system operations within the system boundary. A boundary means a physical barrier, which imposes a diffusion step on the entering/outgoing metabolites.

To apply the concepts of autopoiesis to concrete chemical systems, it is necessary to adhere to the original definition of boundary. Accordingly, the term boundary will be restricted to a three-dimensionally closed structure. This definition eliminates other possible kinds of boundaries, such as an interface between two phases (liquid/liquid or liquid/solid), and the simple surface of a molecule: thus intermolecular binding and recognition of molecules is not *per se* an autopoietic process. Neither are autopoietic boundaries the surfaces of a growing crystal or the contours of cyclic molecules. Rather, vesicles, or micelles, are bounded in the autopoietic sense, in that they form three-dimensionally closed structures with a membrane which is, in principle, able selectively to discriminate among entering/outgoing molecules. Even with such restriction, the definition of autopoietic boundary does allow for cases which are difficult to categorize, but we do not have to discuss these now.

Having defined in more chemical terms the definition of boundary, let us now consider the simplest possible autopoietic system, and its kinetic behavior. This was first proposed by Varela et al.,[57] and more recently discussed by Fleischaker.[20]

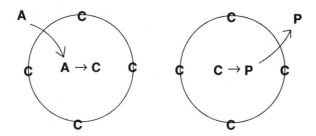

FIGURE 2   The two basic operations in a minimal autopoietic unit.

A *minimal* autopoietic system (Figure 2) would require a boundary composed by at least one component, C; it will be characterized by only one entering metabolite, A, and by two unimolecular reactions, a self-generation reaction leading to C (at the expense of A); and a decomposition reaction which transforms C into a product P, which then goes out of the boundary. These chemical reactions are determined by the bounded unit, i.e., the transformations of A into C and of C into P take place only inside the boundary; we will assume that such spontaneous reactions are chemically irreversible (i.e., that we can neglect the back-reactions).

Let us also assume, for the sake of simplicity, that all diffusion processes throughout the boundary, as well as the rate of self-assembling of C, are very fast with respect to the chemical transformations $A \longrightarrow C$ and $C \longrightarrow P$. The system then produces its own component C, thus regenerating its own structure, with a velocity $v_{gen} = k_{gen}[A]$, and replacing C in a process defined by $v_{dec} = k_{dec}[C]$.

It goes without saying that several other possibilities can be envisaged; for example, P might be transformed back to A or C. Also, in a real system each step can be the result of many chemical events. But this simple system is enough to show a very important point: the actual destiny of the autopoietic unit will depend on the relative magnitude of v_{gen} and v_{dec}. In order to make this point clear, let us use Figure 3.

- If $v_{gen} = v_{dec}$, the system will be self-perpetuating; it is in a steady state (is a homeostatic system) whereby the autopoietic unit is maintained at the expense of the incoming A.
- If $v_{gen} > v_{dec}$, the autopoietic unit will tend to grow; and if the system is monodisperse, i.e., thermodynamically existing in only one size, than we have the necessary conditions for self-replication,[2] as the excessive C molecules will build new bounded structures.

---

[2] The term replication is used throughout this article, without distinction from the term "reproduction." As illustrated by Maturana and Varela,[33] replication implies making identical copies, whereas reproduction implies the division of the parent structure into smaller parts, which being smaller are not identical with the parent structures, and may not be identical with each other.

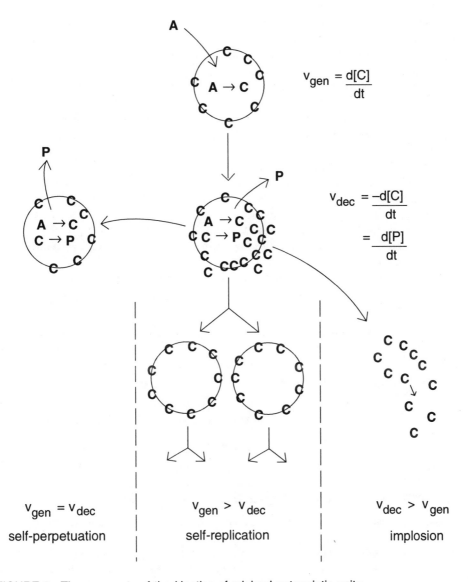

FIGURE 3   Three aspects of the kinetics of minimal autopoietic units.

■   Finally, if $v_{gen} < v_{dec}$, the unit will become impoverished in the C component and will eventually disappear: this will cause implosion of the structure.

On the basis of this simple kinetic scheme of the minimal autopoietic system, we have then reached an important conclusion: that once the system is self-bounded and self-generating, it must be autopoietic, in one of the three dynamic expressions of autopoiesis, namely, self-replication, self-perpetuation, or self-implosion.

The work which will be discussed next is based on the observation that, when self-replication of the components of the bounded structure takes place within the bounded structure, it indicates autopoiesis. The kinetics of autopoiesis deserves a much more complete treatment than the qualitative one I have sketched here. At some future time, this should be done, including more realistic bimolecular reactions as well as all back reactions; and also the exchange rates of the reagents throughout the boundary, as well as the assembly rate of C molecules into the boundary.

A mathematical model of autopoiesis has been presented by Schwegler and collaborators.[2,54] This was in the form of a "Stefan condition" giving rise to a nonlinear feedback of surface motion to the reaction and diffusion processes inside the protocell. It would be certainly very useful to integrate these physical concepts into chemical systems.

For the time being, the application of the notion of autopoiesis to concrete chemical systems permits us to modify slightly the previously presented criteria of autopoiesis in the following way:

1. self-boundary: the system must be tridimensionally closed by a self-assembling boundary structure which selectively filters incoming/outgoing molecules.
2. self-generation: the components of the system (including those of the boundary structure) must be produced by reactions taking place within the boundary.

Any of the three kinetic consequences of self-generation (self-replication, self-perpetuation, and implosion) can be taken as an expression of autopoiesis.

THE MICELLAR SELF-REPLICATING SYSTEMS

The arguments developed in the previous section bring one, almost naturally, to look at micelles, vesicles, and liposomes as candidates for autopoietic structures. (In this article we will use the classic nomenclature according to which the term vesicle is inclusive of the term liposome, the latter being a particular class of vesicles, in which the surfactant is a glyceride.) In fact, micelles and vesicles have the capability of spontaneous formation, or, as is commonly said, self-assembling, which is a thermodynamically controlled creation of order. This capability of self-organization alone would suffice to make these structures very interesting in relation to the chemistry of life.

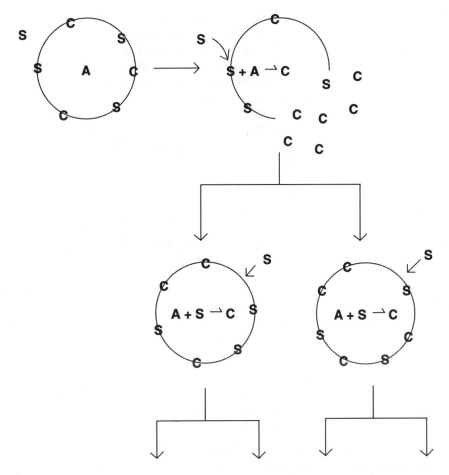

FIGURE 4 The principle of micellar self-replication, which is at the basis of the reactions illustrated in the experimental systems in Figure 5. C is the surfactant and S the co-surfactant, respectively; A is a reagent in the micellar center. By reacting A with S, the surfactant C is produced.

Furthermore, micelles and vesicles exemplify the principle of micro compartmentation: these structures have an interior which is chemically and physically different from the bulk solvent in which they are formed. The chemistry which takes place in the core of these structures, or at the interface is quite particular, i.e., dictated by the structure itself.

It is important to bear in mind that both aqueous and reverse micelles, as well as vesicles, are dynamic systems: not only do the surfactant monomers move in and out of the micelle domain, but they exchange continuously from one micelle to another, due mostly to collisions between micelles in solution. The same holds for vesicles and liposomes. The actual rate of these processes depends on the structure of the aggregate and on the external conditions. For more information about the chemistry of micelles, vesicles, and liposomes, the reader is referred to the specialized literature, e.g., Fendler and Fendler[16] for aqueous micelles, Luisi et al.[27] for reverse micelles, and Fendler[17] for vesicles and liposomes.

The general principle by which micelles and vesicles can be converted into self-replicating minimal autopoietic structures has been presented by Luisi and Varela.[28] Conditions are sought under which a reaction takes place within the boundary of the micelle or vesicle, which leads to formation of the very surfactant which makes the micelle (or vesicle). If conditions are such as to warrant also monodispersity of the bounded structure, then the fresh surfactant leads to formation of more bounded structures, and since these are generated within the boundary of the parent structures, the process can be properly defined as self-replication.

The experimental implementation of this principle in the first series of self-replicating micellar systems is based on a reaction between a reagent, which is strictly localized in the micellar interior, and a co-surfactant. The co-surfactant is present in large excess in the bulk, from where it is continuously delivered to the interface of the micelle. Reaction proceeds until the reagent in the micellar pool is depleted. A schematic representation is offered in Figure 4. The analogy with the principles illustrated in Figures 2 and 3 is apparent.

The first experimental system which we found[4] consisted of reverse micelles formed by sodium octanoate in isooctane/octanol, in a ratio of 9:1 $v : v$. The reaction is the hydrolysis of octyloctanoate, as schematized in Figure 5. The hydrolysis agent LiOH is strictly localized in the water pool. The figure also illustrates two other micellar systems[5] which utilize octanoate micelles, with octanol acting as co-surfactant and/or co-solvent. In these cases, self-replication is based on oxidation of octanol by permanganate. The same reaction system has been utilized both for reverse and aqueous micelles, as octanoate forms micelles of both types.

In the case of reverse micelles, permanganate ions are localized strictly in the water pool, so that the octanol oxidation takes place on the interface of the reverse micelle; in the case of the aqueous micelles, permanganate is localized in the bulk solvent (water) and the reaction of the water-insoluble octanol is also localized on the micelle interface. In addition to the systems illustrated on Figure 5, others have been described,[5] e.g., an enzymatically driven, self-replication, reverse micellar system. The reverse micelles with permanganate represent the "record" of self-replication: a factor of an approximately tenfold increase in the number of micelles.[5] For the oxidation of octanol in aqueous micelles, an increase of the number of micelles by about 50% was obtained.

In all these examples, the reaction takes place at the interface of the micelles; the chemical reactivity is a direct result of the presence of a boundary. We have then a classic autopoietic situation, in which the product of the reaction is the consequence of the boundary constraints of the original structure, and the surfactant product

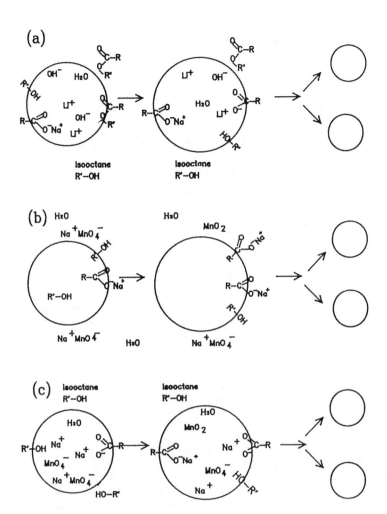

FIGURE 5 The first self-replicating micellar systems. (a) System 1: base-catalyzed ester hydrolysis in reverse micelles. $CH_3\text{-}(CH_2)_6\text{-}CO\text{-}O(CH_2)_7\text{-}CH_3 + LiOH = CH_3\text{-}(CH_2)_6\text{-}COO^- \ Li^- + HO\text{-}(CH_2)_7\text{-}CH_3$. (b) System IIA: 1-octanol oxidation in aqueous micelles. $3\ CH_3\text{-}(CH_2)_7\text{-}OH + 4\ NaMnO_4 = 3\ CH_3\text{-}(CH_2)_6\text{-}COO^- \ Na^+ + 4MnO_2 + NaOH + 4\ H_2O$. (c) System IIB: 1-octanol oxidation in reverse micelles. $3\ CH_3\text{-}(CH_2)_7\text{-}OH + 4\ NaMnO_4 = 3\ CH_3\text{-}(CH_2)_6\text{-}COO^- \ Na^+ + 4\ MnO_2 + NaOH + 4\ H_2O$.

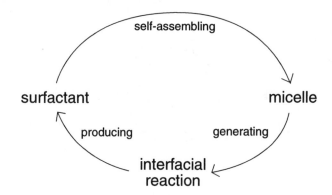

FIGURE 6   The circular organization in an autopoietic micellar system.

then assembles spontaneously in the boundary itself. The situation can be depicted as in Figure 6; the analogy with Figure 1 is again apparent.

Concerning the comparison of reverse versus aqueous micelles, one may think that aqueous micelles offer a greater biological relevance. Reverse micelles, however, owing to their more stringent monodispersity, seem to be able to replicate "better." The reader is referred to the original literature for a discussion of the experimental problems, for example, the determination of the absolute number of micelles and their dimensions; see also a review of our experimental work.[62]

FROM MICELLES TO VESICLES

In the experiments described in the previous section, micelles were initially present in the reaction mixture, i.e., the microcompartmentation for the reaction was given with the initial conditions. A more interesting situation would be one in which the particularly reactive micro-environment is itself created by an initial chemical reaction.

One system of this type has been recently described[6]: the hydrolysis of the water-insoluble ethyl caprylate (EC) produces caprylate ions which are soluble in water, where they build micelles above the critical micelle concentration. As soon as micelles are built, the hydrolysis of EC increases exponentially. This is due to a micellar catalytic process. Micelles, once formed, rapidly bind ethyl caprylate, and its hydrolysis in the micellar domain is very fast. The formation of caprylate ions brings about new caprylate micelles. We have thus a case of self-replication and autocatalysis at the same time, as the fresh micelles originate from a reaction taking place in previous micelles. The more micelles that are present, the higher the velocity of micelle formation.

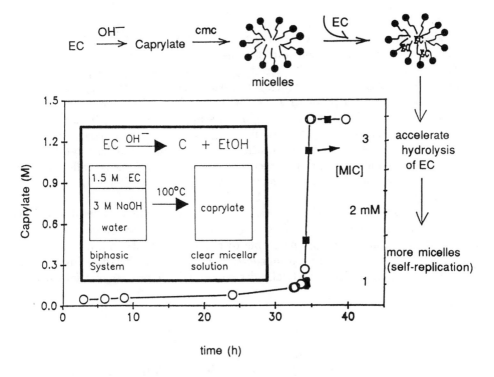

FIGURE 7   Time progress of the hydrolysis of ethyl caprylate (EC) to yield caprylate ions (C) and aqueous caprylate micelles. The concentration of formed caprylate is shown on the left-hand side of the ordinate; the resulting concentration of micelles on the right-hand side.

Thus, the hydrolysis of a simple organic molecule, ethyl caprylate, has given rise to an autopoietic entity. The possible relevance of this finding is increased by the observation that aqueous caprylate micelles can be transformed into vesicles, although not very stable ones, by lowering the pH to about 6.5. An illustration of the EC cycle is depicted in Figure 7.

Thus, in more than one way, the simple experiment with ethyl caprylate has some bearing on prebiotic chemistry. We are dealing with a process that is related to the origin of a basic structure for minimal life but, mark well, without the presence of DNA, RNA, or proteins. As mentioned earlier, the question of the origin of life emerges as a consequence of the logic of autopoiesis, although this is not included in its explicit definition.

These results reinforce our initial stand that, by taking the "cell-centric" autopoietic view, we can arrive at a different way to look at the problem of the origin of life. Of course, at some point the question of the DNA/protein interaction, so essential for life as we know it today, must arise. But it does not have to be the first important event; perhaps we can start elsewhere in order to look for the beginnings.

Generally, vesicles are more interesting than micelles from the point of view of cellular models and/or precursors. As already mentioned, the hypothesis that liposomes are a form of prebiotic cells, has been advanced for some time[14] and, therefore, the idea of self-replicating liposomes appears particularly challenging. One such system has been in fact described, although in a preliminary form.[53]

The basic idea is to construct lecithin liposomes which host the synthesis of lecithin. To accomplish that, the four enzymes which catalyze the so-called salvage synthesis of lecithin are bound to such lecithin liposomes. It could be shown, in fact, that this system is able to synthesize fresh lecithin, which is the prerequisite for liposome replication. One problem with this kind of proteoliposome lies in the fact that replication is not perfect, as it actually yields liposomes which do not contain enzymes and are therefore no longer able to replicate further; and, actually, only those liposomes which contain all four enzymes can be considered as autopoietic units. The possibility of a catalytically driven autopoietic process was foreseen by Varela.[58]

THE TEMPLATE-BASED SELF-REPLICATIONS SYSTEMS

After this discussion of the self-replication of a bounded structure, it is quite interesting to review briefly the self-replication of short, linear, nucleotide-based structures, for here self-replication takes place on quite a different basis. We will focus attention on the papers by von Kiedrowski and collaborators[60,61] and by Rebek and collaborators.[39,50,51,56]

The underlying philosophy in that work is DNA-centric: the simulation of DNA replication in simpler systems is the principle which may bring us closer to understanding the basic replicative mechanism of life processes and, thus, the origin of life. From this approach comes the idea of a nonenzymatic, template-induced synthesis of short oligonucleotides. Starting from work by Naylor and Gilham,[38] and by Orgel and collaborators,[12,24,25,26,45,64] this notion has gained wide recognition and has been clearly related to the prebiotic chemistry and origin of life. In the work of Günter von Kiedrowski, a former collaborator of Orgel, self-replication of oligonucleotides is clearly implemented.[60]

The general principle is illustrated in Figure 8, in which two molecules, say, A and B, (or entire oligomeric structures) are able to form specific complexes with single B and A, respectively (for example, through base-pairing), and can also be linked to each other in a covalent structure A---B. The template A---B, by properly binding and positioning the monomeric reagents B and A, assists in

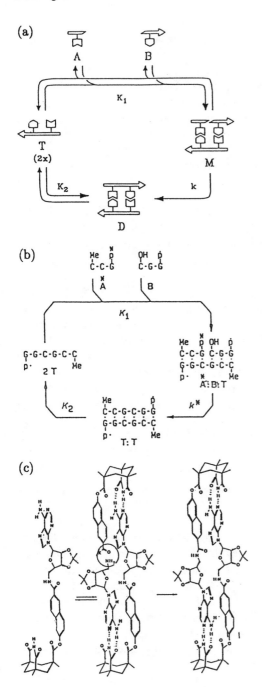

FIGURE 8 The self-replicating template systems: (a) the general principle—a simple schematization of the template-directed synthesis of complementary and palindromic sequences, the basis of the work of von Kiedrowski and of Rebek. Also shown are (b) the actual system used by von Kiedrowski[60] and (c) the system used by Rebek and collaborators.[39,50,51,56] Modified with permission from *J. Am. Chem. Soc.* Copyright ©1991 American Chemical Society.

the chemical synthesis. Thus, the rate of A–B coupling will be higher than in the absence of the template.

Note also that the cyclic process is autocatalytic, since in each cycle the number of template molecules doubles. This illustrates three fundamental points: (i) the importance of molecular complexation and recognition (which is actually not so central in the work with micelles and liposomes); (ii) the notion of palindromic sequence: A---B and B---A are the same, once they are read in opposite directions; and (iii) the fact that in this kind of work, one will be dealing with two competitive reactions: template-assisted coupling and normal coupling. This introduces a difficulty from the analytical point of view, as the rate of the template-assisted reaction must be read against a background.

The bulk of the work of Günter von Kiedrowski has been accomplished with a hexadesoxyribonucleotide, as illustrated in the Figure 8. Von Kiedrowski concluded his 1986 paper[60] by saying that "...we could show that the inherent capability to self-replication of nucleic acids is already present at the level of chemical—i.e., no enzyme being necessary-model systems...."

Let us consider now the work of Rebek and collaborators.[39,50,51,56] The working principle is similar, in the sense that self-replication is based on a template reaction, complementary recognition, and hydrogen bonding. They departed, however, from the straight nucleotides and started, rather, from a triacid, converting it to the imide (see Figure 8) which, as they state, provides a hydrogen-bonding edge similar to that of thymine. This self-replicating compound can be seen as a structure of the type A---B, in which A and B are the imide and the adenine components, respectively. Based on his studies, Rebek remarks that these model reactions can aid one to understand the basic mechanisms of prebiotic life, and how the condensation of catalytically active RNA may have come about.[50,51]

CONCLUDING REMARKS

Self-replicating bounded structures can be realized with commonly known laboratory techniques. What are the implications of this finding? For synthetic chemistry at large, the work on micelles and the work by von Kiedrowski and by Rebek can be taken to indicate a novel field within organic chemistry: molecules or molecular systems with the capability of making copies of themselves, namely, an organic synthesis which is caused and directed by the structural properties of the parent molecular systems.

But it is probably in the field of the chemistry of life that the implications are even more interesting. As Maddox, the editor of *Nature*, commented recently[29] "...what will that mean? Nothing decisive, of course. Merely that another step in the origin of life has been shown to be feasible...." In this chapter, I have compared two different approaches to the chemistry of the living, one based on the question "how did life start?" and one based on the question "how can one define life?" This second question has introduced the notion of autopoiesis. I take this stand mostly for heuristic reasons. In fact, the shift in perspective from one to the other of the two questions above has shown to be very useful in focusing and crystallizing certain arguments. We have seen, for example, that according to the autopoietic view, it is possible to define "minimal life" without invoking macromolecules, or advancing the question of whether "living" entities existed in primordial times prior to the development of RNA or DNA.

It was also interesting to see that properties such as replication do not need to be included *a priori* in a definition of the living—but derive as its consequence. Likewise, with respect to another fundamental property of the living—evolution. Although not discussed in this article, it is easy to conceive how an autopoietic entity can evolve, once particular experimental conditions are set. It appears feasible, for example, to have two reacting micellar systems which compete against one another for an intermediate metabolite, so that one of the two micellar systems will eventually "die" and disappear from the solution.

One of the important concerns of this article was self-replication, and it is proper to make a remark on this. If one takes cell reproduction as a standard, this being characterized by a replication of the outer membrane as well as of the cell content, it appears that the micellar self-replication described until now corresponds to a "shell replication," whereas the template replication of nucleotides corresponds rather to a "core replication." In a way, then, the two self-replication schemes are complementary. Eventually, the two kinds of replication should be put together in a more complex autopoietic unity: it appears feasible to build a shell replication with micelles or liposomes which contain self-replicating nucleotides, DNA or RNA.

I would like to end by going back to the initial statement, that one cannot work on the chemistry of life without having a clear operational definition of minimal life. The autopoietic approach has the advantage of offering a simple and operative definition of the minimal living being—although, as pointed out, each definition automatically means a self-imposed restriction. Within the limits of this restriction, the notion of autopoiesis suggests that life must be based on autopoietic units, and, conversely, any structure which is autopoietic must correspond to the minimal form of life. Accordingly, the self-replicating bounded structures illustrated in this article should be considered as minimal synthetic life. Such a statement may possess an unappealing flavor, but I believe one should not be afraid of it. This feeling of unappealingness probably arises for psychological reasons, but it should not cloud the scientific issue. The scientific issue is whether or not a certain set of of self-imposed definitions and criteria are implemented by a given experimental system. Even if we give a positive answer to these criteria, certainly nothing is taken away from the mystery of life at large.

ACKNOWLEDGMENTS

The collaboration and criticism of a number of colleagues has been very useful. I like to mention in particular Francisco Varela, Peter Walde, Pascale Bachmann, Ulrich Müller-Härold, Gail Fleischaker, and Marco Maestro.

REFERENCES

1. Altman, S. "Ribonuclease P: An Enzyme with a Catalytic RNA Subunit." *Adv. Enzymol.* **62** (1989): 1.
2. An der Heiden, U., G. Roth, and H. Schwegler. "Principles of Self-Generation and Self-Maintenance." *Acta Biotheoretica* **34** (1985): 125.
3. Aw, S. E. *Chemical Evolution.* Singapore: University Education Press, 1976.
4. Bachmann, P. A., P. Walde, P. L. Luisi, and J. Lang. "Self-Replicating Reverse Micelles and Chemical Autopoiesis." *J. Am. Chem. Soc.* **112** (1990): 8200.
5. Bachmann, P. A., P. Walde, P. L. Luisi, and J. Lang. "Self-Replicating Micelles: Aqueous Micelles and Enzymatically Driven Reactions in Reverse Micelles." *J. Am. Chem. Soc.* **113** (1991): 8204.
6. Bachmann, P. A., P. L.Luisi, and J. Lang. "Autocatalytic Self-Replicating Micelles and Models for Prebiotic Structures." *Nature* **357** (1992): 57.
7. Bernal, J. D. *Theoretical and Mathematical Biology*, edited by T. H. Waterman and H. J. Morowitz. New York: Blaidswell, 1965.
8. Bernal, J. D. *The Origin of Life*, 168. London: Weidenfeld and Nicolson, 1967.
9. Cairns-Smith, A. G. "A Case for an Alien Ancestry." *Proc. Roy. Soc. Lond. B* **189** (1975): 249.
10. Cech, T. R., and B. L. Bass. "Biological Catalysis by RNA." *Ann. Rev. Biochem.* **55** (1986): 599.
11. Cech, T. R. "The Chemistry of Self-Splicing RNA and RNA Enzymes." *Science* **236** (1987): 1532.
12. Chen, C. B., T. Inoue, and L. E. Orgel. "Template-Directed Synthesis on Oligodeoxycytidylate and Polydeoxycytildylate Templates." *J. Mol. Biol.* **181** (1985): 271.
13. Crick, F. *Of Molecules and Men.* Seattle, London: University of Washington Press, 1966.
14. Deamer, D. W. "Role of Amphiphilic Compounds in the Evolution of Membran Structure on the Early Earth." *Origins of Life* **17** (1986): 3.
15. Eigen, M. "Self-Organization of Matter and Evolution of Biological Macromolecules." *Naturwissenschaften* **58** (1971): 465.

16. Fendler, J. H., and E. J. Fendler. *Catalysis in Micellar and Macromolecular Systems.* New York: Academic Press, 1975.
17. Fendler, J. H. *Membrane Mimetic Chemistry.* New York: John Wiley & Sons, 1982.
18. Fleischaker, G. R. "Autopoiesis: The Status of Its System Logic." *BioSystems* **22** (1988): 37.
19. Fleischaker, G. R. "Origin of Life: An Operational Definition." *Origins Life Evol. Biosphere* **20** (1990): 127.
20. Fleischaker, G. R. "The Models of a Minimal Cell." In *Prebiological Self-Organization of Matter*, edited by C. Ponnamperuma and F. R. Eirich, 235. Hampton: DEEPAK, 1990.
21. Fox, S. W. *The Origin of Prebiological Systems.* New York: Academic Press, 1965.
22. Fox, S. W., and T. Nakashima. "The Assembly and Properties of Protobiological Structures: The Beginnings of Cellular Peptid Synthesis." *BioSystems* **12** (1980): 155.
23. Fox, S. W. "Oparin: The First Cells, and Selection Processes." In *Prebiological Self Organization of Matter*, edited by C. Ponnamperuma, and F. R. Eirich, 183. Hampton, USA: DEEPAK, 1990.
24. Inoue, T., and L. E. Orgel. "A Nonenzymatic RNA Polymerase Model." *Science* **219** (1983): 859.
25. Inoue, T., G. F. Joyce, K. Grzeskowiak, L. E. Orgel, J. M. Brown, and C. B. Reese. "Template-Directed Synthesis onto the Pentanucleotide CpCpG-pCpC." *J. Mol. Biol.* **178** (1984): 669.
26. Lohrmann, R., and L. E. Orgel. "Efficient Catalysis of Polycytidylic Acid-Directed Oligoguanylate Formation by PB2+." *J. Mol. Biol.* **142** (1980): 555.
27. Luisi, P. L., M. Giomini, M. P. Pileni, and B. H. Robinson. "Reverse Micelles as Hosts for Proteins and Small Molecules." *Biochim. Biophys. Acta* **947** (1988): 209.
28. Luisi, P. L., and F. J. Varela. "Self-Replicating Micelles—A Chemical Version of a Minimal Autopoietic System." *Origins Life Evol. Biosphere* **19** (1990): 633.
29. Maddox, J. "Towards Synthetic Self-Replication." *Nature* **354** (1991): 351.
30. Margulis, L., and R. Guerrero. "From Origins of Life to Evolution of Microbial Communities: A Minimalist Approach." In *Prebiological Self Organization of Matter*, edited by C. Ponnamperuma and F. R. Eirich, 261. Hampton: DEEPAK, 1990.
31. Maturana, H. R., and F. J. Varela. "De Máquinas y Seres Vivos." In *Una Teoría Sobre la Organizacíon Biológica.* Santiago, Chile: Editorial Universitaria S.A., 1973.
32. Maturana, H. R., and F. J. Varela. *Autopoiesis and Cognition—The Realization of the Living.* Dordrecht, Holland: H. D. Reidel, 1980.
33. Maturana, H. R., and F. J. Varela. *The Tree of Knowledge.* Boston: New Science Library, Shambala Publications Boston, MA, 1987.

34. Miller, S. L. "A Production of Amino Acids Under Possible Primitive Earth Conditions." *Science* **117** (1953): 528.

35. Miller, S. L., and L. E. Orgel. *The Origins of Life on the Earth.* Englewood Cliffs, NJ: Prentice-Hall, 1974.

36. Morowitz, H. J. "Biological Self-Replicating Systems." In *Progress in Theoretical Biology*, Vol. 1, edited by F. M. Snell, 35. New York: Academic Press, 1967.

37. Nagyvary, J., and J. H. Fendler. "Origin of the Genetic Code: A Physical-Chemical Model of Primitive Codon Assignments" *Origins of Life* **5** (1974): 357.

38. Naylor, R., and P. T. Gilham. "Studies on Some Interactions and Reactions of Oligonucleotides in Aqueous Solution." *Biochemistry* **5** (1966): 2722.

39. Nowick, J. S., Q. Feng, T. Tjivikua, P. Ballester, and J. Rebek, Jr. "Kinetic Studies and Modeling of a Self-Replicating System." *J. Am. Chem. Soc.* **113** (1991): 8831.

40. Olomucki, M. *La Chimie du Vivant.* Paris: Hachette, 1991.

41. Oparin, A. I. *Origin of Life.* New York: MacMillan, 1938.

42. Oparin, A. I. *Genesis and Evolutionary Development of Life.* New York: Academic Press, 1968.

43. Oparin, A. I., and K. L. Gladilin. "Evolution of Self-Assembly of Probiants." *BioSystems* **12** (1980): 133.

44. Orgel, L. E. *The Origins of Life.* London: Chapman and Hall, 1973.

45. Orgel, L. E., and R. Lohrmann. "Prebiotic Chemistry and Nucleic Acid Replication." *Acc. Chem. Res.* **7** (1974): 368

46. Oro, J., and A. Lazcano. "A Holistic Precellular Organization Model." In *Prebiological Self Organization of Matter*, edited by C. Ponnamperuma and F. R. Eirich, 11. Hampton: DEEPAK, 1990.

47. Pattee, H. H. "The Problem of Biological Hierarchy." In *Towards a Theoretical Biology*, edited by C. H. Waddington, 268. Edinburgh 8, North America: Edinburgh University Press, 1969.

48. Perret, J. "Biochemistry and Bacteria." *New Biolog.* **12** (1952): 68.

49. Polanyi, M. "Life's Irreducible Structure." *Science* **160** (1968): 1308.

50. Rebek, J., Jr. "Molekulare Erkennung mit konkaven Modellverbindungen." *Angew. Chem.* **102** (1990): 261.

51. Rebek, J., Jr. "Molecular Recognition and the Development of Self-Replicating Systems." *Experientia* **47** (1991): 1096.

52. Sagan, L. "On the Origin of Mitosing Cells." *J. Theor. Biol.* **14** (1967): 225.

53. Schmidli, P. K., P. Schurtenberger, and P. L. Luisi. "Liposome-Mediated Enzymatic Synthesis of Phosphatidylscholine as an Approach to Self-Replicating Liposomes." *J. Am. Chem. Soc.* **113** (1991): 8127.

54. Schwegler, H., and K. Tarumi. "The Protocell: A Mathematical Model of Self-Maintenance." *BioSystems* **19** (1986): 307.

55. Spiegelman, S. "An Approach to the Experimental Analysis of Precellular Evolution." *Qtr. Rev. Biophys.* **4** (1971): 215.
56. Tjivikua, T., P. Ballester, and J. Rebek Jr. "A Self-Replicating System." *J. Am. Chem. Soc.* **112** (1990): 1249.
57. Varela, F. J., H. R. Maturana, and R. Uribe. "The Organization of Living Systems, Its Characterization and a Model." *BioSystems* **5** (1974): 187.
58. Varela, F. J. *Principles of Biological Autonomy.* New York: North Holland, 1979.
59. Varela, F. J. "Describing the Logic of the Living: The Adequacy and Limitations of the Idea of Autopoiesis." In *Autopoiesis: A Theory of Living Organization*, edited by M. Zeleny, 36. New York: North Holland, 1981.
60. von Kiedrowski, G. "Ein Selbstreplizierendes Hexadesoxynucleotid." *Angew. Chem.* **98** (1986): 932.
61. von Kiedrowski, G., B. Wlotzka, and J. Helbling. "Sequenzabhängigkeit Matrizengesteuerter Synthesen von Hexadesoxynucleotid-Derivaten mit 3'-5'-Pyrophosphatverknüpfung." *Angew. Chem.* **101** (1989): 1259.
62. Walde, P., P. A. Bachmann, P. K. Schmidli, and P. L. Luisi. "Chemical Autopoiesis: Self-Replication of Micelles and Vesicels." In *Membrane Mimetic Chemistry and Its Applications*, edited by T. F. Yen et al. New York: Plenum, in press.
63. Zeleny, M., G. J. Klir, and K. D. Hufford. "Precipitation Membranes, Osmotic Growths, and Synthetic Biology." In *Artificial Life*, edited by Christopher G. Langton. Santa Fe Institute Studies in the Sciences of Complexity, Proc. Vol. VI, 125–140. Redwood City, CA: Addison Wesley, 1988.
64. Zielinski, W. S., and L. E. Orgel. "Autocatalytic Synthesis of a Tetranucleotide Analogue." *Nature* **327** (1987): 346.

Peter T. Saunders
Department of Mathematics, King's College, Strand, London WC2R 2LS, England

The Organism as a Dynamical System

One of the most characteristic properties of the developmental process is that it is stable. An embryo does not need an absolutely perfect environment and it can survive many small disturbances and even some large ones. Two embryos do not have to be clones to turn into very similar adults. The stability of development is, however, not just the simple sort of stability that we observe in such familiar examples as a ball at the bottom of a cup. An embryo that is perturbed will not return to the state that it was previously in. If it can recover at all, it will continue to develop, eventually reaching more or less the state it would have attained had it been left alone. What is stable is not the state of the embryo at any one time but its pathway of development.

In biology, the property of returning to the state a system was in before it was disturbed is called homeostasis, from the Greek words meaning similar and standing. C. H. Waddington, one of the first to stress the role of stability in development, introduced the words homeorhesis (similar flow) to describe a system which returns to a trajectory and chreod (necessary path) for the trajectory itself.[13,15,17] He also used the term canalization to describe the property that development typically can proceed to one or more of a restricted number of alternative end states rather than to a broad spectrum. And as an aid to understanding the role of these phenomena in development and evolution, he devised the epigenetic landscape, a metaphor which is illustrated in Figure 1.

Thinking About Biology, Eds. W. D. Stein and F. J. Varela, SFI Studies in the
Sciences of Complexity, Lect. Note Vol. III, Addison-Wesley, 1993

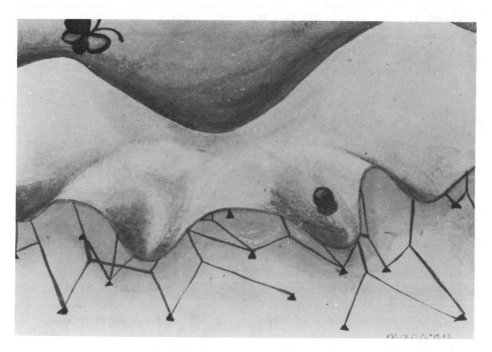

FIGURE 1   The epigenetic landscape. The butterfly symbolizes the role of catastrophe theory, among other things.

Waddington arrived at his view from his experience in embryology. Homeorhesis, chreods, and canalization describe features which developing organisms do possess, and recognizing their role leads to insights into both development and evolution. It is, however, possible both to see why existing developmental systems have these properties and, also, to realize that it would be hard to imagine a developmental system that did not. The properties arise from the fact that organisms, whatever else they are, are dynamical systems.

Now an arbitrary system may or may not be stable. If it is not stable, it will not persist, and we probably would not think of it as a system. Consequently a typical property of those entities that we recognize as systems is that they are stable.

When applied to systems in general, this argument is not very enlightening. It tells us more about how we view things than about how things are. But an organism, and in particular its developmental system, is not just any kind of system. It is a complex, nonlinear dynamical system. And such a system, if it is stable at all, generally has not just stability about a single equilibrium point but the richer kind of stability that Waddington identified in development. This allows us to replace the weak statement "The systems we observe are stable" by the much stronger "The complex nonlinear systems we observe, and this includes the developmental process in organisms, have a special kind of stability which gives them a number of important properties including those depicted in the epigenetic landscape." It may be surprising that there are organisms, but once we know that there are, we should not be surprised to find that they have these properties.

While the qualitative study of dynamical systems goes back at least to Poincaré at the turn of the century, it is only comparatively recently that it has become a major focus of mathematical research, largely on account of advances in topology and in computing. The aim of this chapter is to show how modern techniques, even just the modern point of view, can lead to progress in directions that Waddington first indicated many years ago.

THE EPIGENETIC LANDSCAPE

Waddington imagined the developmental system as a mountainous terrain (Figure 1) stretching from the heights, where the process begins, to the lowlands below, where it culminates. The valleys represent possible developmental pathways. The precise shape of the landscape depends on a network of guy ropes beneath it; the ropes stand for the effects of the genes and the complexity of the network is to remind us that while genes certainly affect development they do so in a very complicated way.

A ball rolls down the landscape, and the path that it follows stands for the actual development of the organism. An environmental perturbation is represented either by the ball being pushed up the side of the valley or by a small temporary change in the shape of the valley itself, as if the landscape were a tent and someone had kicked it. In either case, if the disturbance is not too great, then the ball returns to the valley bottom not at the point where it was disturbed but somewhere further down. The system thus exhibits homeorhesis.

If the ball is disturbed so much that it is forced out of the valley, it will probably reach a dead end, but it is also possible that it will pass over a watershed and then continue down a different valley just as stable as the original one. This, too, illustrates a property which is actually observed: sometimes when an embryo is perturbed, it neither dies nor returns to normal development but switches to an alternative pathway which leads to the production of a viable, though significantly different organism.

A mutation is represented by a change in the tension or position of a guy rope which may (though on account of the complexity need not) bring about a permanent change in the shape of a valley. Most of these will be relatively minor and the ball will generally return to its original course.

Mutations are most likely to have significant effects if they disturb the landscape near where one valley divides into two. At such a point even a small alteration in the topography can be enough to send the ball down a different path. Thus a small genetic change can bring about a large change in the phenotype, and without necessarily affecting the action of any other gene.

The model also suggests that mutations and environmental perturbations can have similar effects: they are simply different ways of diverting development into the same alternative pathway. This is observed in the common phenomenon of phenocopying. If, for example, genetically normal *Drosophila* embryos are immersed for a short time in ether vapor, some of them will develop into adult flies that resemble the *bithorax* mutant.[3]

Phenocopy is largely ignored by evolutionists, who seem content to treat everything between the genome and the adult as a black box whose properties can safely be disregarded. It is, however, very important once we start thinking in terms of dynamics. Its significance for the epigenetic landscape is that it provides a clear example of the existence of alternative pathways, i.e., that the neighboring valleys in the landscape really do exist. For if they did not, it would be very hard indeed to explain how it is that so often a mutation and an environmental perturbation can have the same effects.

THE EPIGENETIC LANDSCAPE AND EVOLUTION

The stability of development is very important for evolution. In the picture, the valleys are deep enough that neither minor changes in topography nor small random disturbances to the ball are likely to divert it from its normal course. Since the complexity of the network of guy ropes means that changing the position or tension of any one rope is unlikely to alter the landscape very much, most such changes will have little or no influence on the end result. The majority of exceptions will be near the end of the process, where the sides of valleys are flattening out and where the ball may not have enough time to return to its original trajectory.

Many perturbations, whether mutational or environmental, have little or no effect on the phenotype. There is typically very great genetic variability in populations which have very nearly identical phenotypes. The model (or, if you prefer, the recognition of the role of chreods) makes it easy to understand why this should be so.

Like most theories, the model has no difficulty with small evolutionary changes, but we have to ask what it suggests about large evolutionary changes. In principle, of course, they might occur by long sequences of small changes. This would require a succession of small deformations of the landscape, each moving the end point a little bit until it had reached a significantly different position. While this is possible in principle, it seems unlikely. If pulling a little on a rope shifts the landscape in the appropriate direction at first, after a while the tensions from the other ropes to which it is linked will stop it from continuing the same effect. And at the same time, the complexity of the network means that other parts of the landscape may also be affected, and in ways that were not intended.

Again, what the model illustrates are real properties. Even artificial selection seems inevitably to run up against limits. It is also generally accepted that both pleiotropy (one gene affecting more than one character) and genetic linkage (the fact that genes that are physically close together on a chromosome tend to be passed on together) will oppose strong directional selection. A sequence of advantageous changes in one character is likely also to bring about a sequence of random changes in others, and if this is carried on for some time, the net effect will almost certainly be deleterious.

But we have already noted that a small change in the landscape can have a significant effect on the organism if it occurs just where it can divert the ball down a different path. No other part of the landscape would have to be affected significantly, because any concomitant changes would also be small and all that is needed is that none of them is at a critical point. What is more, since it is only a matter of a diversion into an existing path, not the creation of a new one, we would expect that any one of a number of different small alterations in the network of ropes would do.

The model thus suggests not only that a single mutation can bring about a large phenotypic change but also that different single mutations can have the same effect. This could be because different mutations led to the production of the same protein, though we shall see later that this is not the only way. This helps us to understand how evolutionary change can occur without requiring that the same mutation occur more or less simultaneously in a large number of organisms. (As Haldane[2] showed many years ago, a mutation that occurs in only one individual in a large population is unlikely to survive, even if it confers a significant selective advantage.)

The model therefore predicts that large changes will happen rapidly and will probably not be related to any minor ones that are going on at the same time. Thus the mode of evolution suggested by the epigenetic landscape is precisely that of punctuated equilibria.

This is a very significant difference between the genetic and epigenetic views of evolution. From Darwin to the present, most evolutionists have insisted that evolution is gradual. Some are now prepared, albeit reluctantly, to accept that large changes might occur, and they see them as caused by mutations in regulatory genes that affect the action of several other genes. Such mutations can no doubt occur, but it is very hard to see how the effect of all the changes they cause will be to produce a viable organism, let alone one that will be in any sense fitter than the normal one. The genetic theory does not *predict* large changes. At best it suggests how they might possibly be accommodated within the theory.

THE DYNAMICS OF DEVELOPMENT

The epigenetic landscape illustrates remarkably well some important properties of developmental systems. It does not, however, explain why they should all have these properties, when there are so many differences between organisms or even between different regions of a single complex organism. The answer is that the properties are common not just to developing organisms but to most nonlinear dynamical systems, certainly to those of any marked complexity.

Waddington was aware that one should think of development in dynamical terms and he even wrote down some equations as an indication of how this might be done. In fact, he was closer to the modern approach when he drew the epigenetic landscape. For we can define a dynamical system as a manifold with a vector field defined on it. In other words, we describe a dynamical system by specifying the complete set of possible states, the phase space, and then providing rules that tell us at each point in the space where to go next. The rules can be given as differential equations (as they will be in the rest of this chapter), but they could also be, for example, difference equations or cellular automata.

In the case of the epigenetic landscape, the manifold is the Euclidean plane. The vector field is provided by the landscape: its slope at any point indicates the direction in which the ball will be accelerated. If we think of the system in more than two dimensions, as we should, the manifold is R^n, n-dimensional Euclidean space, but the situation is otherwise much the same, except that it is harder to visualize and impossible to draw.

Actually, it is not quite correct to say that the epigenetic landscape is a dynamical system. Figure 1 is not meant to depict a particular developmental process, only to illustrate some typical features of development. So it would be better to say that the epigenetic landscape represents a class of dynamical systems which share a number of important properties which are typically found in developmental systems. The mathematical problem is to determine the properties of this class.

In the rest of this chapter, it will be assumed that many processes of development can be modeled by one particular kind of dynamical system, differential equations. This not a difficult hypothesis to defend, and besides, it is part of practically all conventional modeling in development. The results will mostly be consequences of the assumption that the differential equations are nonlinear and, therefore, have many properties that are not found in the linear systems with which we are more familiar.

NONLINEAR DIFFERENTIAL EQUATIONS

It should not come as a surprise that nonlinear differential equations are significantly different from linear ones. Remember what happens with algebraic equations. If a and b are real and a is not zero, then the equation

$$ax + b = 0$$

has the unique solution $x = -b/a$, which is a real number. Even the more complicated case, a system of n linear algebraic equations in n unknowns, still usually has a unique solution, and any solution must be real.

On the other hand, the quadratic equation

$$ax^2 + bx + c = 0\,,$$

with a, b and c all real and a not zero, has two roots, and these can be complex numbers. So the transition from linear to nonlinear introduces multiple solutions and a completely new kind of solution as well.

A quadratic term is about the smallest amount of nonlinearity you can have, so you might ask what happens if we make the equation much more nonlinear, with higher-order terms or even functions that are not polynomials. The answer is: not a lot. There may be more roots, and they may be harder to find, but that's about all. Even if the coefficients in the equations are complex, nothing worse happens: there is no new kind of number lurking beyond complex. The most important differences between linear and nonlinear equations appear right away. Once we understand the quadratic case, we know the most important consequences of nonlinearity. (To the sort of accuracy to which Nature works, we may ignore the distinction between rational numbers on the one hand and irrational and transcendental numbers on the other.)

Things are much the same with ordinary differential equations. A linear ordinary differential equation, or a set of them, typically has a unique critical point. If this is stable, the system remains at equilibrium at the so-called attractor or, if disturbed, returns to it. If it is unstable, then the system does not persist. And that is pretty much the full repertoire of a linear system.

The behavior of nonlinear systems is much more interesting. They can have more than one equilibrium point, so that depending on the initial conditions or perturbations during their course, they can end up in different final states. These states typically form a discrete set, not a continuum. Besides stable points, however, they can also have stable trajectories. If the system is moving along one of these trajectories and is perturbed not too far away from it, it will return. Because there can be more than one stable trajectory, if a system undergoes a large disturbance, it may not return to its original trajectory but may be drawn to another stable one instead.

If a system has more than one stable equilibrium or stable trajectory, and almost all but the very simplest stable nonlinear systems are bound to, we may expect transitions from one to another. These will typically happen on a time scale that is short compared with the rest of the behavior of the system. Relatively small changes in parameters can, by causing attractors to disappear, bring about large changes in the state of the system. Thus large effects do not have to have large causes, as they generally do in linear systems. What is more, in nonlinear systems different causes can have the same effect. In particular, when a large change is possible, it can usually be initiated by altering the value of any one of a number of different parameters. This is very important for evolution, because it means that not only are large changes possible, they do not depend on one particular mutation nor even on a small set of mutations, each with very much the same effect.

Another characteristic of nonlinearity is that we cannot add solutions: the whole may be greater than the sum of the parts or less, but it is seldom the same. This, too, is familiar from elementary algebra: for example, the square of the sum of two numbers is not just the sum of the squares of the same numbers.

All these properties of nonlinear systems are well known, and have been for many years. What is more, as with algebraic equations, only a small amount of non-linearity is needed to make them appear: for instance, the famous Lorenz attractor has only two nonlinear terms and even they are only products of two variables.[12] Unfortunately, most of our intuition is based on linear systems because they are easier to analyze and are, therefore, more common in undergraduate courses. Even when modeling leads to nonlinear equations, the first thing we do is generally to linearize the system so that we can study its behavior near the critical points.

Fortunately, the ready availability of powerful computers is bringing about a change in the situation. Mathematicians are devoting a great deal of attention to nonlinear systems and even non-mathematicians are becoming acquainted with their remarkable behavior. When you look at one of the beautiful fractals that can be generated from simple mathematical formulae,[7] you are seeing just one of the many consequences of nonlinearity.

There is nothing mysterious about the properties that Waddington identified as typical of developmental systems: they are likely to be found in any complex system that can be modeled by nonlinear differential equations. There are also more such properties than Waddington found. For while nonlinear differential equations—and so the systems that can be modeled using them—typically have all the properties of development that are illustrated in the epigenetic landscape, the converse is not true. Ingenious though the picture is, it is only a picture and cannot be expected to capture all the properties of nonlinear systems.

MULTIPLE SPECIATION

One of the shortcomings of the epigenetic landscape is that it is drawn as a two-dimensional surface. It has to be, because that is the most that can be shown in a picture. Waddington acknowledged that this was a simplification, but it actually makes much more difference than he thought, because two dimensions is a very special case in dynamics. This is largely because we are concerned with trajectories, which are one-dimensional, and there are many things that are true only when the dimension of what you are studying is precisely one less than the space that it is in. One important difference is that in the landscape each valley has only two neighbors, but in higher dimensions a path can have many neighbors: imagine a multicore electric cable. This means, for instance, that there can be a very large number of dead ends and still a few viable pathways, which is not obvious from the picture.

Note how the crucial difference again appears at once. It does not matter very much how many dimensions there are, as long as it's more than two. Another peculiarity of differential equations in two dimensions, incidentally, is that the only attractors are points and limit cycles. In three or more dimensions, there can also be strange attractors, i.e., chaos.

If developmental pathways can have many neighbors, we may ask if multiple speciation can occur. The dynamical systems approach suggests that it should. To see why, it is perhaps easiest to think of a simple physical analogue. If you gradually increase the loading on a pillar, for a long time nothing much happens. This is convenient, because otherwise your house would sag appreciably as soon as you moved the furniture in.

If you increase the load beyond a certain critical value, however, the pillar will buckle. The direction in which this happens depends on either an asymmetry in the pillar which makes it weaker in one direction than in others, or else an imbalance in the load. Sometimes the bias is obvious, as when a lumberjack cuts a notch to ensure that a tree falls in the right direction, but often it is not, and then it can be impossible to predict in which direction the failure will occur.

Now imagine a number of symmetric pillars, each carefully made to the same specification and with no obvious flaws, and suppose each is given the same, gradually increasing load. For a while, the pillars will appear not to respond. Eventually, however, and more or less together, they will all buckle. But they will not all buckle in the same direction, because the minute biases that determine the direction of buckling will be different. When the stability of a system breaks down, previously unimportant distinctions can become crucial.

In the same way, the stability of the epigenetic system, which is necessary for normal development, means that individual organisms will develop in the same way even though there are considerable genetic differences among them. What is more, so long as evolution involves only minor phenotypic changes, canalization ensures that most individuals change in the same way. For speciation to occur, the stability has to break down, and when that happens these previously unexpressed genetic differences may come into play, diverting development into different developmental pathways. We may, therefore, observe the almost simultaneous appearance of a number of new forms. This will not always happen, but we should not be surprised when it does.

The idea that genes that previously had no effect on the phenotype can later come into play is not new, but it is usually seen as a matter of silent genes being turned on. Here we are supposing that the genes are already on, but that the stability of the epigenetic system has been preventing differences in them from affecting the phenotype. When this stability breaks down, the previously unimportant (but not *unexpressed*) genes now have the opportunity to influence development, since the relatively unstable system is susceptible even to small genetic differences. In this way a number of different genes can become significant simultaneously without having either to appear, or first be expressed, simultaneously.

DIRECTION IN DEVELOPMENT AND EVOLUTION

We now demonstrate that where transitions between distinct states can occur, they are more likely to happen in one direction than in the other. This lack of symmetry, which cannot be inferred from the picture, arises from the details of a physical or chemical process, but can be passed up through the dynamical system to affect processes with quite different time scales, including ultimately evolution.

To make the argument easier to follow, and to show that we do not have to postulate a complicated and implausible mechanism, it is convenient to work with a particular equation. The result does not, however, depend on this choice, but is true for almost any model which can produce a sharp transition between states.

The model that will be used as an illustration was proposed by Lewis, Slack, and Wolpert,[4] who wanted to show how a sharp frontier could form in a region of tissue. The states of cells in the region were supposed to be specified by the concentration of a gene product, g, which is activated by a "signal substance" S, and the rate of change of g was given by the equation

$$\frac{dg}{dt} = K_1 S + \frac{K_2 g^2}{K_3 + g^2} - K_4 g \,, \tag{1}$$

where the K_i are all constants. Lewis, Slack, and Wolpert set K_2 and K_3 equal to unity and K_4 equal to 0.4. Without loss of generality we can choose the units of S such that K_1 is also equal to unity.

Suppose that both g and S are initially zero and that S is then gradually increased. Then dg/dt will be positive, so g will increase as well. If the increase in S is slow enough, g will always be close to the equilibrium value (for the current value of S) which can be found by solving the equation $dg/dt = 0$ for g. With the given values for the constants, this equation is

$$2g^3 - 5g^2(1 + S) + 2g - 5S = 0\,,\tag{2}$$

and for $S = 0$ it has three real roots: 0, 0.5, and 2. There are thus three possible steady states, and it is not hard to show that those at $g = 0$ and $g = 2$ are stable and that at $g = 0.5$ is unstable.

Thus, g will remain zero until S begins to increase. When S is greater than zero but small, Eq. (2) still has three real roots and so there are still three equilibrium points, two stable with an unstable one in between. The smallest equilibrium value of g increases with S, which means that as S increases slowly, so does g.

If, however, S is increased above a critical value S_c, which is approximately 0.0418, a significant change occurs. Eq. (2) now has only one real root, together with a complex conjugate pair. The two steady states (one stable, one unstable), corresponding to smaller values of g, have coalesced and disappeared. The system will, therefore, move rapidly to the higher equilibrium; i.e., there will be a rapid increase in g. Thus, at this point a very small change in S from just below S_c to just above it, will cause a large change in g.

This gave Lewis, Slack, and Wolpert what they required, because if there is a smooth gradient in S throughout a region then, at the position where S takes on the value S_c, there will be a discontinuity in g, i.e., a definite frontier between two sub-regions. Thus the continuous gradient has given rise to a sharp division. Here we are not so much interested in the formation of a frontier as in an abrupt change in evolution, but the model will provide this too, if we imagine that S has the same final concentration throughout the region but that a mutation increases this value from below S_c to above it, or vice versa. The principle is the same, but the important variable is time rather than space.

It turns out, however, that there is more to the model than first appears.[10] Equation (1) has more properties than we have seen so far. What is more, these properties are likely to be found in almost any system which is capable of producing a transition from one steady state to another.

We know this as a result of catastrophe theory, which, in its noncontroversial role as a theory about classes of systems of differential equations, tells us that a system that can produce two steady states should have at least two parameters. A model with only one parameter is structurally unstable, which means that almost any other model only very slightly different from it makes different predictions. If we believe that Eq. (1) with all but one parameter fixed is *precisely* the right model for a particular process, that is one thing, but it cannot tell us how a typical process of this kind will behave.

Fortunately, it is both necessary and sufficient to allow two parameters to vary. Here we shall choose K_3 (which we shall write simply as K) because varying a saturation constant seems physically plausible. Allowing the other parameters to vary as well would not change the general pattern of the behavior, only the details.

Other, quite different models, providing that they were not unnecessarily complicated, would also exhibit the same general pattern. We shall see later on that the chief result of this section applies to more complicated models as well, even though Eq. (1) does not describe their entire repertoire of behavior.

With this change, the equilibrium condition is

$$2g^3 - 5g^2(1+S) + 2Kg - 5KS = 0.\tag{3}$$

Figure 2 is a diagram of the "control space" for the system, i.e., the S–K plane. The idea is that the system is controlled by variations in the parameters S and K, and responds by adjusting the gene product g to an equilibrium value. This value will naturally depend on S and K. Since Eq. (3) is a cubic, it has three roots, but they are all real only for certain combinations of the parameters. For others, there is one real root and a complex conjugate pair. Hence, for some values of S and K, there are two stable equilibrium values of f (and one unstable one in between) whereas for others there is only one. The cusped curve in Figure 2 is the bifurcation set, i.e., the boundary between the two regions. (Those familiar with catastrophe theory will have recognized Eq. (3) as a cusp catastrophe, though not in its canonical form.[6])

In general, a small change in either S or K will cause only a small change in g. The exceptions are when the change takes the system from the region of the control space in which there are two equilibria to that in which there is only one. If the equilibrium, that the system is in, is the one that disappears (by coalescing with the unstable equilibrium to form a saddle point), then there will be a sudden change in g as the system moves rapidly to the other equilibrium.

Suppose that S is initially zero. Then $g = 0$ is a (stable) equilibrium, so if the gene product is initially at zero concentration, it will remain there. If S is then increased slowly enough that the system is not disturbed appreciably from steady state, g will also increase slowly. This will continue until the increase in S takes the system across the right-hand branch of the bifurcation set: for $K = 1$ this is at $S = 0.0418$. When that happens, the low-value equilibrium disappears and g will will increase rapidly. (For the values used in this illustration, the change is from $g = 0.227$ to $g = 2.165$.)

If S is then decreased, nothing significant will happen as the trajectory crosses the right-hand branch because the high-valued equilibrium is still present. The sudden jump back to a low value occurs only when the trajectory crosses the *left* branch of the cusp, which for $K < 1.563$ would require S to be negative. Hence, for all values of K below 1.563, even if S returns to zero the concentration of the gene product, g will remain at a significantly nonzero value; for $K = 1$, this is 2.0.

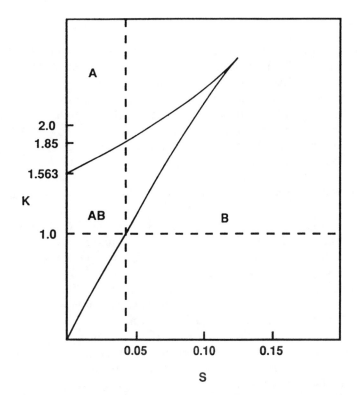

FIGURE 2   The control space for Eq. (3), showing the bifurcation set. The regions in which only the A phenotype, either the A or B, or only the B phenotype are possible are marked by A, AB, and B, respectively. The lines $K = 1$ and $S = 0.0418$ are relevant to the two models described in the text.

This is an example of the very important phenomenon of hysteresis. When a system can move from one state to another, even if the reverse transition is possible, the system will not simply retrace its path in the opposite direction. As the name hysteresis (from the Greek word for delay) suggests, systems tend to remain close to whichever end state they happen to start in. Because of this, even reversible transitions are typically not symmetric, and this, as we shall see, can contribute to irreversibility.

We now consider two different ways in which the model we have been describing might be involved in development. For simplicity we will suppose that whether g has a low value or a high one is directly observable in the phenotype, and we will refer to individuals with low and high values of g as A and B, respectively.

MODEL I

Suppose that during development K is fixed and S rises slowly from 0 to a maximum value S_m. Both K and S_m are assumed to be under genetic control. Then for K less than about two, combinations of K and S_m that lie in the regions marked A or AB will produce A individuals (since g will never be forced away from the low-valued equilibrium), but those that lie in the region marked B will produce B individuals.

Now suppose that mutations alter the values of K or S_m. Any mutation which did not take the values across the right-hand branch of the cusp, i.e., the boundary separating region AB from region B, would have no observable effect on the phenotype. On the other hand, a mutation that caused only a small change in either parameter but did take the system across that boundary would produce an obvious effect.

Thus, for example, imagine that S_m is fixed at 0.0418 and K varies from 0.5 to 2.0. For all values of K less than 1.0, the result would be B, and for all values of K greater than 1.0 the result would be A. Whether the change in K occurred in a single step or as an accumulation of many mutations, the large change in g and, therefore, in the observable phenotype would occur abruptly. So the mathematical model, like Waddington's picture, predicts punctuated equilibria.

Suppose that after K and S have reached their normal values, there is a change in the concentrations of one or both of them. If $K < 1$, then there will be no net effect. If $K > 1$, however, so that the normal phenotype is A, then a temporary reduction in K or increase in S which carried the point across the right-hand branch of the cusp would cause g to move to the high equilibrium. Even if K and S returned to their original values, g would remain at the high equilibrium, and so the phenotype would be an B. This phenocopying is most likely to happen for K not very much greater than 1.0, because as can be seen from the diagram, as K increases so does the size of the perturbation required to have a significant effect.

If the gene product g is perturbed, it will return to an equilibrium value for the given values of K and S. If $K < 1$, there is only one equilibrium and so the organism will still be a B. If K lies between 1.0 and 1.85, there are two equilibria. The value of g will normally be at the lower equilibrium, but if g is increased too far, it will not return to this equilibrium. Instead, it will stabilize at the higher one, i.e., the organism will become a B instead of an A. As K increases the two equilibria move further apart, so that the transition becomes less likely, and finally when $K > 1.85$, the higher equilibrium disappears and the phenotype must be A.

Thus, as K increases, there is a range within which the phenotype must be B, a range within which the normal phenotype is A but B phenocopies can occur, and finally a range in which the phenotype must be A. Phenocopying never occurs in the direction B to A.

MODEL II

Model I follows Lewis, Slack, and Wolpert's original idea, that S rises from zero to a final value S_m where it remains.[4] Saunders and Ho,[10] however, suggested that a more reliable way to create a boundary would be to establish a gradient in K and then have S rise to a large but not precisely specified value, say, 0.1 or thereabouts, and then fall back to zero. The boundary would then occur at a particular value of K. The advantage would be that since for any K, g remains zero as long as $S = 0$, there would be plenty of time to set up an accurate gradient in K. To set up an accurate gradient in S is harder, because as soon as S starts to increase, so does g. Here we are not concerned with boundaries, but it is still of interest to see how this different dynamic behaves.

Because the maximum value attained by S is no longer critical, we may assume that the only variable under close genetic control is K. If there are no perturbations, then all individuals with $K > 1.563$ are A and all those with $K < 1.563$ are B.

Now suppose that K undergoes a perturbation, which we again take to be a small and temporary change from its normal value K_0. If this happens before the rise and fall in S, it will have no effect because so long as $S = 0, g = 0$ is a stable equilibrium for all values of K. If it happens during or after the rise, however, the equilibrium reached when S falls back to zero will be that corresponding to S_t, the temporary value of K. The only interesting cases are, of course, when one of K_0 and K_t is less than 1.563 and the other is greater. If $K_0 > 1.563$ and $K_t < 1.563$, as S falls, g will remain at the high equilibrium, but when K returns to K_0, this is no longer available, g will fall to the low equilibrium, and the organism will be an A, as it should. In the reverse situation, however, as S falls, g will go to the lower equilibrium, and since this does not disappear when K returns to K_0, g will remain there. So the genetically B organism can be perturbed to an A phenotype.

Alternatively, since for $K < 1.563$ there are two equilibria but for $K > 1.563$ there is only one, a perturbation of g can cause a permanent change in the former case but not in the latter. So again, phenocopying can occur only in the direction B to A.

The two models differ in the direction in which phenocopying is possible. This is not important, because if Eq. (1) described a real process in development, either model I or model II would have to apply, not both. What is significant is that there is always an asymmetry.

Like the epigenetic landscape of which it is a mathematical version, the model predicts that phenocopies can occur. It also predicts that during evolution from one form to another, there may be a limited period during which phenocopying is possible, but that the system may later stabilize so that it is not.

The most interesting prediction, and one which cannot be seen in the epigenetic landscape, is that we expect phenocopying in one direction only. We know that we can readily produce a *bithorax* phenocopy from a genetically normal *Drosophila* embryo; this should not lead us to expect that we can produce a phenotypically normal *Drosophila* from a *bithorax* mutant embryo.

When an evolutionary change occurs, there may be a period just before the change in which the form that is about to take over appears from time to time as a phenocopy. Alternatively, there may be a period shortly after the change in which the old form sometimes appears as a phenocopy. But we should not expect to see both patterns of behavior.

Finally, the model illustrates the point made earlier, that the mutation required to cause a large change is unlikely to be unique. Model I has only two parameters, K and S_m, and yet any mutation that altered either of them in the right direction would do. And in the original model given by Eq. (1), the same effect could be brought about by a mutation affecting any of four different parameters. The magnitudes of the changes caused by different mutations would not have to be the same because the magnitude of the change in gene product is a property of the dynamic, not the mutation. In the numerical example given above, almost any mutation that changed g significantly would increase it from about 0.2 to about 2.1.

Dynamical systems typically have more parameters than are necessary to produce the entire range of behavior of the system. In the model we have been discussing here, there are four physical variables but only two that are in this sense mathematically independent. Since Eq. (1) was chosen as an illustration rather than proposed as a mechanism for a particular process, we may expect that in many real situations there would be still more physical parameters, even though the behavior was, in general terms, the same.

Mathematicians naturally prefer to work with model systems that have no such redundancy. The first step in analyzing a system is often to identify and eliminate nonessential parameters, as we have done here with Eq. (1). The value of this strategy is obvious, but it does tend to give the misleading impression that each mode of behavior of a dynamical system can be attributed to a single physical parameter. Not every parameter is necessarily capable of causing every kind of change, but the real situation is usually far more complicated than the idealized models in textbooks.

Biologists now recognize that neither "one gene/one character" nor "one gene/one polypeptide" is a valid assumption. Since the polypeptides ultimately act through the parameters of chemical reactions, we can add the third side to the triangle: "one polypeptide/one character" is also false.

GENETIC ASSIMILATION

The beginning of a major transition often poses a problem for Darwinian evolutionary theory. However great the selective advantage of the final result, it can be very hard indeed to imagine what possible use the first step towards it could have been.

A famous example, due to Mivart,[6] concerns flat fish, like plaice and flounder, which swim on their sides near the bottom of the sea and have both eyes on the same side of the head. An eye that pointed downwards would presumably not be of much use, so let us agree that there is a significant selective advantage in having both eyes on top. Let us also suppose that the change from the usual arrangement occurred gradually, by a number of separate mutations. Then each of these mutations must have individually conferred a significant selective advantage, and this is not so easy to understand. Above all, what benefit could there have been in having the bottom-facing eye just marginally further forward than before? Yet if there was no benefit, how did the process of moving the eye ever get started?

Darwin's own suggestion,[1] incidentally, was that the first stages of the transition "may be attributed to the habit, no doubt beneficial to the individual and to the species, of endeavoring to look upwards with both eyes, while resting on one side at the bottom." This may have been an adequate response at the time, since what is now called Lamarckism was then widely accepted, but most modern evolutionists would not be happy to rely on it.

Suppose, however, that the asymmetric form could occur as a phenocopy, and that each of the mutations not only moves the eye slightly but also increases the probability that the phenocopy will occur. Then each mutation would increase fitness by increasing the chance that the individual that possessed it has both eyes on top. A small fraction of an advantageous trait may be of no use; a small probability of having the trait is another matter.

Whether genetic assimilation plays a significant role in evolution is not known, though Waddington[16] and others have demonstrated it in the laboratory. But the simple example here illustrates in concrete terms how it could come about. Suppose that K is initially 1.4 so that the normal phenotype is A, and suppose that the B form is fitter. Suppose also that any point mutation that affects K reduces it by 0.1. A single such mutation would not change the phenotype, and so would appear to have no selective advantage. But it would increase the chance of the individual that possessed it becoming a B phenocopy, and this would give a selective advantage. As a result, we would expect genes that reduced K to increase and that the B form would eventually replace the A.

Our results suggest that phenocopying will frequently be observed when a significant evolutionary change is about to occur. They also show how genetic assimilation can contribute to irreversibility in evolution, because if it assists a change in one direction, it will not help in the other. This could help to explain Dollos's law, which states that a feature that is once lost is unlikely ever to reappear.

MORE COMPLEXITY

The results of the preceding sections depend on two slightly different assumptions about simplicity. The first was that the transition from one state to another occurred in a way that is relatively easy for nature to bring about. An electric light switch may operate without a hysteresis loop, but it is an artificial device. What is easy for us is not always easy for nature.

We also assumed that the mechanism was the simplest natural one that can produce a transition between states, and here simplest was used in the sense of producing such a transition and nothing more. The next simplest class are those with three stable equilibria, and as a concrete example we can use the following extension of Eq. (1):

$$\frac{dg}{dt} = S + \frac{ag^2}{K + g^2} + \frac{bg^2}{L + g^2} - Dg. \tag{4}$$

This seems a plausible example, but its real justification is that we know from catastrophe theory that Eq. (4), like the simpler Eq. (1), captures the essential behavior of a very large number of systems, many of them far more complicated and with many more variables.

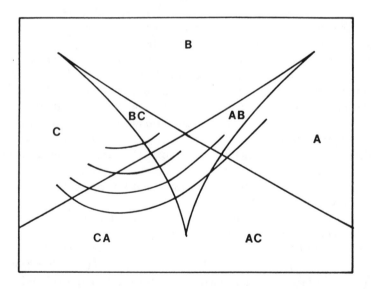

FIGURE 3   A qualitative projection of the control space for Eq. (4), showing the bifurcation set and control trajectories for $L = 0.55, 0.60, 0.65, 0.75$ as S rises from 0 to 0.5 or above. The letters A, B, and C indicate the possible phenotypes as in Figure 2. All three are possible in the unlabelled diamond-shaped region in the center of the figure.

For most values of the parameters, there are only one or two stable equilibria, and so most of the behavior of this equation is very much like that of Eq. (1). The results of the previous section therefore apply. If, however, we take $K = 36.6, A = 1, B = 0.14, D = 0.11$, and L greater than zero but less than about 2.5, then for small values of S there are three equilibria, which allows more possibilities than before.

Some of these are illustrated in Figure 3. The axes have been neither scaled nor even labelled on the figure because it is to be interpreted only qualitatively. This is necessary because the analogue of Figure 1 would require four dimensions, and so the most that can be shown is a projection. (A full description of the figure and how it is to be interpreted is given by Saunders and Kubal.[11])

There are now three possible equilibria, one at low g, one at an intermediate value, and one at high g. These have been labelled A, B, and C, respectively. The four curves illustrate what happens for different values of L as the signal substance S is increased from 0 to a large value. Because we are concerned only with the sequence of equilibrium states that the system will pass through under various conditions, a figure that is only topologically correct is good enough. If we took the model seriously and wanted to compute the critical values of the four parameters, we could always do this starting from Eq. (4).

As in Eq. (1), if $S = 0$, then $g = 0$ is a stable equilibrium, so all four systems start in state A, the low g equilibrium. What happens after that depends, as before, on the role of S. If we suppose it rises to a precise maximum value, and remains there, the behavior is almost exactly as in model I of the previous case. The system will remain in state A until the corresponding equilibrium disappears, which occurs at the value of S at which the trajectory crosses the part of the bifurcation set that separates the regions marked CA and C. If the normal value of S is below that value but S is temporarily raised above it, the system will move to C and stay there; if the normal value is above the critical value and falls below it for a short time, nothing much happens. Unless S is very small, if g is at the low equilibrium, it can be perturbed to the higher equilibrium, but if S is large enough so that g has moved to the higher equilibrium, there is no longer a lower one for it to be perturbed to. Hence, phenocopying can occur in the direction A to C only.

State B cannot be reached by mutation or by perturbations in L or S_m, but it can be reached by a perturbation of g when L and S_m have appropriate values. At such values, the normal state is A, so phenocopying will be in the direction A to B, not C to B. Thus the only phenocopies are A to B and A to C.

The absence of normal B phenotypes arises out of a distinction between mathematical systems and the real phenomena they are being used to model. Large regions of the phase space of a mathematical model are often inaccessible in reality. The most obvious reason for this is that many physical variables cannot take on negative values.

An important consequence of this is that some regions of the phase space which correspond to permitted values of the parameters are for the most part inaccessible, because they can only be reached from normal starting points by passing through a forbidden region. In effect, the phase space for the real process may not be connected. It might be, however, that a perturbation could force the system into such a region. This would correspond to forms that can occur only through environmental disturbance, never by normal development or mutation.

As a general point, it is worth noting that while the assumption that a phase space is a smooth, simply connected manifold is seldom stated explicitly, many results depend on it. For example, a number of authors have discussed evolution in terms of fitness landscapes with species striving to reach adaptive peaks, i.e., phenotypes with (locally) maximum fitness. In principle there is no reason why one cannot imagine a phase space for phenotypes. On the other hand, its topology (loosely speaking, its structure) would be very different from that of a Euclidean space. At any point we can move only in a very restricted number of directions, and we cannot go from almost any point to almost any other point by a continuous curve. (Translated into biology: the number of possible variants is much fewer than the number of conceivable alterations in the phenotype, and not every transition from any one phenotype to any other can be accomplished by infinitesimal changes.) That there can be large phenotypic changes implies that there are points in the phase space that are in one sense far apart and yet in another sense close together. We should not assume that just because it is possible to define a space, it, and therefore the system or process it is supposed to model, will have all the nice properties of the Euclidean plane.

Returning to the model, let us consider what happens if we assume that development involves S becoming quite large, say 0.5, and then falling back to zero, like Model II in the first example. It is then easy to track what happens on the figure. For $L = 0.55, 0.60, 0.65$, and 0.75, the final states are C, C, B, and A respectively. Temporary changes in L or in g can cause permanent transitions from C to B or A or from B to A, but not the reverse. Note that these are all in the opposite direction to those permitted when we assumed S remained at its maximum value. Thus we have the same result as before: the direction in which phenocopying is possible depends on the details of the process, but it is only possible in one direction.

In the model based on Eq. (1), we could see all the interesting behavior using obviously plausible values of K and S_m. In Eq. (4), on the other hand, it takes a fair amount of trial and error to find parameter values that give anything new. This might be an artifact, but it probably reflects a general rule that even if complex systems do have the potential for complex behavior, most of the time they act like simpler ones. For example, a complex elastic structure may have many different failure modes, but it will still usually behave in much the same way as the pillars mentioned above, showing little response to loading until the critical load is reached, only then buckling. If, however, the parameters have just the right values necessary to make the critical loads for two failure modes equal, the behavior will be quite different. The imperfection sensitivity will be very great, which means that the maximum safe load may be much lower than expected, and if the structure is overloaded, it will collapse without warning.[14]

While the consequences are unlikely to be as disastrous as if we were building bridges, we too should remember that a complex system can have hidden within it a capability for much more complicated behavior than it usually displays. The potential is likely to be realized only rarely, but then significant steps in evolution do not occur very often.

TIME SCALES

In *The Strategy of the Genes*, Waddington[17] pointed out that a living thing is involved in at least three different types of temporal change, all going on simultaneously. These are, in order of time scale, physiology, life history, and evolution. Development is part of the medium time scale and interacts both with the short-term physiological processes that bring it about and also with the long-term process of evolution. Biologists, faced with the immense complexity of organisms, have naturally tended to look at the three time scales separately, and physiology, developmental biology, and volution exist as separate disciplines. We might call this a sort of epistemological adiabatic approximation, drawing on a term that is sometimes used in physical science to describe the mathematical decomposition of a dynamical system into sub-systems with different time scales.

This device is often very useful in mathematics, and it is certainly hard to imagine how biology could progress if we had to take absolutely everything into account all the time. The subject is hard enough as it is. All the same, we must not forget that the levels do interact, and in both directions. It is not simply a matter of the slowly varying quantities acting as parameters for the fast processes, of evolution setting the framework for development. Here we have seen how a fast process can profoundly affect the nature of a slow one. The origin of the asymmetry was that the gene product g started at zero rather than a high value. This created a lack of equivalence between two equilibrium values, i.e., between two developmental states. This, in turn, meant that an evolutionary increase in a certain parameter would not be simply the mirror image of a decrease in the same parameter. The asymmetry was thus passed upwards from the shortest time scale to the longest.

CONCLUSION

By considering organisms as dynamical systems, we are able to understand why they have the properties that Waddington portrayed in the epigenetic landscape. We can also make a number of inferences about development and evolution. Both large changes and multiple speciation should occur. Large changes, even more then small ones, are likely to be capable of being initiated by a number of different mutations. This greatly increases the likelihood that they will occur, certainly by comparison with regulatory gene mutations, which presumably involve one specific gene. Where a transition between two states is possible, it will be more likely to occur in one direction than in the other. In particular, phenocopying should occur in one direction only. This lack of symmetry will contribute to the irreversibility of evolution.

In biology, organisms are traditionally arranged into a hierarchical classification. A butterfly has certain properties, say, its color and markings, because of the species and variety it belongs to, whereas other properties, such as being segmented and having three pairs of legs, are common to all insects. Still others, like the triplet code a butterfly uses to translate its genetic information, are properties of almost all organisms. We may feel we understand how DNA is transcribed and translated in a butterfly, but very little of our knowledge of this process was gained from work on butterflies.

Conversely, we often study a particular organism more because of the light it can throw on general problems than for its own sake. The aim of the vast amount of research that has been done on *Drosophila* has been to learn about genetics and development, not to satisfy an apparently limitless curiosity about fruit flies.

Now an organism is also a dynamical system. Hence, even if biologists do not generally include dynamical systems as a level of the Linnaean hierarchy, it is still an appropriate level for studying biological phenomena, especially the most general ones that are shared by almost all organisms—and also some systems that we do not usually think of as organisms, like the Earth.[5] And, just as biologists do, we can also use particular dynamical systems as examples to help us in our work. Equations (1) and (4) are what we might call mathematical fruit flies. We are interested in them not because they model important chemical reactions—it may even be that they do not model any real reactions at all—but because they are convenient to work with and yet share important properties with large classes of dynamical systems which almost certainly include many of those that occur in development.

Waddington saw the value of this approach half a century ago, which is why he devised the epigenetic landscape. Since then, mathematics has provided some of the techniques needed to make his idea work. We know far more about the behavior of classes of dynamics, as distinct from individual equations.[9] We are also better able to judge which equations are typical of large classes and which are not. It may be a new departure that mathematics can play such a fundamental role in biology, but then it is a rather new kind of mathematics that is involved.

ACKNOWLEDGEMENT

I am grateful to Mae-Wan Ho for helpful comments and for drawing Figure 1.

REFERENCES

1. Darwin, C. *The Origin of Species*. 6th edition. London: John Murray, 1875.
2. Haldane, J. B. S. "A Mathematical Theory of Natural and Artificial Selection. Part V. Selection and Mutation." *Proc. Cambridge Phil. Soc.* **23** (1927): 838–844.
3. Ho, M. W., E. Bolton, and P. T. Saunders. "*Bithorax* Phenocopy and Pattern Formation I. Spatiotemporal Characteristics of the Phenocopy Response." *Exper. Cell Biol.* **51** (1983): 28–290.
4. Lewis, J., J. M. Slack, and L. Wolpert. "Thresholds in Development." *J. Theor. Biol.* **65** (1977): 579–590.
5. Lovelock, J. E. *The Ages of Gaia*. Oxford: Oxford University Press, 1988.
6. Mivart, St. G. *On the Genesis of Species*. London: Macmillan, 1871.
7. Peitgen, H. O., and P. H. Richter. *The Beauty of Fractals: Images of Complex Dynamical Systems*. Berlin: Springer-Verlag, 1986.
8. Saunders, P. T. *An Introduction to Catastrophe Theory*. Cambridge: Cambridge University Press, 1980.
9. Saunders, P. T. "Mathematics, Structuralism and the Formal Cause in Biology." In *Dynamic Structures in Biology*, edited by B. C. Goodwin, G. C. Webster, and A. Sibatani, 107–120. Edinburgh: Edinburgh University Press, 1989.
10. Saunders, P. T., and M. W. Ho. "Primary and Secondary Waves in Prepattern Formation." *J. Theor. Biol.* **114** (1985): 491–504.
11. Saunders, P. T., and C. Kubal. "Bifurcations and the Epigenetic Landscape." In *Theoretical Biology: Epigenetic and Evolutionary Order from Complex Systems*, edited by B. C. Goodwin and P. T. Saunders, 16–30. Edinburgh: Edinburgh University Press, 1989.
12. Stewart, I. *Does God Play Dice? The Mathematics of Chaos*. Oxford: Blackwell, 1989.
13. Thom, R. "An Inventory of Waddington Concepts." In *Theoretical Biology: Epigenetic and Evolutionary Order from Complex Systems*, edited by B. C. Goodwin and P. T. Saunders, 1–7. Edinburgh: Edinburgh University Press, 1989.
14. Thompson, J. M. T., and G. W. Hunt. *A General Theory of Elastic Stability*. London: Wiley, 1973.
15. Waddington, C. H. *Organizers and Genes*. Cambridge: Cambridge University Press, 1940.
16. Waddington, C. H. "Genetic Assimilation of the *Bithorax* Phenocopy." *Evolution* **10** (1956): 1–13.
17. Waddington, C. H. *The Strategy of the Genes*. London: George Allen & Unwin, 1957.

Jay E. Mittenthal, Bertrand Clarke,† and Mark Levinthal‡

Department of Cell and Structural Biology, 505 S. Goodwin St. (and Center for Complex Systems Research, Beckman Institute; and College of Medicine), University of Illinois, Urbana, IL 61801, U.S.A.; †Department of Statistics, Purdue University, W. Lafayette, IN 47907, U.S.A.; and ‡Department of Biology, Purdue University, W. Lafayette, IN 47907, U.S.A.

Designing Bacteria

The genome of an *E. coli* cell contains codes for about 2000 different proteins. About half of these have now been characterized to some extent. We can imagine that within the next 50 years the structures and functions of every one of these proteins will have been determined and that the entire genome will have been mapped in the greatest possible detail, so that we will know the exact position of each *E. coli* gene, how the expression of each gene is regulated, and the exact chemical nature of both the gene and its products. It will then be possible to write an *Encyclopedia of E. coli Life Processes*. If you were to ask someone the question "What is life?" and your respondent handed you the *Encyclopedia of E. coli Life Processes*, would you be satisfied?

—J. D. Rawn,[30] p. 22

1. INTRODUCTION

The experimental study of biological systems has emphasized analysis at ever-finer levels. This approach is essential for understanding and manipulating organisms, but it is unlikely to show high-level patterns of organization. Although in principle such patterns could be inferred from detailed knowledge of molecular organization, in practice the detail obscures the patterns; one can not see the forest for the trees.

Several authors[24,25,31,44,45] have sought a theory of the organization of organisms that is independent of molecular details, just as there is a theory of computation that does not depend on the particular materials and circuits in computers. To find such a theory, a deliberate search is necessary. This is a task for theoreticians in biology, to assimilate knowledge gained from experimentation in order to provide general theories of biological organization.

Here we sketch a theory for the organization of the bacterium *Escherichia coli*. While it is essential eventually to develop a detailed predictive model, that enterprise will be protracted and laborious. Here we offer merely a conceptual framework, which is intended to provide a rationale for what is observed rather than to make new predictions.

We view a bacterium as a network of coupled biochemical reactions.[1] Two reactions are coupled if an output of one is an input to another. The pattern of coupling among all reactions is the structure or connectivity of the network. The theory aims to understand the structure of biochemical networks, by answering two questions. (1) What characteristic structures are evident in biochemical networks? (2) Why do these structures evolve?

The theory we propose can be summarized as follows. A bacterium embodies two complementary networks, or nets, of processes. Each net is a distributor; it allocates limited resources to perform diverse tasks. The metabolic net is a distributor that converts diverse substrates to a standard set of key metabolites. The macromolecular net polymerizes key metabolites into diverse species of macromolecules. Because the enzymes for the metabolic net are a subset of the outputs from the macromolecular net, the characteristic structure of a biochemical network is a hierarchy of distributors.

We argue that a hierarchy of distributors is likely to evolve, under selection for high performance using limited resources. Criteria for performance include reliability, flexibility, economy, and speed. In particular, limits on internal resources—on the coding capacity of the genome and on the solubility of macromolecules—favor the evolution of distributors.

These arguments suggest answers to the preceding questions. (1) The characteristic structure of a biochemical network is a hierarchy of distributors. (2) This

[1]Abbreviations: ATP, adenosine triphosphate; DNA, deoxyribonucleic acid; RNA, ribonucleic acid; mRNA, messenger RNA; rRNA, ribosomal RNA.

structure evolves because distributors achieve high performance with limited resources. These answers suggest an appropriate use for the *Encyclopedia of E. coli Life Processes*. It contains information about the hierarchy of distributors, information which can be used to understand the dynamical modes of operation of a bacterium.

This paper is organized as follows: The remainder of this section contrasts the generative and design approaches to a theory of organization. After justifying our use of a design approach, we relate general performance criteria to the specific tasks that a bacterium must perform. Section 2 characterizes the ensemble of possible metabolic nets as a collection of directed graphs. Imposing the performance criteria selects a unique graph: The optimal metabolic net is a distributor in which branches extend from a central hub of reactions to the key metabolites. Section 3 identifies the possible macromolecular nets as autocatalytic sets of polymers. We suggest that the optimal macromolecular net is again a distributor with a hub-and-branch structure, but now the hub and the branches have more complex structure than in the metabolic net. Finally, Section 4 discusses the relations between the two nets, and points toward elaboration and testing of the theory.

1.1 GENERATIVE AND DESIGN APPROACHES TO A THEORY OF ORGANIZATION

There are two complementary approaches to understanding a complex structure. The generative approach derives structures from basic principles and building blocks. Starting from an initial population of organisms, it allows specified modes of mutation to generate other kinds of organisms. All organisms compete for available resources and proliferate differentially according to their success in the competition. A generative simulation can produce novel structures not anticipated *a priori*.[3,21] However, if this approach simply evolves structures blindly, it need not reveal why some structures persist, rather than others. By contrast, an approach through design allows one to test hypotheses about why structures persist.

The design approach asks what problems an organism is solving, and what structures can best solve those problems. The problems imply criteria of performance. The preceding criteria—reliability, flexibility, economy, and speed—are a useful general set, though this list may not be comprehensive. Some conceivable structures are disallowed because they do not meet constraints that are imposed by natural laws (for example, a limit on solubility), by the present organization of the system (as is the capacity for error correction), or by the environment. The remaining structures constitute the ensemble of possibilities; they differ in adjustable variables on which their performance depends. An evaluation procedure, which typically involves an objective function, is used to find structures that perform well.

1.1.1 JUSTIFICATIONS FOR USING A DESIGN APPROACH TO THE ORGANIZATION OF BACTERIA. There are at least two justifications for using a design approach to understand the organization of molecular pathways in bacteria. First, any systems that undergo selection for a long time in competition with alternatives will often show favorable features of design because they meet performance criteria better than the alternatives. This argument has limitations, as discussed in the following section. However, bacteria have survived and reproduced for at least a billion years, so it is plausible that they will show good design. As Riedl[31] has emphasized, in a multilevel organism high levels of organization may have evolved before lower levels achieved optimality. The burden of compatibility between the higher and lower levels might arrest the optimization of the lower levels. This problem is least severe in bacteria, which have the fewest levels of any free-living organism.

Second, plausible arguments suggest that since bacteria perform metabolism and self-reproduction, they meet performance criteria. Processes must be sufficiently reliable if bacteria are to survive. Bacteria, such as *E. coli*, that live in the gut of a host animal must be flexible, that is, be able to proliferate using diverse substrates with variable abundances. Because coding capacity, solubility, and available substrates are limited, a bacterium must operate economically; it should use few enzymes at low concentrations, to make products without side reactions. A type of bacterium that proliferates faster than its competitors will tend to predominate in a population.

1.1.2 OBJECTIONS TO USING A DESIGN APPROACH TO THE ORGANIZATION OF BACTERIA. The design approach raises several questions. A biologist might ask whether this approach is applicable to organisms, which have, in truth, evolved. The answer is yes, because what is called design is merely a special kind of natural selection. Humans design machines; that is, humans generate variant types of machines and preferentially make the ones that perform well. In biological evolution, the variations are generated within organisms that proliferate differentially.

A biologist might also point out that many features of organisms are special tricks, peculiar to particular groups of organisms. This observation is compatible with a search for favorable designs because a general strategy of design is always realized with particular, historically limited, solutions. Strategies for designing computers, based on the theory of computation, were implemented first with vacuum tubes and then with transistors. Likewise, the idiosyncracies of organisms represent historically attainable realizations of principles of organization. The principles are evident in organisms, though they may be manifest in diverse and peculiar ways.

A design approach is not always successful in predicting what actually occurs. An approach that seeks the optimal design may not characterize the numerically dominant type because an evolving population can become trapped in a suboptimal state through accidents of history; only a generative argument can show this. Also, organisms may use special tricks that were not included in the theoretician's trial ensemble of candidate structures. This is especially likely to occur in coevolution, when the performance criteria are not fixed but change with the characteristics of the interacting organisms. For example, the type of bacterium that one expects to become most abundant in a population should be that which reproduces fastest. However, a slower growing type may excrete a metabolite that poisons faster growing types, and so may become most abundant. *E. coli* excretes acetate when growing on glucose; the acetate lowers the pH and so removes some important competitors from the gut.

Nevertheless, a design approach can give useful insights, especially if the performance criteria are fixed for an extended time, as is the case with the characteristics of the physical world. This approach has been useful for understanding many systems in biochemistry, physiology, and ecology.[22,28,33,36,39] We have, therefore, approached the organization of bacteria mainly as a problem of design.

1.2 CONSTRAINTS ACT SO AS TO LIMIT ADJUSTABLE VARIABLES

The resources available to a bacterium are limited. Thermodynamics imposes fundamental limits; mass and energy are conserved, and the free-energy changes in chemical reactions are limited. Near thermodynamic equilibrium, the enthalpy and entropy changes in processes constrain the solubility, configuration, and association of molecules, and the stoichiometry and equilibrium of chemical reactions.[30,42] The kinds and quantities of molecules that a cell uses are limited by their stability and solubility.[2]

Chemistry also limits the variety and size of macromolecules in a more subtle way: Errors in the synthesis of macromolecules limit the capacity of the genome to encode the synthesis of RNAs and proteins. Alternative reactions may occur because, from thermodynamics, the specificity of intermolecular associations is not absolute. Sometimes an atypical base pairing occurs in DNA replication. Because the average number of errors in making a linear polymer such as DNA is proportional to its length, its coding capacity is limited by the accuracy of replication, as Kirkwood, Rosenberger, and Galas[18] have shown. Similarly, transcription of shorter RNAs and translation of shorter proteins are more accurate. However, there are lower limits to size; RNAs and proteins must include sequences that maintain active conformations reliably, bind suitably with ligands, and direct appropriate synthesis and degradation.

Within these limits a bacterium can meet performance criteria through changes in adjustable variables—in the rate constants of individual reactions and in the connectivity of the net of these reactions. Rate constants can change when the amino acid sequence of the corresponding enzyme changes, or when the enzyme associates with other molecules in a multicomponent complex.

1.3 BASIC AND CONTROL REACTIONS

In sections 2 and 3 we offer arguments from design about the connectivity of the metabolic and macromolecular nets. We distinguish two kinds of reactions—basic reactions that generate the outputs of the nets (key metabolites or macromolecules), and control reactions that regulate the allocation of resources among the basic reactions. For example, a metabolic pathway that synthesizes a key metabolite consists of basic reactions, but is often regulated by feedback inhibition. For both nets, we first argue that the optimal connectivity of basic reactions is a hub-and-branch structure, and then infer the connectivity of control reactions. We offer formal proofs for the optimality of a hub-and-branch structure in a simplified basic metabolic net, but the rest of the arguments are heuristic.

2. THE ORGANIZATION OF METABOLISM

The metabolic net transforms the substrates that are available in the medium into the metabolites, cofactors, and energy supply for the biosynthesis of macromolecules. Here we propose that good design of the metabolic net to meet performance criteria produces the basic and regulatory connectivity that is observed.

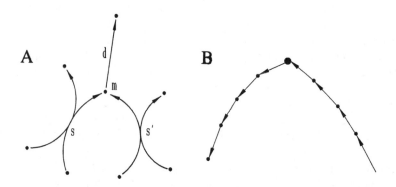

FIGURE 1   Directed graphs representing reaction networks. (a) A metabolite (m) can be synthesized in some reactions (s, s′) and consumed in others (d). (b) A key pathway, in which a key metabolite (large dot) is synthesized and degraded through linear sequences of unimolecular reactions involving intermediate metabolites (small dots).

Within a framework of simplifying assumptions, we define an ensemble of possible nets. Under specified performance criteria, a net with a hub-and-branch structure (that is, a hub net) is optimal. Relaxing some of the assumptions allows us to examine the structure of the hub and branching pathways. These basic pathways are regulated; a hierarchy of regulatory pathways extends from the control of flux in individual branches to the allocation of precursors among competing pathways. The biochemistry texts by Rawn[30] and Stryer[42] summarize the empirical observations used here.

2.1 ASSUMPTIONS

To specify an ensemble of possible metabolic nets, we shall use seven assumptions about basic connectivity and two more about performance. Here we list and discuss these assumptions.

2.1.1 STRUCTURAL ASSUMPTIONS ABOUT THE BASIC CONNECTIVITY OF THE METABOLIC NET.

1. *Unimolecularity:* The reactions of the metabolic net can be approximated as unimolecular.

   In a typical reaction two or more input (substrate) molecules are transformed into two or more output (product) molecules. Such reactions can be shown as in Figure 1(a); a dot represents each metabolite. For each reaction that synthesizes a metabolite, an arrow points toward the corresponding dot; for each reaction that consumes a metabolite, an arrow points away from the dot.

   The main action of a step is often, though not always, to transform a particular input to an output. The other inputs and outputs are often widely available auxiliary molecules such as water, energy suppliers, and oxidizing or reducing agents. We assume that these auxiliary molecules are widely available and relatively abundant, so that variations in their concentration can be ignored in inferring the connectivity of the metabolic net. Thus each reaction will be regarded as transforming one input to one output. Enzymes convert substrates to products in a sequence of reactions because of the magnitude of the free energy changes accompanying the making or breaking of interatomic bonds (Figure 1(b)).

   While the assumption of unimolecularity may be adequate for treating the connectivity of the basic net, this assumption is invalid for regulation: The relative abundance of oxidizing and reducing agents helps to regulate the relative rates of catabolism and biosynthesis.

2. *Digraph:* The metabolic net can be represented by a directed graph of reactions. We are assuming that every enzyme-catalyzed reaction (step) is unidirectional and is catalyzed by a unique enzyme. In fact, each step is reversible to some degree, but many steps are strongly biased in one direction. A metabolite is usually synthesized by different enzymes than those that degrade it.

3. *Connectedness*: The metabolic net can convert any metabolite to any other; it is a connected net.

   Because the substrates from the medium that are available to a bacterium vary with time, it will fare better if it can convert any metabolite to any other.

4. *Key metabolites*: It is possible to distinguish a set of K key metabolites from other metabolites.

   Key metabolites are used as monomers in the synthesis of macromolecules. In the figures key metabolites are represented as large dots. There are about 60 key metabolites—about 20 amino acids, 8 nucleotides, 25 sugars, and 8 fatty acids.[13] Note that ATP is treated as a key metabolite for the synthesis of RNA, but as an auxiliary molecule when it is used as an energy supplier for other reactions.

   Small molecules other than key metabolites occur as intermediates in the synthesis or degradation of key metabolites; some, such as succinate and oxaloacetate, may be available to a bacterium as substrates in the medium. Thus the set of key metabolites overlaps, but is not identical with, the set of substrates. Many bacteria can use diverse molecules as substrates.[11,41]

5. *Key pathways*: Each key metabolite is synthesized and degraded by a unique unbranched sequence of steps, called a key pathway.

   The number of enzymes will be reduced if each key metabolite is synthesized and degraded through a unique pathway. As Figure 1(b) shows, a key pathway starts at an intermediate metabolite, and synthesizes its key metabolite through an unbranched sequence of steps. The key metabolite is degraded through an unbranched sequence of steps, leading to an intermediate metabolite.

   In fact, two or more key pathways may share steps, forming a single branched pathway. Key pathways often share steps if they synthesize or degrade key metabolites that are related (in chemical type, carbon skeleton, and polarity). Branching pathways are discussed below. Furthermore, there may be more than one pathway to synthesize some key metabolites. Such redundancy would increase reliability.

6. *Junctions*: Key pathways are connected to each other at junctions.

   Junctions are represented in the figures as open circles. The metabolic net consists of K key pathways and the junctions between them. A junction connecting only two key pathways will be called a simple junction; a junction connecting three or more key pathways will be called a compound junction. The junctions between key pathways may be arranged in diverse ways, illustrated in Figure 2. In a rim net (Figure 2(a)), the key pathways occur in sequence around a loop, all in the same orientation. The end product of catabolism in one pathway is transformed through a simple junction into the starting metabolite for biosynthesis in the next pathway. In a hub net (Figure 2(b)) all key pathways extend from a hub, a compound junction that interconverts the starting metabolites for all biosyntheses and the end products of catabolism for all key pathways. Many other nets are possible, with many combinations of simple and compound junctions; Figures 2(c) and 2(d) show examples.

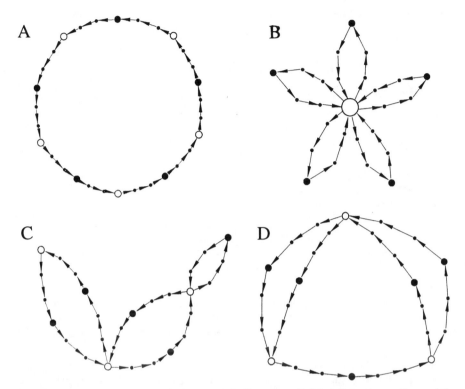

FIGURE 2 Examples of candidate metabolic nets satisfying the assumptions of the model. Each net has five key pathways. (a) A rim net. The key pathways occur in sequence around a loop, with the same polarity. A simple junction connects the starting metabolite of each pathway to the end metabolite of the adjacent pathway. (b) A hub net. All key pathways join a hub, a single compound junction. (c, d) Other candidate nets from the ensemble. (d) is an example of a graph that cannot be obtained by pulling junctions of a rim net together.

7. *Standardization*: All key pathways have equivalent kinetic behavior. Each key metabolite is synthesized in S steps and degraded in S steps. The average number of junctional steps that connect each key pathway to the others is J; J is much less than $2S$.

Two key pathways are kinetically equivalent if, when supplied with the same concentration of the starting metabolite, they generate and degrade their key metabolites at the same rate. The assumption that all key pathways are kinetically equivalent simplifies analysis, in that nets of the ensemble differ only in the connectivity among key pathways. Therefore, the optimization is independent of the actual differences among key pathways in the number of steps, in the amount of enzyme per step, and in the rate constants for a step. In a more realistic model, the optimization would depend on these variables.

2.1.2 ASSUMPTIONS ABOUT PERFORMANCE.

1. *Conversion*: Conversion from any key metabolite to any other is equally likely. This assumption allows us to prove the theorems in the Appendix, although it is unrealistic. Since glucose is the most likely substrate, the transition from glucose to ATP occurs more often than the reverse transition. Below we suggest that the key pathway for glucose has become part of the hub.
2. *Stability*: The connectivity of the net should tend to produce the same pattern of key metabolite concentrations, regardless of the substrate used.

   Presumably bacteria can synthesize macromolecules more reliably if the concentrations of key metabolites are stable against variations in the supply of metabolites, the demand for metabolites, and kinetic parameters of enzymes. Of course, media for the culture of bacteria differ in the degree to which they supply key metabolites, cofactors, and energy; correspondingly, the intracellular concentrations of key metabolites will be higher in a richer medium.[13,14]

   The metabolic flexibility of a bacterium is increased if the key metabolites are equivalent, in the sense that supplying any one of them will yield roughly the same pattern of concentrations of all. That is, the variance of the pattern over the set of possible substrates is a minimum. Because all the key pathways are kinetically equivalent, variations in the pattern with substrate depend only on the connectivity of the net, which determines path lengths. The length of the path from one key metabolite to another in a net is the minimal number of steps between them. Therefore, the variance of the pattern of concentrations with substrates is minimized if the variance of the path lengths is minimized. As discussed below and in the Appendix, we shall show that a hub net minimizes the mean and variance of the path lengths.

2.1.3 SUMMARY OF ASSUMPTIONS.

The first seven assumptions characterize the set of directed graphs admitted to the ensemble of nets. With the standardization assumption, we oversimplify the biological problem to obtain a formally tractable problem. The last two assumptions specify the performance criteria: A favorable candidate net is flexible, in that it can use any key metabolite as a substrate, and reliable, in that any substrate produces a roughly equivalent pattern of concentrations of key metabolites. These assumptions suggest that favorable nets will be economical, in that the mean and variance of the minimum path length between any two key metabolites will tend to be minimal. If so, a favorable candidate net will mediate metabolism with speed, converting any key metabolite to any other along a short path.

2.2 THE CONNECTIVITY OF THE METABOLIC NET: INFERENCES FROM THE ASSUMPTIONS

Next we shall see what these assumptions imply about the optimal connectivity of the metabolic net. Note that our aim is only to see which of the alternative possible connectivities best meets the performance criteria. We are not inferring the number of steps per key pathway, the rate constants of steps, or the relative sizes of pools of key metabolites.

2.2.1 A HUB NET MEETS THE PERFORMANCE CRITERIA OPTIMALLY. In the Appendix, in Proposition 1, we show that a hub net minimizes the mean and the variance of the path lengths between key metabolites. This conclusion can be understood heuristically as follows:

The rim net provides equivalence poorly because the key pathways are linked in series. As Figure 3(a) shows, a rim net tends to produce metabolites in lesser concentration when these are more steps away from the substrate (because intervening metabolites are consumed in other pathways). Also, the key metabolites do not respond at the same rate to a change in the availability of a substrate; key metabolites that are more steps away from the substrate will respond with a greater delay. Thus, in a rim net the pattern of metabolite concentrations is relatively sensitive to the identity of the substrate.

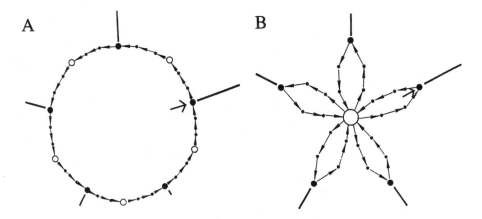

FIGURE 3 Concentrations of key metabolites in a rim net (a) and a hub net (b) are designated by the lengths of bars. The open arrow designates the substrate externally supplied. Each key metabolite is consumed in synthesizing macromolecules. In the rim net this consumption reduces the concentration of key metabolites that are more steps away from the substrate, so that the pattern of metabolite concentrations depends strongly on the identity of the substrate. In a hub net the concentration of the substrate is highest, but all other key metabolites are equidistant from the hub.

The variability in distance between two key metabolites can be reduced greatly by converting the rim net to a hub net. One can do this by pulling all of the junctions together at the center of the loop and uniting them in a compound junction. In a hub net each key metabolite is the same distance from every other key metabolite. All key metabolites have similar concentrations, since they are equidistant from the hub, although the substrate has the highest concentration (Figure 3(b)). The exact pattern of concentrations depends on the pattern of supply and demand for the metabolites.

In Proposition 2 and the subsequent discussion in the Appendix, we show that of all the candidate networks in the ensemble, a hub net is least sensitive to mutation. In any net, deletion of a step in a key pathway prevents either biosynthesis or catabolism of its key metabolite. In a hub net, the mutation does not affect transitions among the other key metabolites. However, in other networks this need not be so.

In the remainder of Section 2 we shall confine our attention to properties of the hub net—to the structure of the hub, branching pathways, and regulation.

2.2.2 THE STRUCTURE OF THE HUB. The hub is a net of reactions, in which neither the connectivity nor the reactants have been specified. Although it is unclear at present how to specify the connectivity and reactants in the hub, certain considerations are evident. Biosyntheses should start from several hub metabolites, which should be readily interconvertible. If all key pathways started from a single intermediate metabolite, competition among them would drain the pool of this metabolite, slowing the generation of key metabolites. If each key pathway started from a different intermediate metabolite, more coding capacity would be required for the corresponding enzymes than is essential, since key pathways can be amalgamated into branching pathways, as discussed below.

The composition and connectivity of the hub will probably depend on the relative availability of various substrates. If all key metabolites are equally likely to occur in the environment, all key pathways should have an equivalent relation to the hub, as diagrammed in Figure 2(b). However, if one substrate is most abundant, the paths from it to the other key metabolites will be shorter if the other key pathways connect closer to the abundant substrate. That is, the abundant substrate, and reactions leading to and from it, should be part of the hub (Figure 4).

In bacteria the pathways that synthesize and degrade glucose—the reactions of glycolysis, the pentose phosphate pathway, and the citric acid cycle—form the hub; key pathways for the synthesis and degradation of amino acids, nucleotides, and lipids are external to the hub (Figure 5). The presence of the glucose-metabolizing reactions in the hub suggests that glucose is the optimal substrate for *E. coli*; diverse lines of evidence support this inference.[23] Note that even though glucose has been incorporated into the hub, it is still a key metabolite, in the sense that it is a monomer for the synthesis of polysaccharides.

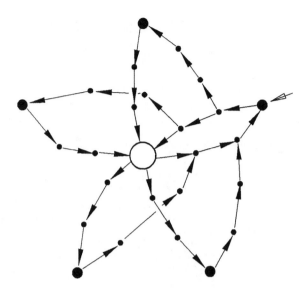

FIGURE 4   Connectivity in the hub if one key metabolite (open arrow) is the most common substrate, and its key pathway becomes part of the hub. The distance from this primary substrate to other key metabolites is particularly short, even though its key pathway is a hub loop.

The hub is a microcosm of the entire metabolic net; it should be a connected net that can rapidly convert any hub metabolite to any other. A loop of reactions is suboptimal because in the hub, as in a rim net, a loop increases the variance in the minimum path length between metabolites. Small loops have persisted in the hub—the citric acid cycle and the pentose phosphate pathway—perhaps because of constraints or historical circumstances neglected in our optimality argument. For example, the degradation of glucose to carbon dioxide and water is constrained to occur through a sequence of reactions that extracts a large fraction of the energy and reducing power from glucose. The citric acid cycle provides a pathway with greater energetic efficiency for oxidizing acetate to carbon dioxide than does an alternative linear pathway.[42]

2.2.3 BRANCHING PATHWAYS. So far we have assumed that each key pathway uses a distinct set of reactions; that is, steps are not shared between key pathways. However, pathways that synthesize or degrade related key metabolites might share steps in branching pathways. A treelike, branching, biosynthetic pathway may extend from one metabolite in the hub to several key metabolites; similarly, a catabolic tree may degrade several key metabolites to the same metabolite (Figure 6(a)).

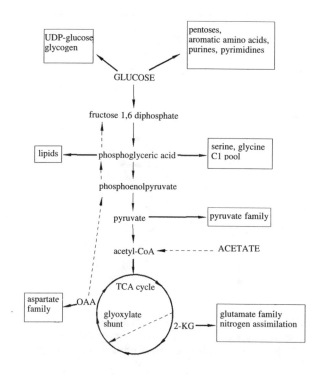

FIGURE 5 The metabolic net in *E. coli* resembles a hub net. The hub includes glycolysis and the tricarboxylic acid (TCA) cycle. Thin solid arrows indicate reactions specifically used when glucose is the carbon source. Dashed arrows indicate reactions specifically used when acetate is the carbon source. Heavy arrows designate reactions common to both carbon sources; these include the treelike pathways that synthesize key metabolites, indicated by arrows pointing toward the periphery.

Sharing of steps between pathways improves the performance of the net by reducing the number of different enzymes needed. To support an adequate flux of metabolites, the concentrations of the enzymes shared between pathways should increase. Branching does not change the lengths of the pathways, so the hub net remains the most favorable organization for metabolism.

Now, consider the factors influencing the number and geometry of trees and their connection to the hub. To minimize the number of steps in a tree, the branch points to distinct key metabolites should occur as far as possible from the hub. Many small trees, rather than few large trees, should be used because the smaller a tree, the less time that is required to traverse it. Hence a network with smaller trees will respond the more rapidly to changes in the concentrations of metabolites.

A small tree, bearing a cluster of biochemically similar key metabolites, connects to the hub at a metabolite that can be regarded as a minihub. Thus, the transition from one small tree to another occurs via a transition between the corresponding minihubs, rather than directly (Figure 5). An analogy to air routes is useful: Routes extend from a geographic cluster of small towns (class of key metabolites) to a major city (minihub). To travel between small towns near different major cities, one must travel from town to city, between cities, and from city to town.

These expectations are fulfilled in the metabolic net. Each biosynthetic tree generates a family of biochemically similar end products. For example, the twenty amino acids required for protein synthesis are synthesized in six biosynthetic families[30] (Figure 5). Members of a family share a structural index of similarity; for example, most amino acids of the aspartate family (aspartate, asparagine, lysine, methionine, threonine, isoleucine) have two carbons in the side chain; most members of the glutamate family (glutamate, glutamine, proline, arginine) have three.

2.3 THE CONNECTIVITY OF METABOLIC REGULATION

Although the basic connectivity of the metabolic net tends to maintain stable pools of key metabolites, regulation can improve its performance. To see this, consider how the unregulated metabolic net would perform. Without regulation, the macromolecular net would synthesize metabolic enzymes at a constant rate. The metabolic net would process metabolites at a rate dependent on the levels of enzymes, their rate constants, the availability of substrates, and the consumption of key metabolites by the macromolecular net. As these variables changed, the pools of key metabolites would change, even in a hub net. Regulation of both nets can increase the stability of the pools.

Here we consider the regulation of enzyme activity by metabolites within the metabolic net; regulation of the synthesis of enzymes is discussed in Section 3. As we shall discuss, there is a hierarchy of regulatory mechanisms; mechanisms operate on one branch of a tree, on entire trees, and on the allocation of metabolites to competing pathways.

2.3.1 REGULATION OF INDIVIDUAL PATHWAYS. Feedforward and feedback are the classic ways to stabilize the output of a process against variations in the input and in the rate constants. In a biochemical net, these controls occur mainly through allosteric interactions, in which a regulator protein acts as an adaptor that can bind both to a metabolite and to an enzyme that would not itself bind the metabolite. Savageau[33] has analyzed the feedback regulation of a linear sequence of steps. If only one feedback path is allowed, then the input-output relation of the sequence is maximally stabilized against variations of rate constants, if the end product inhibits the first enzyme in the sequence. Use of a single feedback path through one regulator is sufficient, and economical in terms of coding capacity. Thus feedback promotes reliable and economical performance.

Feedforward inhibition can promote speed: It enables a pathway that synthesizes an amino acid to respond more quickly to changes in the substrate concentration.[37] If the product of the first enzyme inhibits the consumption of the amino acid by the cognate aminoacyl-tRNA synthetase, an increase in the concentration of substrate for the pathway quickly inhibits the synthetase, increasing the concentration of the amino acid and so, by feedback inhibition, reducing the activity of the first enzyme.

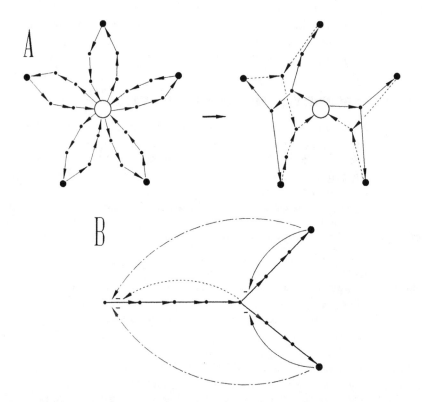

FIGURE 6 Branching in the metabolic net. (a) Branching reduces the number of steps in related pathways. Solid arrows show branching trees of biosynthetic pathways; dashed arrows show trees of catabolic pathways. (b) Regulation of a branching pathway can occur in diverse ways. The demand for metabolites at the tips of different branches is dissociable; correspondingly, feedbacks in distal branches are dissociable. The constraint of maintaining a reasonable level of metabolite at the tip of the mother branch can be met with various feedback pathways. (Short-dashed + thin solid lines are sequential control pathways; long- and-short-dashed + thin solid lines are a hierarchy of control pathways. Modified from Savageau,[33] p. 201).

2.3.2 REGULATING THE ALLOCATION OF METABOLITES TO COMPETING PATHWAYS. Regulation of a branching biosynthetic pathway poses special problems of resource allocation, because the substrate of the pathway must be converted to the several products in a flexible way that depends on the demand for the products. As Savageau[33] discussed, each distal branch has its own negative feedback. Either a sequence or a hierarchy of feedback paths can regulate the flux through the most proximal branch (Figure 6(b)).

Regulation is also needed to adjust the allocation of the medium's carbon sources among competing pathways for the synthesis of key metabolites. To maximize the growth rate, biosynthetic pathways should not be inhibited when the relative concentrations of key metabolites are favorable for the synthesis of macromolecules. Depletion of a key metabolite will slow the synthesis of macromolecules, increasing the pools of other key metabolites, which feed back to inhibit their pathways. Thus more of the carbon source supplying all the pathways will be available for synthesis of the depleted metabolite.

2.4 THE ORGANIZATION OF METABOLISM: DISCUSSION AND CONCLUSIONS

The preceding arguments suggest that the metabolic net has a hub-and-branch structure, the performance of which is improved by regulation. Clearly, regulation would improve the performance of any net in the ensemble. It is an open question whether regulation could improve a non-hub net so much that it would perform better than an optimally regulated hub net.

The hub-and-branch structure is a hierarchy of modules that meets a hierarchy of constraints. A dynamic module is a cluster of reactions with few inputs that generates an output. As a whole, the net is a high-level module, with substrates from the medium as inputs and the key metabolites as outputs; its task is to maintain stable pools of key metabolites. A biosynthetic or catabolic branch of the metabolic net is a lower-level module; feedback stabilizes its performance against variations in the operation of other modules. In a branching biosynthetic pathway, the demand for key metabolites at the tips of different branches is dissociable; correspondingly, the feedbacks in distal branches are dissociable. These features illustrate the matching of modular organization to the correlation among demands that is common in biological nets.[7,26] Our argument is that such matching evolved in response to limitations of coding capacity and of solubility of macromolecules. This argument implements the proposal that a resource limitation can promote the evolution of a hierarchy of modules.[5,26]

3. THE ORGANIZATION OF THE MACROMOLECULAR NET

The macromolecular net uses the key metabolites to synthesize macromolecules, which perform the activities that enable bacteria to survive. Among these activities are processes of the cell cycle, including growth, DNA replication, and cell division; mechanical stabilization of the cell surface and transport of molecules across it; locomotion; and the sensing of, and response to, the physicochemical environment. The macromolecular net must perform all these activities, in appropriate situations, using the limited resources (key metabolites, coding capacity, solubility) available to it. Thus the macromolecular net, like the metabolic net, must act as a distributor for limited resources.

The rates of individual reactions ultimately limit the rates of all the above activities. However, the connectivity of these reactions in the macromolecular net greatly affects its performance. We first characterize the ensemble of possible structures for the macromolecular net. In these candidate nets, macromolecules may associate in multicomponent complexes; we argue that such complexes can improve the performance of the net. From the availability of complexes for the synthesis of macromolecules, we conjecture (but do not prove) that, as for the metabolic net, the optimal macromolecular net has a hub-and-branch structure. The differential activity of the branches depends on regulatory processes, which allocate the limited types and quantities of macromolecules among competing activities. Sources for the following discussion include Alberts et al., Rawn, and Stryer.[1,30,42]

3.1 THE BASIC CONNECTIVITY OF THE MACROMOLECULAR NET

Here we look at the basic structure of the macromolecular net, without concern for the way that regulation matches its performance to the demands that are placed on it.

3.1.1 THE ENSEMBLE OF POSSIBLE STRUCTURES FOR THE MACROMOLECULAR NET.
There are two essential prerequisites for a self-reproducing biochemical organism. (1) Its metabolism must convert substrates from the medium to monomers. (2) An autocatalytic set of polymers must use the monomers to generate additional copies of all polymers in the set. Each polymer is synthesized by a sequence of reactions; each of these reactions is catalyzed by a polymer, or a complex of polymers, in the set. Collectively the polymers not only produce themselves, but perform the activities essential for survival of the autocatalytic set. The ensemble of possible structures for the macromolecular net contains all possible autocatalytic sets; these differ in the particular polymers that they contain. Kauffman[16] proposed that self-replicating macromolecular nets evolved from an autocatalytic set of polymers; see also Bagley, Farmer, and Fontana.[3]

For the macromolecular net, as for the metabolic net, it is important that the same limitations on solubility and coding capacity apply to all members of the ensemble. This is the case for the ensemble of autocatalytic sets of polymers. Obviously the solubility of polymers is limited for an arbitrary autocatalytic set. Also, the rate of errors in copying the set will limit the size of the set—the number and length of polymers—for an arbitrary autocatalytic set as for the actual macromolecular net. If the error rate per monomer added is constant, the expected number of errors in the set increases with the total length of polymers that must be copied. This observation generalizes the notion of a limited coding capacity to arbitrary autocatalytic sets.

3.1.2 COMPLEXES OF MACROMOLECULES FORM, AND THESE CAN IMPROVE THE PERFORMANCE OF THE NET.

In any autocatalytic net some polymers can aggregate, forming multicomponent complexes. The formation of a complex from dissociated polymers can help, hinder, or not affect the performance of the net. However, its performance will often be improved if there is clustering between macromolecules that cooperate to perform a process. In the metabolic net, the enzymes that perform a sequence of metabolic reactions may form a complex. As Ovadi[29] pointed out, multienzyme complexes in metabolic pathways reduce the loss of intermediates by diffusion or along extraneous reactions, and allow more rapid transfer of intermediates between enzymes.

Some complexes can synthesize or modify many kinds of macromolecules in a class. These holoenzymes—DNA replicases, RNA polymerases, ribosomes, and spliceosomes—use information in a polynucleotide to specify the sequence of monomers in another linear polymer. Holoenzymes substantially improve the performance of an autocatalytic net, as we shall now see.

An autocatalytic net that uses holoenzymes to synthesize polymers is particularly reliable because the copying of all polymers in the net depends directly only on the performance of the holoenzymes. The performance of other polymers (for example, of enzymes and regulatory proteins in the metabolic net) is indirectly important for copying the net, in that they maintain appropriate levels of key metabolites. By contrast, in an autocatalytic net lacking holoenzymes a large fraction of all polymers might have to perform correctly to copy the net.

A holoenzyme provides flexibility because it can generate any polymer in a class, given an appropriate polynucleotide as input. An RNA polymerase can catalyze the formation of any RNA, given an appropriate DNA sequence. Furthermore, a change in one component of a holoenzyme can make it synthesize particular polymers preferentially, by binding preferentially to the corresponding polynucleotides. An example is the sigma factors of RNA polymerase, which themselves admit specializing modifiers.[19]

Holoenzymes provide economy in at least two ways. Because one holoenzyme can perform diverse syntheses, an autocatalytic net needs fewer species of polymers to accomplish those syntheses. Consequently, to copy itself, the net uses less of its limited total length of polymer. Note also that an investment of polymers in increasing the reliability of a holoenzyme, enhances the reliability of all the syntheses it can perform.

Holoenzymes speed syntheses by reducing the dimensionality of the search space in which to find the site for the next reaction step, from three dimensions to the one dimension along the polynucleotide chain. Furthermore, product polymers are generated faster when several holoenzymes operate in sequence on a polynucleotide. Such pipelining occurs in polysomes, where several ribosomes translate a messenger RNA simultaneously. Pipelining also occurs during rapid growth of *E. coli*, when several DNA replicases copy the circular bacterial chromosome in sequence during one cell cycle.[8]

The availability of separate holoenzymes for making distinct classes of polymers also speeds syntheses by allowing more parallel processing. In an arbitrary autocatalytic set, there may be unavoidable waiting periods while precursors of a given step accumulate. However, the existence of a DNA replicase distinct from a ribosome allows replication of DNA and translation of RNA to proceed in parallel, with negligible interaction. Thus class-specific holoenzymes partition the syntheses into minimally interacting subsets.

The operons of a bacterial chromosome are complexes that increase the speed of DNA handling. In an operon, binding of an RNA polymerase to a single promoter site on DNA can regulate coordinately the synthesis of RNA for several enzymes that cooperate in an activity. The assembly of the resulting proteins into a functional complex may be accelerated through a topographic mapping, in which the sequence of genes in the operon matches the spatial order of proteins that comprise the complex.

In summary, it is reasonable to expect that autocatalytic sets with outstanding performance will use holoenzymes. We now propose that a set of this kind with a particular form should occur in bacteria.

3.1.3 CONJECTURE: THE OPTIMAL MACROMOLECULAR NET HAS A HUB-AND-BRANCH STRUCTURE.

Now we offer a heuristic argument to motivate the conjecture that the optimal macromolecular net has a hub-and-branch structure. Each chemically distinct class of polymers is synthesized by one kind of holoenzyme. Each holoenzyme seems analogous to a junction in the metabolic net. These holoenzymes supply each other with their products: DNA replicase supplies DNA to RNA polymerase, which supplies mRNAs, rRNAs, and tRNAs to ribosomes; ribosomes generate proteins for the holoenzymes and for all other activities of the cell. Therefore, the set of holoenzymes, together with their products (DNA, RNAs, and proteins), that allow the duplication of the holoenzymes, constitute a hub in the macromolecular net (Figure 7). The hub is an autocatalytic subset of the macromolecular net, because it can synthesize all the macromolecules of the hub.

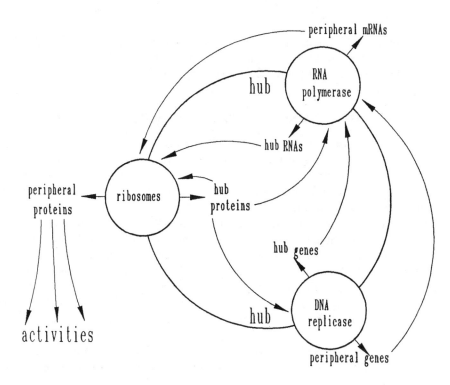

FIGURE 7   The hub-and-branch structure of the macromolecular net.

Peripheral branches, which extend from the hub, synthesize proteins used in other activities, including enzymes for the metabolic net. The peripheral branches copy the genes which code for peripheral proteins, transcribe these genes to make the corresponding mRNAs, and translate these mRNAs to make the peripheral proteins. Evidently this concept of the macromolecular net as a hub-and-branch structure partitions the genes, RNAs, and proteins into two disjoint sets of macromolecules, those in the hub and those peripheral to it. For each of these three classes of macromolecules, there are branches within the hub and branches peripheral to the hub. The operation of these branches depends on holoenzymes, which can copy either hub or peripheral polynucleotides.

There are corresponding elements in the metabolic and macromolecular nets. Key metabolites are analogous to the macromolecules at the tips of branches. For each net, branches that mediate the operation of the hub, such as glycolysis or the synthesis of macromolecular hub components, have been incorporated into the hub. In the macromolecular net, holoenzymes play roles analogous to biosynthetic enzymes in the metabolic net. However, the catabolism of macromolecules involves branching pathways analogous to those in the catabolism of key metabolites. The macromolecular net also includes biosynthetic branching pathways, for DNA repair and for synthesis of polysaccharides and glycoproteins, which are analogous to biosynthetic branching pathways of the metabolic net.

3.2 REGULATING THE PERFORMANCE OF ACTIVITIES: INTRODUCTION

A bacterium needs diverse combinations of macromolecules to perform the activities needed for survival. The limits on coding capacity and on the solubility of molecules restrict the kinds and amounts of coexisting macromolecules. Therefore, regulation must adjust the synthesis and degradation of macromolecules to match the capabilities of a bacterium to its situation.

In this section we consider the regulation of gene expression at increasing levels of complexity. The regulation of operons for the metabolic net shows how relatively simple activities can be controlled. Two further examples, the regulation of catabolism and the synthesis of ribosomes, illustrate more complex problems of control. To see regulation at a higher level, we examine how the allocation of protein-synthesizing machinery changes with the richness of the medium. Finally, we consider how the allocation of resources is regulated for a bacterium as a whole.

3.2.1 REGULATING THE SYNTHESIS OF ENZYMES FOR THE METABOLIC NET. Because the solubility of macromolecules and the availability of key metabolites are limited, a bacterium must suppress the synthesis of all enzymes not currently needed. Maximizing the concentrations of active enzymes relative to inactive ones makes metabolism operate faster, enables a bacterium to use diverse substrates efficiently, and prevents wasteful syntheses. Therefore, a metabolic pathway is typically regulated, not only by feedback inhibition of enzymes within the pathway, but also by regulating transcription of the corresponding mRNAs. As in the metabolic net, regulation typically occurs through regulator proteins. (Regulation by attenuation, discussed by Rawn,[30] is an exception.) Regulator proteins modify the activity of macromolecules in ways that depend on the binding of small molecules to the regulators.

Since coding capacity is limited, a single regulator often regulates the levels of proteins that are active coordinately. In bacteria a single regulator, called a repressor, activator, antiterminator, or proterminator, can regulate an entire operon.[36] Sigma factors regulate clusters of enzymes that cooperate to perform some activities, including nitrogen metabolism, response to heat shock, and replication of bacteriophages.[30]

Savageau[36] has argued that the frequency of demand for an activity should determine the relation between a regulator and an operon that encodes enzymes for the activity. In a given environment the demand is high for some metabolic pathways and low for others. For example, if a substrate is often present in high concentration, there is a high demand for an inducible pathway that degrades it. A cell will often need an inducible biosynthetic pathway for which the end metabolite is used rapidly and, hence, is seldom present in high concentration. If the end metabolite of a biosynthetic pathway is often present in high concentration, the pathway should be repressible, and the demand for enzymes in it will be low.

Now let us look at the factors that influence the relation between demand and the mode of regulation of an operon. The rate of transcription from an operon depends on several factors. The operon has a promoter site that binds RNA polymerase, and an adjacent modulator site that binds the regulator. These interactions, and binding of the regulator with the polymerase and with a metabolite (inducer or repressor), determine the rate of transcription. Transcription is initiated more often from a strong promoter, to which the polymerase binds strongly, than from a weak promoter. If binding of the regulator to the modulator site increases the frequency of transcription, regulation is positive. If binding of a metabolite to the regulator increases its affinity for the modulator site, the operon is inducible. It is important to distinguish positive vs. negative regulation (determined by regulator-polymerase interaction) from inducible vs. repressible regulation (determined by the effect of a metabolite on regulator-modulator interaction). The demand for expression of the operon depends on the availability of the metabolite. Savageau[33,34] has analyzed performance criteria that favor classical vs. autogenous control of the availability of the regulator.

According to the demand theory,[36] a pathway should have positive regulation in a high-demand environment, but negative regulation in a low-demand environment. These conclusions, which observations support, follow from a generative argument: If selection is to maintain a control mechanism reliably, the unregulated system should not perform well; performance should only be good with regulation. For a pathway in a high-demand environment, selection maintains positive regulation: If the promoter is weak, then the regulator must be used to activate the gene. If the promoter were strong, the regulator would be irrelevant; the gene would continue to be expressed even if mutations degraded the binding of the regulator to the modulator site. A similar argument favors negative regulation in a low-demand environment.

3.2.2 REGULATING THE SYNTHESIS OF RIBOSOMES.

A multicomponent complex is more likely to assemble correctly if the relative concentrations of its component macromolecules are appropriate. Economy also favors regulating the stoichiometry of components, since excess production of some components would be wasteful. That is, for the reliable and economical assembly of a complex, as for the synthesis of a polymer, the pool sizes of the components should be stabilized against variations in the rates of transcription and translation, and in the lifetimes of molecules.

For a complex with few components, the relative sizes of pools can be regulated by synthesizing the components from a single operon. However, this strategy may not work for a large complex, for at least two reasons. (1) The stoichiometry of components in the complex may not match the relative rates with which they would be produced from one operon. This difficulty would prevent the synthesis of a ribosome by a single operon. A ribosome contains 52 proteins ($r$proteins) and three ribosomal RNAs (rRNAs). Each rRNA is used directly in a ribosome, whereas the mRNAs for rproteins are translated several times into rproteins. Thus the rates of synthesis of rRNAs and rproteins from a single operon would not match their stoichiometry in a ribosome. (2) Replication and transcription may interact adversely, reducing the rate or reliability of both processes. Such interaction is more likely in a larger operon, and in faster-growing bacteria, because the number of replication sites per chromosome increases with the growth rate.

Synthesis of rproteins and rRNAs from several operons circumvents problem (2), but does not show how the operons should be regulated so as to preserve a favorable stoichiometry. Negative feedback from the end product of a pathway to an early reaction can stabilize the concentration of the product, as discussed for feedback inhibition in metabolic pathways. Analogous feedback can stabilize the pool sizes of components of a complex encoded by several operons, if one product of each operon feeds back to inhibit an early stage of synthesis from that operon. Whichever operon generates product most slowly will limit the rate of assembly of complexes, because free components produced by the other, faster-running operons will accumulate and feed back to reduce synthesis by their operons. Thus feedback inhibition within each operon can preserve the stoichiometry of all components.

The synthesis of ribosomes in *E. coli* supports these expectations.[27] The rprotein genes are in several operons. One rprotein from each operon feeds back to repress translation of its mRNA. (The rprotein binds to a site on the mRNA with a secondary structure similar to its binding site on rRNA.) Each rprotein is synthesized from a single gene, but the chromosome contains seven rRNA operons, each having a copy of all three rRNA genes. This difference in gene dosage partially compensates for the different rates of syntheses of rRNA and rprotein. Apparently rRNA feeds back indirectly to inhibit transcription of its operon. According to one hypothesis,[14] abundant rRNA allows assembly of ribosomes that try to translate mRNA at rates higher than the pools of tRNA and amino acids can support. Depletion of these pools slows translation, and indirectly slows transcription, so that RNA polymerase molecules are sequestered on the DNA. The resulting reduction of the RNA polymerase pool preferentially inhibits the transcription of rRNA and other stable RNAs, as we now discuss.

3.2.3 REGULATING THE ALLOCATION OF PROTEIN-SYNTHESIZING MACHINERY AMONG ACTIVITIES. Three classes of pathways of synthesis compete for RNA polymerases and ribosomes. (1) These holoenzymes must be synthesized. (2) The enzymes that catalyze reactions in the metabolic net and the regulators for these enzymes must be synthesized. (3) The enzymes and structural proteins that are used in growth (that is, duplication of the cell's contents and cell division) must be synthesized. These three classes of pathways compete for the monomers, energy, and reducing power in the pools of key metabolites, because the substrates available to a bacterium limit the rates at which the pools can be replenished.

Consider first the allocation of holoenzymes within (2), among the pathways of the metabolic net. When several substrates are available, a bacterium should use the substrates that yield the most carbon skeletons, energy, and reducing power per substrate molecule. In *E. coli* such differential usage of substrates occurs, at least in part, through a competition among catabolic operons: The operons that degrade richer substrates bind RNA polymerase more strongly. If glucose is present, poorer substrates are not used, because RNA polymerase binds poorly to the promoters of the operons for enzymes that degrade these substrates. If a bacterium is starving, then a regulatory mechanism allows it to use some poorer substrates: Cyclic adenosine monophosphate increases in concentration and binds to a regulator, the catabolite activator protein, which enhances the binding of RNA polymerase to the operons for poorer substrates.

Now consider the allocation of holoenzymes among (1), (2), and (3)—that is, to making holoenzymes, metabolic enzymes, and proteins for growth. To maximize the rate of growth, this allocation should depend on the richness of the substrate: A poorer substrate replenishes the pools of key metabolites more slowly. So, with a poorer substrate fewer holoenzymes should operate, draining the pools more slowly and leaving more key metabolites available to synthesize proteins for metabolism and growth.

Jensen and Pedersen[14] have proposed that this regulation occurs through competition among promoters for a limited amount of free RNA polymerase. According to their model, the promoters for making RNA polymerase and ribosomes are weak—they become saturated only at a high concentration of free RNA polymerase. Thus transcription from these promoters decreases considerably when little RNA polymerase is free. By contrast, promoters for the mRNAs for metabolic enzymes and growth-related proteins are presumably strong; they saturate with little free RNA polymerase. When a shift from a rich to a poor substrate occurs, synthesis of these mRNAs drains the pools of key metabolites and of free RNA polymerase. The reduced availability of key metabolites slows transcription and translation in general, as bound RNA polymerase and ribosomes stall for lack of activated monomers. The reduced availability of free RNA polymerase preferentially deprives the promoters for RNA polymerase and ribosomes, slowing their synthesis and so reducing their concentrations. Consequently, the pools of key metabolites increase, so that transcription and translation accelerate. With suitable rate constants, this mechanism might maximize the growth rate for diverse substrates.

3.2.4 ALLOCATING MACROMOLECULES TO ACTIVITIES IN A BACTERIUM AS A WHOLE.

To conclude this discussion of regulation of the macromolecular net, we ask how a bacterium should allocate species of macromolecules to activities in general. We have assumed that a small number of macromolecules is likely to perform each activity, given the limited coding capacity of the genome. This assumption seems valid for the activity and regulation of pathways in the metabolic net. In general, activities will also be performed more reliably, the smaller the set of macromolecules that is required, because the performance is less susceptible to damage by mutations. If q is the probability that mutation does not reduce the activity of a macromolecule, then the probability that mutation does not reduce the activity of M cooperating macromolecules is q^M, which decreases with increasing M.

Contrary to the idea that reliability decreases as more units are used, multicomponent complexes often contain many kinds of macromolecules. Several factors may favor this anomaly. Holoenzymes can synthesize many species of macromolecules, so the average number of components in the holoenzyme used per synthesis of a species is small. Replication and translation are particularly intricate processes, requiring the close coordination of many parts and the occurrence of many processes in close contiguity. DNA replicases and ribosomes operate as elaborate mobile scaffolds, in which supporting structures move the substrates reliably through a sequence of configurations and reactions. Reliability is enhanced by the activity of error-correcting components in the complex; these tend to prevent the addition of inappropriate monomers, by kinetic proofreading. Thus, contrary to the preceding argument, reliability can actually be increased by using additional macromolecules.

The performance of a bacterium depends not only on the allocation of genes to activities, but also on their positions on the chromosome. In *E. coli* the circular chromosome replicates in both directions from the origin of replication to its terminus. A gene nearer to the origin replicates earlier than a gene further away, so transcription can proceed from more copies of an earlier replicating gene, for more of the cell cycle. This difference in copy number is greater, the faster bacteria grow, because pipelining of replication occurs. The gradient of copy number can speed growth, if gene products needed in greater abundance are encoded by genes nearer to the origin of replication.

At least two biosynthetic systems show this correlation. The more that an amino acid is needed during the cell cycle, the closer to the origin is the operon encoding the enzymes for its synthesis.[40] Four of the seven rRNA genes are closely adjacent to the origin, and the other three are located within the half of the chromosome that replicates earlier.[27] This distribution of rRNA genes enhances the ratio of rRNA to mRNA for ribosomal proteins, and so accelerates the synthesis of ribosomes, especially at high growth rates.

I. DISCUSSION AND CONCLUSIONS

Our goal in this inquiry was to understand the operation of a bacterium as an integrated unit. We have proposed that a bacterium is organized as a hierarchy of distributors (Figures 5, 7, and 8). Here we first discuss relations between the distributors, considering basic structure and then regulation. Then, we examine how the theory of design, as sketched here, can be implemented in a detailed theory for the molecular organization of a bacterium. With integrative models, the detailed theory might be used to understand the operation of a bacterium as an integrated unit.

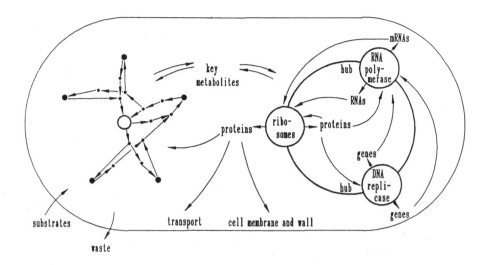

FIGURE 8 Summary diagram for the organization of a bacterium, represented as metabolic and macromolecular distributors. The metabolic net interconverts key metabolites, including substrate from the medium, through its hub. The macromolecular net synthesizes and degrades macromolecules that perform activities, using information stored in DNA.

4.1 RELATIONS BETWEEN THE METABOLIC AND MACROMOLECULAR NETS

The metabolic and macromolecular nets are complementary. The metabolic net stabilizes pools of small molecules against external and internal fluctuations. From these pools the macromolecular net synthesizes macromolecules, and then disassembles useless macromolecules, returning their monomers to the pools. This turnover proceeds on a spectrum of time scales that matches the time scales of demands. The spectrum extends from rapid turnover of metabolic enzymes (for rapid response to a changing environment) to the slow turnover of ribosomes, cell wall, and DNA involved in growth and repair.

In both nets, a hierarchy of controls allocates limited resources among competing activities. Controls in the metabolic net regulate the linear sequences of enzymes within a tree of reactions, entire trees, and relations among trees and the hub. In the macromolecular net, controls allocate holoenzymes among syntheses of holoenzymes, of enzymes for metabolism, and of proteins for growth. Some controls operate within each of these three categories; examples are the competition among catabolic operons when several substrates are present, and the coordinated expression of operons for components of a ribosome.

Regulation plays a more important role in the macromolecular net, than in the metabolic net. Regulation in the metabolic net modulates the activity of existing enzymes. The metabolic net can act as a distributor even without regulation, but regulation increases the rate and specificity of conversions among metabolites. However, in the macromolecular net, the regulation of gene expression controls the presence or absence of enzymes, and so adjusts the competence of the cell to perform activities at all.

4.2 UNDERSTANDING THE OPERATION OF A BACTERIUM AS AN INTEGRATED UNIT: ANALYZING THE MOLECULAR ORGANIZATION

The key to our analysis is the idea that biological structure compensates for limited resources by increasing the diversity of activities that an organism can perform. Indeed, if resources were not limited, any self-replicating dynamical system would suffice, so one could not predict the occurrence of particular structures. Only when molecular processes compete for resources will differential survival favor self-replicating structures with particular patterns of resource allocation.

To design an organism, one must identify an ensemble of candidate organisms that could survive and reproduce in an ecological niche. Each candidate consists of a set of parts with allowed interactions among them. Limits on resources restrict the diversity and amounts of parts. We now ask, how can the global problem of designing an organism be resolved into problems of designing its parts and their interactions? This resolution depends on identifying parts and interactions, and on specifying performance criteria for them. After candidate organisms have been evaluated, the candidates that perform well must be compared with candidates that evolve through a generative approach, and with empirical observations.

4.2.1 IDENTIFYING PARTS. The parts of an organism cooperate to perform activities. To identify parts we can examine the patterns in which structures are coordinately active. Two structures that characteristically are active together belong to the same part. Often a part mediates a cluster of processes. For example, an enzyme is a part; its own parts are amino acids. They cooperate in a cluster of processes—binding a small set of substrates and cofactors, catalyzing a reaction, and releasing the products. In a bacterium, higher level parts include a branch of a metabolic tree, an entire tree, the entire metabolic net, a single operon, and the cluster of operons that synthesizes a ribosome.

If an organism can perform two activities with a variable coordination between them, the two activities are dissociable. Dissociable activities may be performed by distinct parts, or by one part that is specialized or used in different ways. RNA polymerase illustrates the latter case; the enzyme is used with various sigma factors to transcribe different classes of RNAs preferentially.

Organisms perform activities to meet constraints on survival. To discern the parts of an organism, it may be useful to look at the correlations among constraints, because the coordination among activities (which defines parts) tends to match the correlation among constraints. For example, the demand for each key metabolite is a constraint on the metabolic net. The demands for different key metabolites may vary independently. Correspondingly, the regulation of a biosynthetic tree must adjust independently the supply of each key metabolite that the tree makes. Thus the coupling among processes (supplies) matches the correlation among constraints (demands). At a higher level, the constraint that a bacterium should be able to convert one key metabolite to another can be dissociated from the constraints that it should locomote and reproduce; correspondingly, the metabolic network is distinct from the parts that mediate locomotion and reproduction. The statement that the correlation among constraints corresponds to the coordination of activities is the principle of matching.[7,26]

The parts of an organism constrain each other, and they must meet these constraints as well as constraints on the organism as a whole. Parts must match adequately at the interfaces between them; each part must provide outputs that other parts need as inputs. For example, at the interface between the metabolic and macromolecular nets the stabilized pools of key metabolites provide such matching.

The idea that dissociable processes meet dissociable constraints can be extended to make a generative argument that the designs of two parts are dissociable. For example, the basic connectivity of the metabolic and macromolecular nets may be dissociated from their kinetics and regulation. It seems reasonable to assume that the basic pathways evolved to give minimally adequate survival.[43] These pathways then were refined, by the evolution of regulation and by adjustments of kinetic parameters. To the extent that the constraints of getting a minimally adequate performance and a nearly optimal performance are temporally dissociable, the principle of matching suggests that it is reasonable to separate the design of the basic connectivity of a net from the design of its regulatory pathways and kinetic parameters. The theory that we have presented uses this separation.

4.2.2 SPECIFYING PERFORMANCE CRITERIA FOR PARTS.

Once parts are identified, several considerations help to define the performance criteria for each part. Parts compete for limited resources—coding capacity, solubility of macromolecules, and pools of metabolites. The use of resources by any part reduces the resources available to other parts. From a design perspective, no one part should get a disproportionate share of resources. In such cases the amount of a resource used by each part is likely to be the minimum compatible with the performance criteria for the system. For example, the limited coding capacity of DNA can encode proteins for more processes if each process uses fewer, shorter proteins.

Limits on resources may imply tradeoffs in the allocation of resources. For instance, increasing the number of components in a holoenzyme increases its vulnerability to mutation, but still can increase the reliability of its synthesis, as discussed above. As a further example, note that specializations for speed in an organismal network often include short pathways operating in parallel, encoded by a small genome. The price of attaining speed in this way can be a reduction in the reliability and economy of processes, as Karr and Mittenthal[15] discuss.

An adjustment of other variables may resolve a conflict between design criteria. For example, Brown[6] has suggested that a limit on the volume fraction of protein in a cell was reached early in evolution, and different pathways then competed for this resource. Selection for speed favors increasing the fluxes through pathways, and selection for flexibility favors increasing the number of pathways, but these selection pressures must operate on a limited volume fraction of proteins. Bacteria have responded to these pressures by adjusting the variables that are less subject to resource limitations—the specific activity of enzymes, and the amounts of enzymes present in different situations.

4.2.3 COMPARING DESIGN AND GENERATIVE APPROACHES.

A design approach is useful for understanding the diversity, structure, quantity, and relations of parts in organisms. However, networks of processes were not designed; they evolved. Concepts used as tools for design, such as the principle of matching, must be justified by generative arguments. Although heuristic arguments[26] suggest that dynamic modules should evolve as the mutation of genomic regulatory elements regroups processes, a formal justification of the principle of matching is not yet at hand. Furthermore, it is necessary to see whether structures that are optimal from a design viewpoint are likely to evolve. For example, it is not clear that one can define a biologically plausible ensemble and an evolutionary dynamic that evolves a hub-and-branch structure for both the metabolic distributor and the macromolecular distributor, starting from an arbitrary net in the ensemble.

The generative viewpoint can help to explain discrepancies between predictions from a design approach and observations. For example, we have suggested that coding capacity limits the diversity of proteins. However, during evolution proteins have diversified, through duplication and divergence of genes and through recombination of exons. These ideas are compatible because coding capacity increased during evolution as the accuracy of macromolecular syntheses increased. Increases in accuracy probably occurred in relatively rapid steps as well as continuously, as new components contributed to error correction in a holoenzyme. For a while after a step, coding capacity would not limit diversification of genes, which might then have occurred more freely.

4.2.4 TESTING A THEORY OF DESIGN EMPIRICALLY.

We have sketched broadly a theory of design for bacteria, to show that such a theory is possible and to make progress toward finding it. To be useful a theory must make testable predictions about specific experiments. For this purpose the design approach should be applied to more realistic models. For example, it might be useful to design optimal pathways for a small portion of metabolism—say, for the interconversion of a few amino acids and carbohydrates. The ensemble of possible nets would be constrained to use actual reactions that are physiologically reasonable. The performance of each net would be assessed by calculating an objective function to which reliability, flexibility, economy, and speed contribute. In our work the objective function was defined only using the directed graphs for candidate nets; a more realistic objective function should take account of the concentrations of metabolites, and the kinetics and energetics of reactions. A search procedure would be employed to find the nets with good performance.

If an argument from design predicts that a particular feature of molecular organization is optimal in bacteria, the prediction can be tested by using genetic engineering or natural variation to modify the feature. The performance of the variant can then be assayed, either directly or by allowing it to compete against the wild-type organism. Direct assay has been used to test hypotheses about the arrangement of genes on the chromomosome[32] and about the design of lactose metabolism.[10] Competition experiments allow selection for mutants that perform better than the wild type in an atypical medium.[20]

4.3 UNDERSTANDING THE OPERATION OF AN ORGANISM AS AN INTEGRATED UNIT: SYNTHESIZING THE INTEGRATED BEHAVIOR OF THE MOLECULAR ORGANIZATION

Now we inquire: How could information about molecular organization be used to understand the integrated performance of a bacterium? One could use data about connectivity and kinetics to model in detail the dynamics of pathways. For a whole bacterium the resulting model would be incomprehensibly complex; it would be a dynamical analog of the *Encyclopedia of E. coli Life Processes* proposed at the beginning of this essay. However, a many-variable dynamical system of this kind typically shows only a few dynamical modes, or phases, of behavior. The gaits of a running horse are dynamical modes. In a dynamical mode the dynamics of a few macro-level variables, called order parameters, characterize the collective behavior of many micro-level variables as they follow microdynamical laws. In a bacterium the laws of chemical kinetics and diffusion govern the microdynamics. When a biochemical pathway operates in a steady-state mode, one order parameter, the current of biochemical flux, suffices to describe the rate at which metabolites transform through a linear sequence of reactions, and the sequence behaves as a single reaction. The step from micro- to macrodynamics often eliminates variables through a quasi-steady state approximation, if some processes are much more rapid than others.[12,17,38]

The characterization of large-scale dynamical modes is commonplace in statistical physics, and it is increasingly contributing to the analysis of biological systems.[9,33,35] In organisms, spontaneously occurring dynamical modes presumably are the modules that meet constraints. During evolution these modes vary and differentially survive selection, so that the matching of performance to constraints can improve. Thus the mechanistic analysis of dynamical modes emerging from molecular-level organization should provide a generative basis for understanding how organisms meet demands.

4.4 TOWARD A THEORY OF THE ORGANIZATION OF ORGANISMS

The organization of a bacterium illustrates general features of the organization of organisms in a simple context. An organism uses matter, energy, and information from its environment to perform a set of activities sufficient to deal with the situations characteristic of its niche. These activities—the acquisition of more inputs, the regulation of internal organization, and the production of more organisms—enable its lineage to persist. In bacteria the path from inputs to outputs can be resolved into a sequence of two stages, the metabolic and the macromolecular stages. In each stage, several processes operate in parallel to convert inputs to the outputs that other stages use. If limited resources are allocated among the processes of a stage, it must act as a distributor. In more complex organisms, stages can often be recognized, though they are more numerous and are related in more diverse ways than in bacteria.

Eukaryotes use a hierarchy of distributors, with more levels in the hierarchy than a bacterium has, to match performance, with limited resources, to demand. For example, in vertebrates the circulation is the distributor for the limited volume of blood; the heart is its hub. The lungs distribute air from a hub, the trachea, to the limited area of alveolar-capillary interface available for gas exchange. The brain is the hub of a regulatory distributor, the neuroendocrine system. It distributes regulatory information through branches of the nervous and hypothalamic-pituitary systems. The neuroendocrine distributor activates clusters of processes in distributors at all levels in the hierarchy, and so enables the organism to perform diverse integrated activities.

As in bacteria, structures in the distributors change at time scales that match the time scale of demands. Development generates relatively stable structures—organs with spatially patterned differentiated cells—that are used in various combinations through much of the life cycle. Structures that are used more transiently are made or modified on physiological time scales, with alternative allocations of resources. Thus, the hierarchy of distributors meets demands of situations that are diverse in time scale and in frequency, so that a lineage of organisms can persist.

ACKNOWLEDGMENTS

Our efforts to improve this paper were greatly aided by comments from Arthur Baskin, Tim Karr, Stanley Malloy, Robin Mittenthal, Robin Shealy, Dale Steffensen, and Wilfred Stein. We thank them.

APPENDIX

In this mathematical appendix we formally demonstrate claims made about the collection of graphs that describe possible metabolic nets. We identify two senses in which the hub graph is optimal. First, a hub graph has shorter paths between key metabolites, on average, than any other graph in the collection. Second, a hub graph is among the graphs least sensitive to mutational deletion of enzyme-catalyzed steps.

We begin by characterizing the collection of graphs, which we denote by $\mathcal{G}$. Each graph $G \in \mathcal{G}$ has the following properties:

1. There are K distinct points $x_1, ..., x_K$. Each x_i represents a distinct key metabolite.
2. Each G has a nonvoid set of points, distinct from those in 1. called junctions.
3. Every edge is directed and has a junction at one end and some x_i at the other. Each edge represents a sequence of unimolecular reactions; each reaction is mediated by a unique enzyme.
4. Each x_i has exactly one edge coming into it and exactly one edge leaving it.
5. Each G is connected.
6. Given x_i, x_j there is a path, a sequence of edges, leading from x_i to x_j.

Let G_1 designate the hub graph.

Let X be a random variable taking values in the set of ordered pairs

$$S = \{(x_i, x_j) | i \neq j, \ i, j = 1, ..., K\}.$$

Note that the set S has $2\binom{K}{2}$ elements. We assign uniform probabilities to the values of X, i.e.,

$$P(X = x) = \frac{1}{2\binom{K}{2}}.$$

This assignment corresponds to assuming that each key metabolite is equally likely to be available as a substrate in the environment.

We will compare the mean and variance of functions of X for the hub graph and other graphs.

DEMONSTRATION THAT A HUB GRAPH HAS THE SHORTEST PATHS BETWEEN KEY METABOLITES

Let $d_G(x_i, x_j)$ be the path length—the minimal number of edges that must be traversed—from x_i to x_j in graph G. (By property 6, d_G is well defined for each pair of key metabolites.) The average path length is $Ed_G(X)$, and the variance of path length is Var $d_G(X)$, where the expectation E and the variance Var are taken with respect to the uniform probability.

PROPOSITION 1. Let $K \geq 3$. Then

$$\arg \min_{G \in \mathcal{G}} E d_G(X) = G_1. \tag{i}$$

$$\arg \min_{G \in \mathcal{G}} \mathrm{Var} d_G(X) = G_1. \tag{ii}$$

Note, (i) means that the hub graph G_1 minimizes the average path length; (ii) means that the hub graph minimizes the variation in the path length.

PROOF First note that for a hub graph, $d_{G_1}(x) = 2$ for all $x \in S$. Consequently

$$E d_{G_1}(X) = 2 \text{ and } \mathrm{Var}\, d_{G_1}(X) = 0.$$

To prove the proposition it suffices to show that for any graph $G \in \mathcal{G}$ other than G_1, there are $x, y, z \epsilon S$ such that (a) $d_G(x) > 2$ and (b) $d_G(y) \neq d_G(z)$.

Note that $d \geq 2$ and every graph has some $x' \in S$ such that $d(x') = 2$. Since the hub graph has Var $d = 0$, it suffices to choose y in (b) to be x' and z in (b) to be the x shown to exist in (a). This will give a strictly positive variance. Thus, it is enough to show (a).

To prove (a) the strategy is to show, for graphs other than G_1, that an x can be constructed which has path length at least four. To do this we first note that any graph other than G_1 must have at least two junctions. Denote the two junctions by j_1 and j_2. By condition (6), given any two junctions there is a path from the first to the second. This follows from the fact that each junction must have at least one input edge and at least one output edge. These edges terminate in key metabolites, to which condition (6) applies. Either the two junctions are connected by a path of length strictly greater than two, or by a path of length two. We consider these cases in turn.

Case i. The two junctions are connected by a path of length strictly greater than 2. Since this path length must be an even number, at least four edges are required to go from j_1 to j_2, and there must be a third junction j_3 on a path between them.

Since there must be a key metabolite between j_1 and j_3, and j_2 must have an output edge leading to another key metabolite, the path length between these two metabolites is at least four. Thus (a) has been verified for this case.

Case ii. Junction j_1 goes to j_2 by way of a key metabolite, t. There must be at least one edge into j_1, and one edge out of j_2, with endpoints u and v, respectively. If u and v are distinct, then the path from u to v has length four. If $u = v$, then, by assumption, there must be a third key metabolite w. An edge must lead from w to a junction j_3. If j_3 is distinct from j_1 and j_2, then the path from w to u has length at least four. If $j_3 = j_1$, then again the path from w to u has length at least four. If $j_3 = j_2$, the path from w to t has length at least four. Thus (a) has been verified for this case.

This proposition shows that a hub graph is the only graph in $\mathcal{G}$ in which all paths have the same length, and this length is the smallest it can be, namely two. Thus a hub graph is the only graph that reflects the equivalence of all the key metabolites.

DEMONSTRATION THAT A HUB GRAPH IS RELATIVELY INSENSITIVE TO DELETION OF EDGES

Next, we inquire how removing one edge at a time degrades the convertibility of key metabolites by graphs in $\mathcal{G}$. First, we show that if one edge is deleted, no graph in $\mathcal{G}$ performs better than a hub graph. Other graphs may do as well, so a hub graph is not unique in this respect. However, we shall offer heuristic arguments that if a particular class of additional edges are deleted, a hub graph retains convertibility better than any other graph. After deletion in other classes of edges, a hub graph is no worse than other graphs.

To the hub graph G_1 associate $G'_1, ..., G'_{2K}$, where each G'_i results from deleting one of the edges of G_1. Do this also for the remaining graphs in $\mathcal{G}$. We denote the resulting collection of graphs with one deleted edge by $\mathcal{G}'$.

Let X take values in the set $S = \{(x_i, x_j) | i \neq j, i, j = 1, ..., K\}$, with uniform probabilities, as before. We now define an index of the convertibility of key metabolites. For $G' \in \mathcal{G}'$, let $\rho_{G'}(x) = \rho_{G'}(x_i, x_j)$ take value 1 if there is a path from x_i to x_j in G', and 0 otherwise. We will examine the behavior of the interconvertibility, $E\rho_{G'}(X)$, where G' is an element of $\mathcal{G}'$.

PROPOSITION 2. The maximum convertibility over G' in $\mathcal{G}'$ is $(K-1)/K$, and this maximum is achieved in a hub graph.

PROOF First, we show that the convertibility of any G_i', $i = 1, ..., 2K$, is $(K-1)/K$. If any one edge of G_1 is removed, then there will be $K-1$ ordered pairs of key metabolites for which no path leads from the first to the second. For all other ordered pairs such a path will exist. Since there are $K(K-1)$ ordered pairs, for each i

$$E\rho_{G_i'}(X) = (0)\frac{K-1}{K(K-1)} + (1)\frac{K(K-1) - (K-1)}{K(K-1)} = \frac{K-1}{K}.$$

In any other graph in $\mathcal{G}'$, it is possible that the deletion of an edge will eliminate paths between more than $K-1$ ordered pairs, and so give a lower convertibility.

Now let us consider the effects on convertibility of deleting more than one edge. Every graph in $\mathcal{G}$ has the same collection of edges, so corresponding edges can be deleted in different graphs. Suppose we delete two edges in all possible ways from each graph in $\mathcal{G}$. If we look at a hub graph first, there are three ways to delete two edges: The two edges can be deleted on the same key pathway. Or, the two deleted edges may be on different pathways, either with one edge directed toward the hub and one edge directed away (that is, with opposite orientations), or with the same orientation. It can easily be shown that if both edges are in the same key pathway, or are in different pathways but have the same orientation, the convertibility is $(K-2)/K$. If the two edges are in different pathways and have opposite orientations, the convertibility is slightly larger. Thus to compare the convertibility of two graphs, it is important to delete corresponding edges in them. By reasoning similar to Proposition 2, if we delete two edges in any nonhub graph, its convertibility will be less than or equal to the convertibility of the hub graph after deletion of the corresponding edges. The same argument is valid for deletion of an arbitrary number of edges. Thus with successive deletion of edges a hub graph loses convertibility no faster than any other graph.

Now we argue that it is always possible to delete corresponding edges in a hub graph and a nonhub graph in such a way that eventually the convertibility of the hub graph will be greater. In a hub graph no key pathways are affected by deletions other than the pathways in which deletions occur. Thus passage among key metabolites on undamaged pathways is not affected by deletions. However, this is not the case in a nonhub graph. Consider a junction with at least one edge leading to it and at least two edges leading away from it, which lead to key metabolites m_1 and m_2. By deleting all edges leading to the junction, we prevent transitions between m_1 and m_2. This argument applies to any graph except a rim graph, in which any two deletions separate the graph into two disjoint parts. Correspondingly, if at least two edges lead from metabolites m_3 and m_4 into a junction, then deletion of all edges leading out of the junction prevents transitions between m_3 and m_4.

Although the performance of hub and nonhub graphs under multiple deletions is clearly of mathematical interest, its biological relevance is low, since multiple deletions would be lethal.

REFERENCES

1. Alberts, B., D. Bray, J. Lewis, M. Raff, K. Roberts, and J. D. Watson. In *Molecular Biology of the Cell*, chapters 3, 5, 9, 10. 2nd ed. New York: Garland, 1989.

2. Atkinson, D. E. "Limitation of Metabolite Concentrations and the Conservation of Solvent Capacity in Living Cells." *Curr. Top. Cell Regul.* 1 (1969): 29–43.

3. Bagley, R. J., J. D. Farmer, and W. Fontana. "Evolution of a Metabolism." In *Artificial Life II*, edited by Christopher G. Langton, Charles Taylor, J. Doyne Farmer, and Steen Rasmussen, 141–158. Santa Fe Institute Studies in the Sciences of Complexity, Proceedings Volume X. Redwood City, CA: Addison-Wesley, 1991.

4. Barlow, R. E., and F. Proschan. *Statistical Theory of Reliability and Life Testing*. Silver Spring, MD: To Begin With, 1981.

5. Baskin, A. B., R. E. Reinke, and J. E. Mittenthal. "Exploring the Role of Finiteness in the Emergence of Structure." In *Principles of Organization in Organisms*, edited by Jay Mittenthal and Arthur Baskin, 337–377. Santa Fe Institute Studies in the Sciences of Complexity, Proceedings Volume XII. Reading, MA: Addison-Wesley, 1992.

6. Brown, G. C. "Total Cell Protein Concentration as an Evolutionary Constraint on the Metabolic Control Distribution in Cells." *J. Theor. Biol.* 153 (1991): 195–203.

7. Clarke, B., and J. E. Mittenthal. "Modularity and Reliability in the Organization of Organisms." *Bull. Math. Biol.* 54 (1992): 1–20.

8. Cooper, S. *Bacterial Growth and Division*. San Diego: Academic Press, 1991.

9. DeGuzman, C. G., and J. A. S. Kelso. "The Flexible Dynamics of Biological Coordination: Living in the Niche Between Order and Disorder." In *Principles of Organization in Organisms*, edited by Jay Mittenthal and Arthur Baskin, 11–34. Santa Fe Institute Studies in the Sciences of Complexity, Proceedings Volume XII. Reading, MA: Addison-Wesley, 1992.

10. Dykhuizen, D. E., and A. M. Dean. "Enzyme Activity and Fitness: Evolution in Solution." *Trends in Ecol. & Evol.*, 5(8) (1990): 257–262.

11. Gutnick, D., J. M. Calvo, T. Klopotowski, and B. N. Ames. "Compounds Which Serve as the Sole Source of Carbon or Nitrogen for *Salmonella typhimurium LT-2*." *J. Bacteriol.* 100 (1969): 215–218.

12. Haken, H. *Synergetics*. 3rd ed. New York: Springer-Verlag, 1983.

13. Ingraham, J. L., O. Maaloe, and F. C. Neidhardt. *Growth of the Bacterial Cell*. Sunderland, MA: Sinauer, 1983.

14. Jensen, K. F., and S. Pedersen. "Metabolic Growth Rate Control in *Escherichia coli* may be a Consequence of Subsaturation of the Macromolecular Biosynthetic Apparatus with Substrates and Catalytic Components." *Microbiol. Revs.* 54 (1990): 89–100.

15. Karr, T. L., and J. E. Mittenthal. "Adaptive Mechanisms that Accelerate Embryonic Development in *Drosophila*." In *Principles of Organization in Organisms*, edited by Jay Mittenthal and Arthur Baskin, 95–108. Santa Fe Institute Studies in the Sciences of Complexity, Proceedings Volume XII. Reading, MA: Addison-Wesley, 1992.

16. Kauffman, S. A. "Autocatalytic Sets of Proteins." *J. Theor. Biol.* **119** (1986): 1–24.

17. Kelso, J. A. S., M. Ding, and G. Schoner. "Dynamic Pattern Formation: A Primer." In *Principles of Organization in Organisms*, edited by Jay Mittenthal and Arthur Baskin, 397–440. Santa Fe Institute Studies in the Sciences of Complexity, Proceedings Volume XII. Reading, MA: Addison-Wesley, 1992.

18. Kirkwood, T. B. L., R. F. Rosenberger, and D. J. Galas. *Accuracy in Molecular Processes*. New York: Chapman & Hall, 1986.

19. Kustu, S., A. North, and D. Weiss. "Prokaryotic Transcriptional Enhancers and Enhancer-Binding Proteins." *Trends in Biochem. Sci.* **16(11)** (1991): 397–402.

20. Levinthal, M. "The Evolution of Complexity." In *Biosynthesis of Branched Chained Amino Acids*, edited by Z. Barak, D. M. Chipman, and J. V. Schloss, 163–178. New York: VCH, Weinheim, 1990.

21. Lindgren, K. "Evolutionary Phenomena in Simple Dynamics." In *Artificial Life II*, edited by Christopher G. Langton, Charles Taylor, J. Doyne Farmer, and Steen Rasmussen, 295–312. Santa Fe Institute Studies in the Sciences of Complexity, Proceedings Volume X. Redwood City, CA: Addison-Wesley, 1991.

22. Mangel M, and C. W. Clark. *Dynamic Modeling in Behavioral Ecology*. Princeton, NJ: Princeton University Press, 1988.

23. Marr, A. "Growth Rate of *Escherichia coli*." *Microbiol. Revs.* **55** (1991): 316–333.

24. Miller, J. G. *Living Systems*. New York: McGraw-Hill, 1978.

25. Mittenthal J. E., and A. B. Baskin, eds. *Principles of Organization in Organisms*. Santa Fe Institute Studies in the Sciences of Complexity, Proceedings Volume XII. Redwood City, CA: Addison-Wesley, 1992.

26. Mittenthal, J. E., A. B. Baskin, and R. E. Reinke. "Patterns of Structure and Their Evolution in the Organization of Organisms: Modules, Matching, and Compaction." In *Principles of Organization in Organisms*, edited by Jay Mittenthal and Arthur Baskin, 321–332. Santa Fe Institute Studies in the Sciences of Complexity, Proceedings Volume XII. Reading, MA: Addison-Wesley, 1992.

27. Nomura, M. "The Control of Ribosome Synthesis." *Sci. Am.* 250**(1)** (1984): 102–114.

28. Oster G. F., and E. O. Wilson. *Caste and Ecology in the Social Insects*. Princeton, NJ: Princeton University Press, 1978.

29. Ovadi, J. "Physiological Significance of Metabolic Channelling." *J. Theor. Biol.* **152** (1991): 1–22.

30. Rawn, J. D. *Biochemistry*. Burlington, NC: Neil Patterson, 1989.
31. Riedl, R. *Order in Living Organisms*. Translated by R. P. S. Jefferies. New York: John Wiley & Sons, 1978.
32. Roth, J. R., and M. B. Schmid. "Arrangement and Rearrangement of the Bacterial Chromosome." *Stadler Symp.* **13** (1991): 53–70.
33. Savageau, M. A. *Biochemical Systems Analysis*. Reading, MA: Addison-Wesley, 1976.
34. Savageau, M. A. "Autogenous and Classical Regulation of Gene Expression: A General Theory and Experimental Evidence." In *Biological Regulation and Development*, edited by R. F. Goldberger, Volume 1, 57–108. New York: Plenum, 1979.
35. Savageau, M. A. "Growth of Complex Systems Can be Related to the Properties of Their Underlying Determinants." *Proc. Natl. Acad. Sci. USA* **76** (1979): 5413–5417.
36. Savageau, M. A. "Are There Rules Governing Patterns of Gene Regulation?" In *Theoretical Biology: Epigenetic and Evolutionary Order from Complex Systems*, edited by B. Goodwin and P. Saunders, 42–66. Edinburgh Press, 1989.
37. Savageau, M. A. "Metabolite Channeling: Implications for Regulation of Metabolism and for Quantitative Description of Reactions *in vivo*." *J. Theor. Biol.* **152** (1991): 85–92.
38. Segel, L. A. *Modeling Dynamic Phenomena in Molecular and Cellular Biology*. New York: Cambridge University Press, 1984
39. Sibley R. M., and P. Calow. *Physiological Ecology of Animals: An Evolutionary Approach*. Oxford: Blackwell Scientific, 1986.
40. Snellings, K., and C. W. Vermeulen. "Non-Random Layout of the Amino Acid Loci on the Genome of *Escherichia coli*." *J. Mol. Biol.* **157** (1982): 687–688.
41. Stephenson, M. *Bacterial Metabolism*, 182–192. Cambridge, MA: MIT Press, 1966.
42. Stryer, L. *Biochemistry*. 3rd ed. New York: W. H. Freeman, 1988.
43. Wachtershauser, G. "Before Enzymes and Templates: Theory of Surface Metabolism." *Microbiol. Revs.* **52** (1988): 452–484.
44. Whyte, L. L. *Internal Factors in Evolution*. New York: George Braziller, 1965.
45. Yates, F. E. "Systems Analysis of Hormone Action." In *Biological Regulation and Development*, Volume 3A, edited by R. F. Goldberger and K. Yamamoto. New York: Plenum, 1982.

G. Cocho,† F. Lara-Ochoa,‡ E. Vargas,‡ M. A. Jimenez-Montaño,§ and J. L. Ruis†
†Instituto de Física, Universidad Nacional Autónoma de México, 01000, D. F. México;
‡Instituto de Química, Universidad Nacional Autónoma de México, 04510, D. F. México;
and §Universidad de Las Américas, Cholula, Puebla, Pue., México.

Structural Patterns in Macromolecules

1. ANALYTICAL AND HISTORICAL APPROACHES TO EVOLUTION

At present there exist two conflicting approaches to evolution, one that we might call "historical-accidental" associated with "traditional neo-Darwinism," with Nature as an opportunistic tinkerer with very few restrictions, and the other that we might call the "analytical-rational" approach with its emphasis on the restrictions that physicochemistry and complex system laws impose on the possible evolutionary paths. The "analytical-rational" approach has led to the understanding of important aspects of development,[9,10,18] and to the presention of a rational taxonomy,[5,6,12] more restrictive than the usual Darwinian genealogies. However, the impact of the analytical-rational approach on DNA and RNA structure and dynamics has been rather low, and up to a certain degree, this comment is also valid for protein dynamics. The main purpose of the present work is to show that there exist important restrictions on DNA and on immunological protein structure. However, it is important to mention that the "historical-accidental" and "analytical-rational" approaches might be different aspects and regions of the same dynamical space.

Thinking About Biology, Eds. W. D. Stein and F. J. Varela, SFI Studies in the Sciences of Complexity, Lect. Note Vol. III, Addison-Wesley, 1993

Systems built of parts with low connectivity (a given part interacts with a small number of other parts, very often neighboring ones) and with no dynamical conflicts (it is possible to optimize the system with respect to all the different restrictions) have, in general, one or just a small number of optimal states and the system arrives at them, independent of the initial conditions and perturbations suffered along its history. Field descriptions of morphogenesis usually belong to this class. On the other hand, systems with large connectivity and "frustration" (conflicts among the different optimization constraints) have, in general, a large number of quasi-equivalent states of almost equal optimality. If the connectivity is large enough, these states are separated by regions of low optimality and the particular state reached by the system depends strongly on the initial conditions and on its previous history. Immunological and neural networks are examples of this last type. Therefore, depending on the degree of connectivity and conflictness, we can move from an "analytical" to a "historical" system and we can consider cases where systems of different type coevolve.

The restrictions imposed on the genetic sequences by the "cellular machinery" seem to be of the analytical type. Genes are not amorphous materials on which evolution writes messages: they are the product of a rather complicated cellular machinery and should show the traces of the dynamics of such a machinery. The efficiency and accuracy of DNA and RNA processing steps (replication, transcription, editing, and translation) should depend, to a certain extent, on the local or global physicochemical properties, which, in general, will be functions of the base composition. Modulations in the base composition are, in general, present and one might predict some of these modulations from an analysis of the physicochemical properties. These modulations might be the basis of algorithms for the estimation of the local or semilocal (at domain level) efficiency and accuracy of the different processing steps.

At the protein level, immunoglobulins (Ig) are key elements in the mediation of the immune response. These proteins bind foreign antigens, blocking their ability to bind to receptors on target cells, or mark the invading microorganisms for destruction. The main challenge in immunology has been to understand how the immune system recognizes the presence of foreign macromolecules.

The basic structural unit of an Ig molecule consists of four polypeptide chains, two identical light (L) chains and two identical heavy chains (H). Both L and H chains have a variable region and a constant region. The constant region of an L chain is about 110 amino acids long, and the variable region is of the same size. The variable region of the H chain is also about 110 amino acids long, but the constant region is about 330 amino acids long. An interesting feature of each chain is that the amino acid sequence suggests that it is made up of repeating segments or domains, each of which folds independently to form a compact unit. Accordingly, an L chain consists of one variable V_l and one constant C_l domain; while most H chains consist of a variable domain V_h and three constant domains. On a three-dimensional basis, each one of these domains is roughly a cylinder composed of a sandwich of two extended protein layers. One of these layers is formed by three strands of polypeptide chains, and the other by four. In each layer the adjacent strands are antiparallel and form a β sheet. The two layers are roughly parallel and are connected by a single intrachain disulfide bond. Only part of the variable region in the binding of the antigen participates. These binding sites are restricted to three small hypervariable regions in each chain, called complementary determining regions (CDRs). These CDR segments are located on the loops that connect the two layers of the β sandwiches on one of the sides. The remaining parts of the variable region are known as framework regions and are relatively constant.

It has been proposed in the literature[4] that the number of main chain conformations of at least five of the six CDRs of the Igs seems to be limited to a very few canonical conformations. The adoption of specific backbone conformations is believed to reflect the existence of a few key conserved residues in the loop of the antibody molecule. This suggests that some of the amino acids on the CDRs may have a structural role, while the rest are either involved in the recognition function or irrelevant. It has thus been proposed[13,17] that there must be evolutionarily conserved residues within CDR-1 and CDR-2, and that a few hypervariable positions in the immediate neighborhood of such conserved positions must decide the antigen specificity of an individual V_h. Furthermore, Ohno et al.[17] proposed that if it is possible that some sites on the CDR-1 and CDR-2 could be conserved for the shaping of the primordial antigen-binding cavity, the remaining seven sites can generate 20^7 or 1.28×10^9 varieties of different amino acid sequence combinations. However, a recent analysis[16] suggests that there are some constraints on the wide range of possible and meaningful amino acid sequence combinations, which would imply that amino acids would have nonrandom frequency distributions.

In order to establish the existence and nature of these constraints, an analysis of the frequency of amino acids in 21 sites of the CDR-1 and CDR-2 of the variable heavy domain and of 18 sites of the variable light domain was performed in an alignment of approximately 1500 immunoglobulins.[14]

In section 2 we compare the average enthalpy of exons, introns, and intergenic sequences of mammal globins. Intron values show low dispersion suggesting the presence of physicochemical constraints associated with transcription and/or editing. In section 3 we present patterns of amino acid use in the CDR's of Igs, and discuss how these patterns represent constraints in the wide range of possible solutions of the immune response. These patterns permit keeping general physicochemical properties, or attributes, in specific points of the antibody-epitope binding.

2. DNA AVERAGE ENTHALPY AND INTRON DYNAMICS

DNA molecules are linear polymers built from four different monomers. Thermodynamic measurements in synthetic oligonucleotides have permitted estimates to be made of the enthalpy, entropy, and free energy of nearby base pairs in DNA. In these estimates the basic unit is considered to be two neighboring base pairs which we shall call digrams. In a first approximation the entropy or energy of a given oligonucleotide duplex is equal to the sum of entropies or energies of the digrams. Each digram has a horizontal hydrogen-bonding interaction energy, a vertical stacking energy, and diagonal energy contributions (see Figure 1). Measurements for the enthalpy ΔH and the free energy ΔG were made by Breslauer et al.,[3] and are shown in Table 1.

Perhaps the dominant approach to intron and intergenic sequence structure and dynamics has been to consider them as a rather structureless "junk." However, as we mentioned at the beginning, they are the product of a rather complex nuclear machinery and should show the traces of such a dynamics. This trace might appear as regularities or conserved quantities which should give clues as to the nature of the dynamical restrictions. Zuckerkandl,[26] Soloviev and Kolchanov,[23] and Beckmann and Trifonov[1,2] have commented that perhaps introns are needed in order to render the protein-coding and chromatin sequence requirements compatible. We will present some additional evidence supporting this hypothesis.

TABLE 1 Enthalpy (ΔH) and free energy (ΔG) for DNA digrams[1]

Description	ΔH	ΔG
AA	9.1	1.9
AT	8.6	1.3
TA	6.0	1.9
TT	9.1	1.9
CC	11.0	3.1
CG	11.9	3.6
GC	11.1	3.1
GG	11.0	3.1
AC	6.5	1.3
AG	7.8	1.6
TC	5.6	1.6
TG	5.8	1.9
CA	5.8	1.9
CT	7.8	1.6
GA	5.6	1.6
GT	6.5	1.3

[1] ΔH and ΔG are taken from Breslauer et al.[3] and are expressed in Kcal/mol.

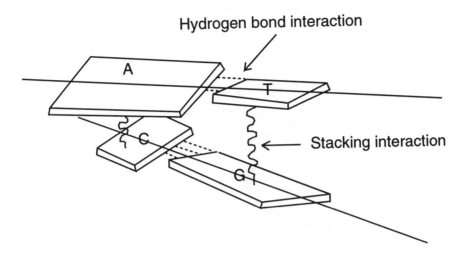

FIGURE 1   Two neighboring rungs (diagram) of a DNA staircase. "Horizontal" H-bond (dashed lines) and "vertical" stacking interactions are shown.

We have computed the average enthalpy for a set of exons, introns, and intergenic sequences (IGS) of mammal globins of the β cluster. This cluster consists of β, γ, δ, and ϵ globins. The structure of these genes is shown in Figure 2. We have taken into account the second exon (222 base pairs) and second intron (between 600 and 800 base pairs) of 14 mammal globins of the β cluster, from man, monkey, lemur, mouse, goat, and rabbit. The average enthalpy is the sum of the enthalpy for each of the digrams of the sequence, taking the values from the work of Breslauer et al.[3] (Table 1), divided by the length of the sequence. We have also computed the average enthalpy for different 700 bp segments of intergenic sequences of the β cluster. In Figure 3 we plot these values (x axis) and compare them with the values corresponding to random sequences, of infinite length, and with the same base composition (y axis). In the figure we see that: (1) Intron average enthalpy is lower than exon

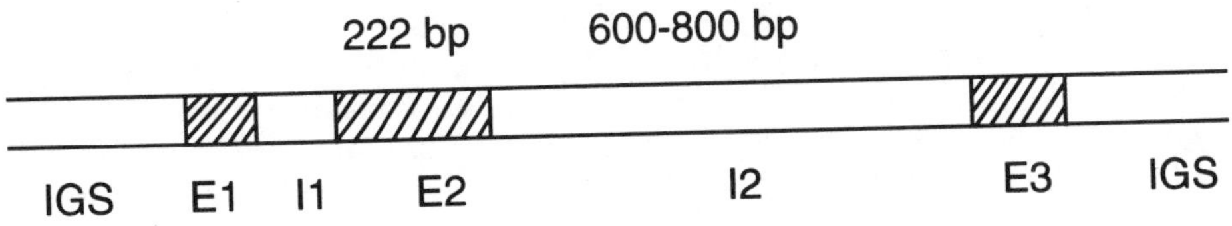

FIGURE 2   Structure of the genes of mammal globin β cluster.

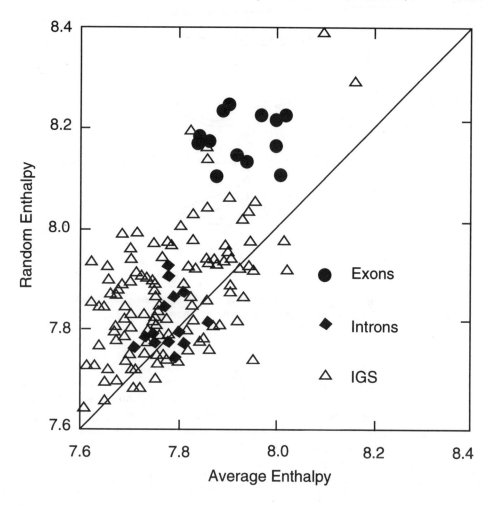

FIGURE 3 Plot of the average enthalpy vs. the enthalpy of random sequences of the same base composition. Exon, intron, and intergenic sequence values are plotted.

average enthalpy. (2) Exons and introns show quite low dispersion while intergenic sequences show larger dispersion. Exon and intron dispersion is, more or less, equal to the standard deviation of the average enthalpy corresponding to sets of random sequences of the same length (700 bp for introns and 222 bp for exons) with a fixed base composition but with different seeds in the random number generator. The dispersion of intergenic sequences is about three times larger than such standard deviation. (3) Intron values are very near to the diagonal, their average enthalpies almost correspond to random sequences with the same base composition. Their values are rather low and, as a matter of fact, they are very near to the minimal value found for random seqences of infinite length and arbitrary composition ($\Delta H = 7.73$ Kcal/mol). (4) Exon values are lower than random sequences with the same composition. This is also true for the average of intergenic sequence points.

The preceding features suggest that introns might be constrained almost as strongly as exons. Since intergenic sequences show a different average enthalpy pattern, it seems that intron constraints might be associated with transcription and/or editing processes. However, intergenic sequences, in spite of their enthalpy dispersion, show average values smaller than those for random sequences, suggesting the presence of additional constraints, perhaps related to replication and chromosome dynamics.

3. CONSERVED AMINO ACIDS AND SKEW DISTRIBUTION OF THE RECOGNITION AMINO ACIDS IN THE HYPERVARIABLE REGIONS OF IMMUNOGLOBULINS

To study the variability of the amino acids in each site of the CDR's, the amino acids distribution table reported by Kabat et al.[14] was examined as the basic source of frequency data. The results of the analysis, showed in Table 2, indicate that some sites are highly conserved (14 of 21, for the heavy chain, and 8 of 18 for the light chain). For instance, Ile in position 51 of the heavy chain, which traditionally is considered as being within the hypervariable region has, by itself, a frequency of 88%. Another case is Gly-26, which has a use frequency of 98%. In some other positions the frequencies of some amino acids may be as low as 43% (for instance, in site 63 of heavy chain). In these cases, a second or third amino acid in order of abundance (residual amino acid) can be found in the same site with almost the same physicochemical properties as those of the more frequent amino acid. That is, in these cases the residual amino acids have the same capabilities for maintaining the general structure of the CDR's, strengthening the idea of a structural role for these amino acids, in agreement with previous suggestions in the literature.[4,13,16,17]

This is a striking finding, since it has been usually considered in the literature[7] that amino acids in sites 26–30 and 53–55 of the heavy domain, and 26–32 and 50–56 of the light domain, being in the loop region, were part of the specificity determining positions. These findings suggest that the concept of CDR should be based not only on the concept of its being on a loop, but on the criterion of how conserved are the amino acids in the site.[25,27]

What arises from the above results is that if the positions discussed are critical in maintaining the canonical structures of the family, the rest of the positions on the CDR may be responsible for the recognition process. Next, the frequencies of use of these amino acids, now considered as having the function of recognition, were

TABLE 2 Frequency of Amino Acids of Invariant Sites of the CDR-1
and CDR-2 of the Heavy and Light Domain

Site	Predominant amino acid	Residual amino acid	Total frequency
Heavy Chain			
26	G (98%)		
27	F (48%)	Y (42%)	90%
28	T (62%)	S (21%)	83%
29	F (75%)	I,L (21%)	96%
30	T (50%)	S (41%)	91%
32	Y (66%)	F (10%)	76%
34	M (60%)	I,V (27%)	87%
51	I (88%)		
55	G (61%)	S (14%)	75%
57	T (76%)		
59	Y (94%)	F (2%)	96%
63	F (43%)	V,L (40%)	83%
64	K (77%)	R (17%)	94%
65	G (60%)	S (23%)	83%
Light Chain			
24	R (47%)	K (15%)	62%
25	A (55%)	S,G (39%)	94%
26	S (86%)	T (2%)	88%
32	Y (72%)	F (5%)	77%
33	L (57%)	V,M (26%)	83%
52	S (78%)	T (7%)	85%
54	L (47%)	R (47%)[1]	94%
56	S (72%)	P,T (18%)	90%

[1] For this case, the second amino acid has different physico-chemical properties from that of the most frequent, but the sum of both frequencies indicate high conservation.

analyzed to determine whether they follow a particular type of distribution or, on the contrary, express the whole repertoire, i.e., have a random distribution.

The most direct form of studying the type of distribution of these remaining sites (heavy chain: 31, 33, 35,...; light chain: 27, 28, 29,...) is by plotting, for each site, the rank of each amino acid versus the log of its frequency. If this plot fits a straight line, the distribution of the objects is skewed.[15,20,21] Thus, twelve sites on

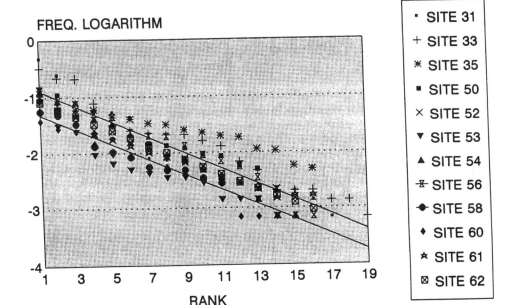

FIGURE 4 Plot of the rank of use of each amino acid versus their relative frequency, for those sites on the CDR-1 and CDR-2 of the heavy chain identified as "recognition" sites. For all the positions, the fit with a straight line was very good, obtaining for all the cases correlation coefficients $r > 0.96$.

the heavy chain and ten on the light chain were analyzed to determine if their use of amino acids follows a particular nonrandom distribution.

The plots obtained are shown in Figures 4 and 5. In all cases the linear fits were very good, with correlation coefficients greater than 0.96 for the sites on the heavy chain, and 0.97 for the light chain. From these data it is evident that a feature of the recognition sites is that they follow a skewed distribution in the frequency of use of the amino acids.

It is interesting to notice that the sites with the best fitting to a straight line ($r > 0.58$) in the heavy chain are the same as those proposed by Ohno et al.[17] as the sites of recognition.

One is then compelled to discard the traditional view that the CDRs are random hypervariable regions, since most of the amino acids selected for each site are constrained as arising from a reduced set of amino acids.

In order to test whether the more frequently used amino acids in each site share some attributes or physicochemical properties, an algorithm normally applied in artificial intelligence was used.[19] This algorithm allows the clustering of elements in a complex system in terms of their common attributes. The algorithm takes objects of a known class, described in terms of a fixed collection of properties or attributes, and produces a decision tree over these attributes, that correctly classifies all the given objects. Each attribute has its own set of discrete attribute values. For the present application, the objects are amino acids; the class is the most frequent amino acid in the site (more than 7%); and the attributes are defined by the properties of the amino acids. By applying the algorithm to the twelve analyzed cases of the heavy chain, using as attributes those physicochemical properties commonly employed in protein structure studies,[11,22] it was found that, in many cases, only a single attribute was necessary to characterize the cluster (Figures 6 and Table 3). In four cases it was not possible to find a reduced set of attributes to characterize the cluster (Figure 7), although this could be due to the lack of identification of an appropriate attribute shared by the set.

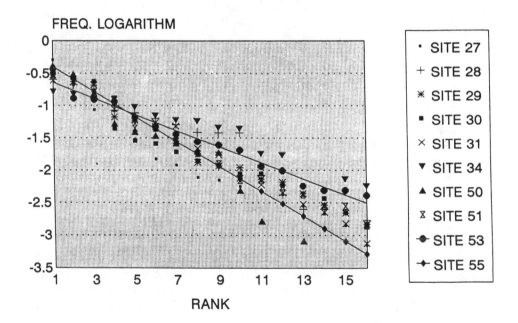

FIGURE 5   Like Figure 1, diagram for those sites on the CDR-1 and CRD-2 of the light chain. Again, the fit to a straight line is very good, with correlation coefficients, for all the cases, $r > 0.97$.

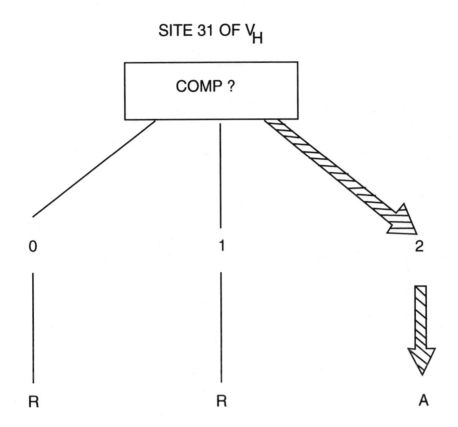

FIGURE 6   Tree constructed with the output data of the induction algorithm of
Quinlan.[19] It was found that most of the recognition sites require only one attribute
to characterize the cluster. As an example of the type of results obtained, site 31 of
the heavy domain is shown. Letters A and R on the figure stand for high (inside the
skewed part of the found distribution) or low use, respectively.

   The above results strengthen the possibility that the recognition process, at
least in certain sites, requires only general properties or attributes, instead of specific complementary amino acids. The general attributes, or physicochemical properties required for each site, are thus satisfied by a reduced set of amino acids
corresponding with the skewed part of the distribution.

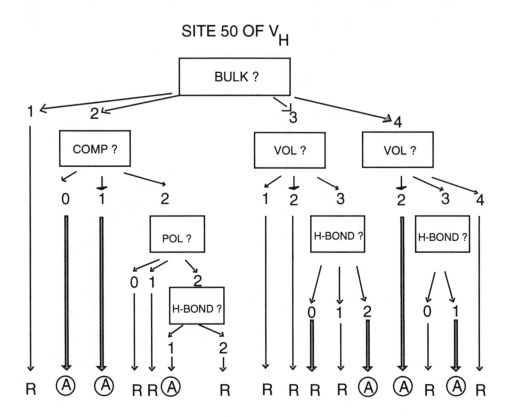

FIGURE 7   Tree constructed with the output data of the induction algorithm of Quinlan.[19] This figure shows site 50 of the heavy chain as an example of an amino acid which was not possible to cluster in terms of these attributes.

The importance of having these general properties in a determined position may be illustrated by the case where large amino acid residues are the more frequently used. These may cover large spaces, facilitating their participation in a wide variety of van der Waals and electrostatic interactions in order to bind a range of epitopes.[16] Another case would be when the common attribute is high flexibility, facilitating the molding of the Igs to different but closely related pathogens and favoring the immune response.[8,24] In this way, amino acids with flexible side chains may generate structurally plastic regions, allowing them to mold themselves around the antigens in order to improve complementarily the interacting surfaces.

TABLE 3 Attributes which identify the cluster of amino acids more frequently used. The sites clustering in terms of one or two common attributes are shown on column 1. For the classification, nine physicochemical properties of the amino acids were considered: composition, polarity, volume, bulk, molecular weight, charge, hydrogen bonding capacity, and aromatic ring. The amino acids were grouped in a discrete way for each one of these attributes. For sites 31, 33, 53, and 60, only one attribute was necessary to cluster the amino acids most frequently used in the site. For the other sites a second attribute was necessary to classify the cluster.

Site	Attributes
31	COMPOSITION
33	VOLUME
35	VOLUME, COMPOSITION[1]
53	BULK
54	VOLUME, COMPOSITION
56	POLARITY, COMPOSITION
58	BULK[1]
60	VOLUME
61	BULK, H-BOND
63	BULK, H-BOND

[1] In these sites the final clusters included an amino acid that was not frequently used.

One question that arises from the above results is about the origin of the observed skewed distributions. Are these indeed a feature of the molecular recognition of Igs, or are they only a result of certain bias existing in the analyzed sample (only 1500 Igs)? In order to elucidate this question, all the reported sequences for the Igs germ-line genes were aligned (Gene-bank, release 65), and the amino acid frequencies for each site were analyzed for the type of distribution. The analyzed frequencies also obeyed a skewed distribution, similar to that found for the protein sample. This seems to indicate that the preferential use of certain amino acids with same physicochemical properties in the recognition sites is indeed a feature of the Igs, suggesting the existence of a general mechanism for the recognition process.

REFERENCES

1. Beckmann, J. S., V. Brendel, and E. N. Trifonov. "Intervening Sequences Exhibit Distinct Vocabulary." *J. Biomol. Struct. & Dynamics* **4** (1986): 391–400.
2. Beckmann, J. S., and E. N. Trifonov. "Splice Junctions Follow a 205-Base Ladder." *Proc. Natl. Acad. Sci. USA* **88** (1991): 2380–2383.
3. Breslauer, K. J., R. Frank, H. Blocker, and L. Marky. "Predicting DNA Duplex Stability From the Base Sequence." *Proc. Natl. Acad. Sci. USA.* **83** (1983): 3746–3750.
4. Chothia, C., A. M. Lesk, A. Tramontano, M. Levitt, S. Smith-Gill, G. Air, S. Sheriff, E. Padlan, D. Davies, W. Tulip, P. M. Colman, S. Spinelli, P. M. Alzari, and R. J. Poljak. "Conformations of Immunoglobulin Hypervariable Regions." *Nature* **342** (1989): 877–883.
5. Cocho, G., R. Perez-Pascual, and J. L. Rius. "Discrete Systems, Cell-Cell Interactions and Color Patterns of Animals. (I) Conflicting Dynamics and Pattern Formation." *J. Theor. Biol.* **125** (1987): 419–435.
6. Cocho, G., R. Perez-Pascual, J. L. Rius, and F. Soto. "Discret Systems, Cell-Cell Interactions and Color Patterns of Animals. (II) Clonal Theory and Cellular Automata." *J. Theor. Biol.* **125** (1987): 437–447.
7. Davies, D. R., and E. A. Padlan. "Padlan Antibody-Antigen Complexes." *Ann. Rev. Biochem.* **59** (1990): 439–473.
8. Dildrop, R. "A New Classification of Mouse Vh Sequences." *Immunol. Today* **5** (1984): 85–86.
9. Goodwin, B. C., and L. E. H. Trainor. "The Ontogeny and Phylogeny of the Pentadactyl Limb." In *Development and Evolution,* edited by B. C. Goodwin, N. Holder, and C. C. Wylie, 75–98. Cambridge: Cambridge University Press, 1983.
10. Goodwin, B. C., and S. A. Kauffman. "Bifurcations, Harmonics and the Four Color Wheel Model of Drosophila Development." In *Cell to Cell Signaling: From Experiments to Theoretical Models,* edited by A. Goldbeter. London: Academic Press, 1989.
11. Grantham, R. "Amino Acid Difference Formula to Help Explain Protein Evolution." *Science* **185** (1984): 862–864.
12. Ho, Mae-Wan. "An Exercise in Rational Taxonomy." *J. Theor. Biol.* **147** (1990): 43–57.
13. Kabat, E. A., T. T. Wu, and H. Bilofsky. "Unusual Distributions of Amino Acids in Complementary-Determining (Hypervariable) Segments of Heavy and Light Chains of Immunoglobulins and Their Possible Roles is Specificity of Antibody-Combining Sites." *J. Biol. Chem.* **252** (1977): 6609–6616.
14. Kabat, E. A., T. Wu, H. M. Perry, K. S. Gottesman, and C. Foeller. *Sequences of Proteins of Immunological Interest,* 5th ed. Bethesda: National Institutes of Health, 1991.

15. Mandelbrot, B. B. *Fractals Form, Chance, and Dimension.* San Francisco: Freeman, 1977.
16. Mian, I. S., A. R. Bradwell, and A. J., Olson. "Structure, Function and Properties of Antibody Binding Sites." *J. Mol. Biol.* **217** (1991): 133–151.
17. Ohno, S., N. Mori, and T. Matsunaga. "Antigen-Binding Specificities of Antibodies are Primarily Determined by Seven Residues of Vh." *Proc. Natl. Acad. Sci. USA* **82** (1985): 2945–2949.
18. Oster, G. F., N. Shubin, J. D. Murray, and P. Alberch. "Evolution and Morphogenetic Rules: The Shape of the Vertebrate Limb in Ontogeny and Phylogeny." *Evolution* **42** (1988): 862–884.
19. Quinlan, J. R. "need chapter title." In *Machine Learning: An Artificial Intelligence Approach*, edited by R. S. Michalski, J. G. Carbonell, and T. M. Mitchell, 463–482, Volume 1. Los Altos, CA: Morgan Kauffman, 1983.
20. Shockley, W. "On the Statistics of Individual Variation of Productivity in Research Laboratories." *Proc. IRE* **45** (1957): 279–290.
21. Simon, H. A. *The Sciences of the Artificial.* Cambridge: MIT Press, 1969.
22. Sneath, P. H. A. "Relation Between Chemical Structure and Biological Activity in Peptides." *J. Theor. Biol.* **12** (1966): 157–195.
23. Solovyov, V. V., and N. A. Kolchanov. "Functional and Evolutionary Role of Intron-Exonic Structure of Eukaryotic Genes." In *Molecular Biology: Structure and Function of Genomes*, edited by V. Ratner and A. Moscow, Vol. 21, 38–73. Nauka, 1985.
24. Tainer, J. A., E. D. Getzoff, Y. Paterson, A. J. Olson, and R. A.Lerner. "The Atomic Mobility Component of Protein Antigenicity." *Ann. Rev. Immunol.* **3** (1985): 501–535.
25. Wu, T., E. A. Kabat. "An Analysis of the Sequences of the Variable Regions of Bence Jones Proteins and Myeloma Light Chains and Their Implications for the Antibody Complementarity." *J. Exp. Med.* **132** (1970): 211–250.
26. Zuckerkandl, E. "Polite DNA: Functional Density and Functional Compatibility in Genomes." *J. Mol. Biol. Evol.* **24** (1986): 12–27.
27. Zvelebil, M. J., G. J. Barton, W. R. Taylor, and M. J. Sternberg. "Prediction of Protein Secondary Structure and Active Sites Using the Alignment of Homologous Sequences." *J. Mol. Biol.* **195** (1987): 957–961.

II. Development and the Individual: Processes of Identity

Brian Goodwin
Developmental Dynamics Research Group, The Open University, Walton Hall, Milton Keynes MK7 6AA, England

Development as a Robust Natural Process

INTRODUCTION

When Darwin contemplated the design of the vertebrate eye, he had very mixed feelings. On the one hand, he was filled with awe at this remarkable result of the evolutionary process; on the other, he found it an enormous challenge to his theory of evolution by natural selection—it made him go cold, he said. How could random variations ever fortuitously conspire to produce a functional eye, that initial step required before natural selection could get a grip on such an apparently improbable organ and subsequently refine it into the sophisticated visual systems found throughout the vertebrates and in invertebrates such as gastropods, cephalopods, crustacea, and insects. What is even more extraordinary is that this organ has evolved independently in at least 40 different taxa.[28] Eyes seem to pop up all over the evolutionary map, and each time they present the same challenge, provoking the same Darwinian shudder: how could random, independent events ever generate such an inherently improbable, coherently organized process as that required to generate a functional visual system in the first place?

In this chapter, I'm going to suggest that the reason why organisms appear to be able to make eyes so easily is that the basic processes involved in embryonic development lead naturally and easily to such organs. This is the meaning of the title: development as a robust natural process. What we see happening during the development of embryos is, I believe, the unfolding of a natural dynamic whose morphological results such as eyes and limbs, hearts and lungs, are high-probability, robust states of the process. Once we understand the principles operating in developing organisms as physical systems of a particular kind, it will become clear why organ sms have the shapes and forms they do. This is rather different from the neo-Darwinian position that organisms generate highly improbable structures, such as the eye, that persist only because they are functionally useful. Natural selection, according to this view, holds organisms in these unlikely states by stabilizing genomes with genetic programs that guide the developing organism through dense thickets of possible states to those that are consistent with survival. I shall be taking a critical look at this perspective and propose an alternative. Nothing of value in the current theory of evolution is lost in the recasting of evolutionary problems that I shall suggest, but there is a significant change of focus from genetic programs and natural selection, towards the principles of development as a major source of the order and organization we see throughout the biological realm. This argument has been developed also in a publication by Goodwin, Kauffman, and Murray,[9] and similar ideas can be found in Newman[23] and Mittenthal.[20]

THE GENERATION OF FORM

The organism whose development I shall now describe and analyze is about as far as you could possibly get from vertebrates, indeed, at the other end of the evolutionary spectrum. All the more interesting, then, will it be to realize that its morphogenesis depends upon the same basic principles of cellular organization as those that operate in higher organisms. The species in question is the unicellular green alga, *Acetabularia acetabulum*, whose life cycle is shown in Figure 1. It has long been a favorite with developmental biologists because this single cell grows into a complex, rather beautiful form several centimeters in size. In doing so, it passes through an interesting sequence of shape changes that are perfectly repeatable—a well-defined developmental sequence that faces us with the same basic problems of morphogenesis as we encounter in the development of multicellular organisms, but in simpler form.

Acetabularia lives in shallow waters around the shores of the Mediterranean. The life cycle is usually described as beginning with the fusion of two gametes (isogametes in this case, there being no sexual differentiation). These produce a more or less spherical zygote which becomes polarized and generates an axis consisting of a rootlike structure (the rhizoid), where the nucleus remains, and a growing stalk or stigma. The stalk extends and, after several weeks of growth when the alga has reached a length of about 1.5 cm, a ring of small elements is produced. These grow into delicately branching structures called hairs or verticils, the whole ring being called a whorl. From the center of a whorl, the tip continues to grow and whorls are produced at irregular intervals of several days. Then, after a number of whorls have been generated, a cap primordium develops at the tip (Figure 2) and grows into the delicately sculpted structure that gives the species its common name, the mermaid's cap. The whorls drop off and we have the adult form of the alga (Figure 3), a giant cell with a stalk 0.5 mm in diameter and 3–5 cm in length, a cap diameter of 0.5–1 cm, and a nucleus that remains in the rhizoid throughout the developmental process. Cytoplasmic streaming carries nuclear products throughout the cell.

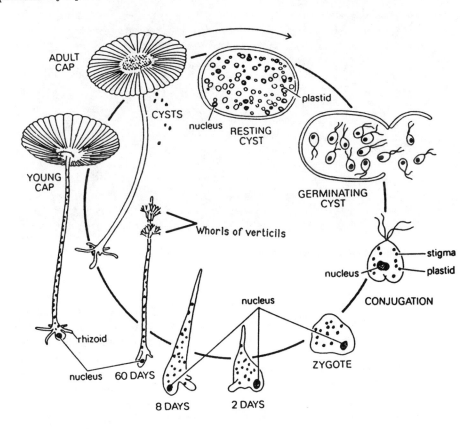

FIGURE 1   The life cycle of *Acetabularia acetabulum.*

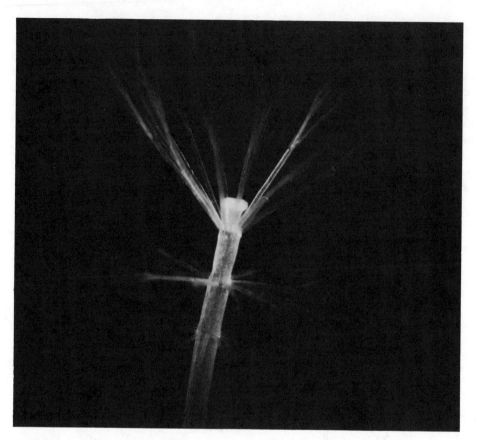

FIGURE 2 A developing cell with three whorls and a cap primordium.

After about three months without further growth or change of form, the alga goes into reproductive mode. The nucleus divides many times, producing thousands of nuclei that are carried up into the cap by the streaming cytoplasm. There they differentiate into gametes, cysts are produced that enclose hundreds of these, and then the cell wall dissolves away and the cysts are released into the seawater. Little "trap doors" open in the cysts, the flagellated gametes swim out and then combine in pairs to start new individuals from zygotes (Figure 1). Plastids, undifferentiated chloroplasts, are included in the gametes, as are mitochondria (not shown), and so are also transmitted from generation to generation with the nucleus.

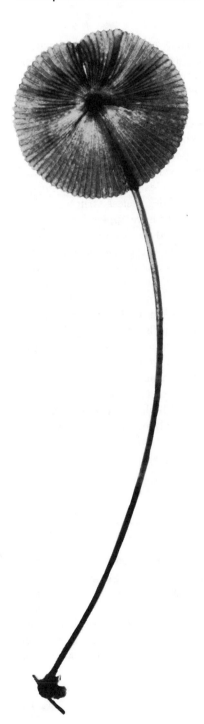

FIGURE 3   The adult form of *Acetabularia.*

Here is a fairly typical life cycle involving a distinct process of morphogenesis leading to a distinctive adult morphology. The cap functions both as a photosynthetic organ and a gametangium, where the gametes are produced. But what is the function of the whorls? Why are they produced when they serve no function in the adult? They certainly are not necessary for photosynthesis. The algae can grow perfectly well without producing whorls or a cap,[12] although they do not reproduce unless caps are made. So how are we to explain the production of whorls by these organisms? The Darwinist perspective appeals to history and function. The following explanatory scenario is typical. Whorls were useful in an ancestor of *Acetabularia acetabulum* and have persisted in this species because of hereditary inertia and a failure of natural selection to eliminate these useless structures during development—like the explanation offered for why we have appendices. To find evidence that supports this interpretation, we are invited to examine other members of the *Dasycladales*, the taxonomic group to which *Acetabularia acetabulum* belongs. Here we do, indeed, find species in which no cap is produced, the whorls functioning as the gametangia in which gametes and cysts differentiate. This is taken as support for the hypothesis. However, there are some serious limitations to this position. First, it fails to explain how whorls are generated in the first place, and so it gives no insight into the process that makes whorls possible consequences of growth and morphogenesis in algal cells. Second, the assumption that natural selection stabilized whorls when they first appeared and were used as gametangia, but failed to eliminate them from the developmental process when they are no longer required, is of virtually no explanatory value since the reasoning is *ad hoc* and arbitrary; it simply redescribes what is observed. Finally, and most seriously, there is no way in which the hypotheses of the historical/functionalist framework of explanation can be investigated systematically and put to direct test. So let's try an alternative approach.

THE DYNAMICS OF MORPHOGENESIS

Let us see if we can find out why *Acetabularia* goes through the particular sequence of shape changes that are observed. Instead of studying algae that are going through their normal life cycle, it is convenient to work with regenerating cells. The cap can simply be cut off at any position along the stalk and a process of regeneration then occurs that follows exactly the same sequence of events as in normal morphogenesis: after healing has occurred at the cut, a tip forms and, after a few millimeters of growth, the tip flattens and a ring of hair primordia is produced which grows into a whorl. The sequence is shown diagrammatically in Figure 4, where only the shape of the wall is shown. The cell membrane (plasmalemma) lies against the inner surface of the wall, with the cytoplasm forming a layer between the plasmalemma and the

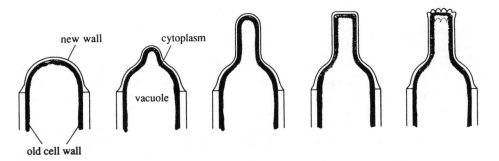

FIGURE 4 Outline of shape changes during tip and whorl formation after cutting through the stalk. The plasmalemma separates the cytoplasm from the cell wall, while the tonoplast separates the cytoplasm from the vacuole.

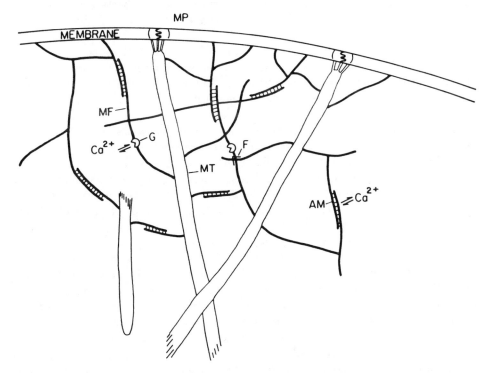

FIGURE 5 The network of elements that make up the cytoskeleton, consisting of microfilaments (MF), microtubules (MT), and actomyosin (AM). Calcium acts on enzymes such as gelsolin (G) that cuts microfilaments and activates actomyosin contraction. Other proteins such as fimbrin (F) strengthen the microfilament mesh, while microtubules are anchored to the plasma membrane by membrane proteins (MP).

tonoplast, another membrane which separates the cytoplasm from the vacuole. The latter makes up the bulk of the cell volume, exerting an osmotic pressure on the cell wall that maintains the turgor of the cell. Tip growth resumes from the center of the whorl and the alga produces a series of whorls before the cap is produced, as in normal growth. The cap grows, the whorls fall off, and the regenerated form is indistinguishable from the original adult alga.

Many workers have studied various aspects of the biochemistry, physiology, and development of *Acetabularia* since Hämmerling discovered in the 1930s that it was a single cell that could be maintained in laboratory culture and lends itself to detailed study because of its large size, its distinctive morphogenesis, and its regenerative powers. With other colleagues, I have been involved primarily in a study of the factors that influence morphogenesis and the process responsible for generating the sequence of shape changes shown in Figure 4, involving the formation of a whorl. Calcium and the cytoskeleton (Figure 5), the network of elements in the cytoplasm of all eukaryotic cells (cells with true nuclei, which excludes bacteria), turn out to be of fundamental importance. So is the behavior of the cell wall and the way it responds to changes of state in the cytoplasm.

In order to try to understand how the interaction of these cellular components results in the integrated process of tip and whorl formation, a mathematical description was developed by Goodwin and Trainor[11] for the basic processes of calcium regulation in the cell and the change of mechanical state in the cytoplasm (elasticity and viscosity). This description was then used for a computer simulation of morphogenesis, adding equations that describe how the cell wall responds to the cytoplasmic state by elastic and plastic changes that describe growth and morphogenesis.[5] These equations define a morphogenetic field. This gives a model that is mathematically quite complex but is biologically extremely simple—a shell of cytoplasm surrounded by an elastic wall. The osmotic pressure exerted by the vacuole was described by equal pressure throughout the inside of the cell wall. The shell of cytoplasm is about 20–30 μm thick and the wall about 5 μ in thickness. The model describes a kind of universal plant-type cell.

A crucial feature of any model of morphogenesis is that it should have the capacity to spontaneously initiate shape formation starting from a spatially uniform state. Otherwise, structure is being put into the model initially, and this is a kind of historical preformationism that begs the question of where this structure comes from in the first place. Organisms have to be self-organizing, self-generating systems. A study of the dynamics of the calcium-cytoskeleton system showed that this system has the properties of an "excitable medium"—it spontaneously generates spatially nonuniform states, described by free calcium concentration and mechanical strain (the degree of stretching or compression of the cytoplasm), from an initially uniform state under random perturbation. The reason for this behavior is as follows.

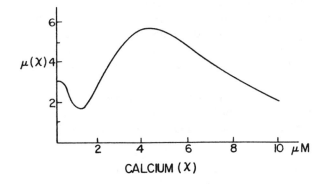

FIGURE 6 Variation of the elastic modulus, $\mu(\chi)$, as a function of free Ca^{2+} concentration, χ (micromolar). The elastic modulus is expressed in arbitrary units, and the graph is a qualitative description only.

Cytosolic free calcium is regulated in eukaryotic cells at concentrations of 100 nM or so by plasmalemma pumps, by sequestration mechanisms involving the endoplasmic reticulum, vesicles, or vacuoles, and by binding to cytoplasmic proteins and to chelating agents such as calcitonin and calmodulin. Studies of actin gels have shown that as calcium rises above 100 nM, it induces gel breakdown and solation, by activation of enzymes such as gelsolin. At higher concentrations, calcium initiates contraction of actomyosin filaments so that the cytoplasm becomes more resistant to deformation.[24] At calcium concentrations above about 5 μM, depolymerization of filaments and microtubules, and the progressive action of gelsolin, have the consequence that the cytoplasm becomes progressively solated. A qualitative description of this behavior in terms of changes in the elastic modulus of the cytoplasm as a function of calcium is shown in Figure 6. This describes how calcium affects the mechanical state of the cytoplasm, measured by the elastic modulus, $\mu(\chi)$. We deduce that there is also a reciprocal action of the mechanical state of the cytoskeleton on free calcium concentration. It is assumed that strain or deformation of the cytoplasm results in the release of calcium from the bound or sequestered state to free ions. Therefore, regions that happen to have elevated strain will also have elevated free calcium levels. But increased free calcium causes gel breakdown and solation. This results locally in more strain (deformation) since the cytoplasm is assumed to be under tension, hence in further calcium release. The result is a positive feedback loop in the regions of calcium concentration where the slope of the elastic modulus curve as a function of calcium is negative (see Figure 6): a local, random increase of calcium above the steady-state level initiates a run-away calcium release and increase of cytoplasmic strain. However, this is stabilized by the effects of diffusion which tends to reduce the calcium gradients and also by the opposing effects of calcium on actomyosin contraction, which increases the elastic modulus and so decreases the strain (region of positive slope, Figure 6).

The argument also works in reverse: where calcium levels are decreased, the strain is also reduced since the cytoplasm is more gel-like (higher elastic modulus) and so free calcium will be bound or sequestered, decreasing it still further. In terms of reaction-diffusion dynamics calcium plays the role of short-range activator while mechanical strain is like the long-range inhibitor. The result of the interactions is that spatial patterns of calcium concentration and strain can arise spontaneously from initially uniform conditions when the equation parameters are in the bifurcation range. The model is qualitatively similar to that discussed by Oster and Odell.[27]

All models have parameters, and the one used for *Acetabularia* morphogenesis has plenty—more than 20. However, the behavior of the model is sensitive to only about six of these; changes in the others produce small effects. Parameter values were chosen to satisfy two criteria:

1. The calcium-cytoskeleton system was in its bifurcation range so that patterns would form spontaneously.
2. Localized growth should occur, initially limited to the top of the dome.

These imposed some general constraints on the model, but nothing that could be called fine tuning. The procedure was then to let the morphogenetic process unfold without parametric change. The process is known mathematically as a moving boundary problem: the dynamics of the calcium-cytoskeleton cell wall system generates a geometrical form which then acts back on the dynamics which, in turn, generates a new geometry. Not much is known about such systems, but they are the category of process to which developing organisms belong. It was expected that many patterns would be possible, and that it would take a long time to find the parameter values that stimulated *Acetabularia* morphogenesis. There was also the distinct possibility that our simple model was lacking basic ingredients and would fail to get even close to the observed sequence. As it turned out, the results were remarkable.

MORPHOGENETIC ATTRACTORS

The model was studied on a computer using a finite-element procedure in which the geometry of the cell is represented by a regular network of finite elements. Figure 7 shows the initial shape of the dome as generated by the model, representing the regenerating tip in the first diagram of Figure 4. Once the parameters were set, the model followed its own course. The first event was the spontaneous formation of a gradient of calcium and mechanical strain in the cytoplasm with a maximum at the pole of the dome, starting from uniform initial conditions. This led to a reduction in the elastic modulus of the wall towards the pole due to the

WALL

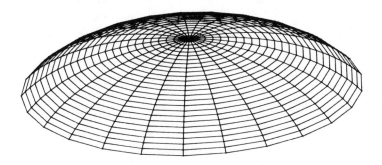

FIGURE 7   The initial shape of the regenerating tip, represented by finite elements that obey the equations of wall deformation and growth defined in the model.

WALL

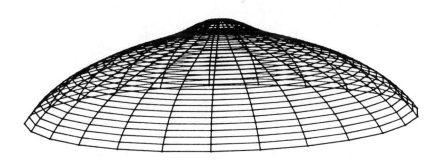

FIGURE 8   The initiation of tip growth in the model.

effect of increased cytoplasmic strain which was assumed to activate pumps that secrete wall-softening agents, in accordance with experimental evidence. The result was the formation of a tip due to elastic formation of the wall under the action of the constant turgor pressure exerted by the vacuole (Figure 8). Growth followed, resulting in plastic deformations of the wall, and then a complex series of shape and state changes occurred. Something that we had never understood was why and how the initially conical tip flattens just before whorl formation (see Figure 4). The model gave us an explanation. The gradient of calcium with a maximum at the pole becomes unstable as growth proceeds, and transforms into an annulus with a ring of elevated calcium and increased strain (see Figure 9). This results in a ring where the wall is softened because of the coupling between cytoplasmic strain and wall elasticity, so the wall develops maximum curvature in the region of the annulus and flattens towards the tip. This also gave the clue to whorl formation. There was a possibility that the calcium-strain annulus would itself become unstable and break up spontaneously by a secondary bifurcation into a ring of peaks that initiate the ring of hairs in a whorl. Perturbation of the annulus led precisely to this

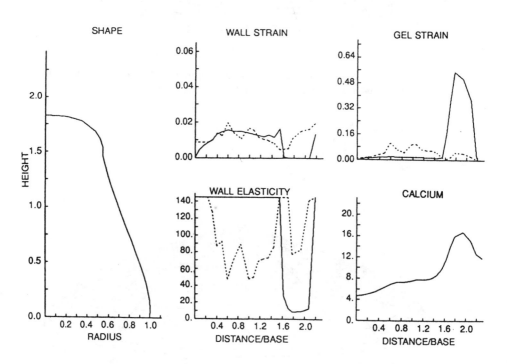

FIGURE 9 Behavior of the model during tip growth. The shape of the tip is shown as a half-section. The variation in cytoplasmic strain (gel strain) and free calcium along the length of the regenerate from base to tip are shown on the right, both rising to a maximum in the region of maximum curvature near the tip and falling again at the pole where the wall flattens. Wall strain and elastic modulus are likewise shown, both falling where gel strain and calcium are maximal. Dotted lines describe properties of the longitudinal finite elements of the model, while the accompanying solid lines describe the properties of the latitudinal elements.

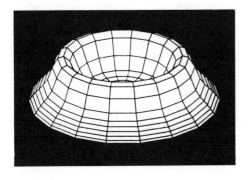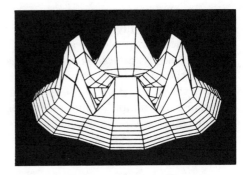

FIGURE 10   A three-dimensional image of the calcium annulus at the tip as in Figure 9 before (left) and after (right) perturbation, showing the formation of a ring of six calcium peaks corresponding to the initiation of a whorl of six verticils.

result, as shown in Figure 10. This provided a natural explanation for the sequence of events leading to whorl formation. Experimental evidence on calcium distribution by Harrison et al.[16] was consistent with this result, though these investigators use a reaction-diffusion model to explain *Acetabularia* morphogenesis. They observed the distribution of bound calcium by using a fluorescent probe, chlorotetracycline, and showed that calcium is maximal towards the tip of a growing cell. This distribution changes into an annulus in association with tip flattening and then breaks into a set of peaks that coincide with the hair primordia.

The behavior of our model suggests that the events leading to whorl formation are a robust and natural consequence of the dynamics of the calcium-cytoskeleton cell wall morphogenetic field; i.e., once the dynamic principles of the system are defined, including the interactions of the components, one of the natural consequences is the generation of a whorl. Technically, a whorl is a consequence of the morphogenetic field equations and so may be a morphogenetic attractor in the space of possible forms of the moving boundary process. The reason why all members of the *Dasycladacles* make whorls may be because they are a natural form that arises from the dynamic principles embodied in the organization of the cell. Some species use them as gametangia and others do not, but all generate them.

The model gave further results. The annulus is not stable as cell growth proceeds, but transforms into a gradient which then changes again into an annulus, this transition occurring with an irregular periodicity—like the irregular formation of successive whorls during algal growth. As yet no structure has arisen from the model that has the precise form of a cap. Cap formation appears to require more anisotropy than is in the model (e.g., more latitudinal than longitudinal strain, leading to differential growth). However, there is no doubt that the model, simple as it is, encourages the view that biological forms are a result of what comes naturally, inevitably, from the dynamic properties at work in developing organisms. My objective here is not to prove that this is so, but simply to demonstrate that it is a distinct possibility, and that there is a clear research programme for the study of biological morphologies as natural forms, as attractors in the space of morphogenetic field dynamics. These solutions must be stable states, part of a stable life cycle if the species is to survive, so natural selection is automatically satisfied. But clearly the explanation of the form comes from an understanding of morphogenetic dynamics, not simply from a recognition that the process must be dynamically stable in a given environment. So natural selection clearly doesn't explain morphology; and neither does gene activity, as we shall now see.

FITTING IN THE GENES

What about the role of genes in these generative processes? Their obvious influence is on parameter values. For growth and morphogenesis to occur in the model, the parameters must be in a particular range. Within this range, parametric variation results in modifications of morphogenesis: larger or smaller tip diameter, changes in the wavelength of the whorl pattern, failure of annulus and whorl formation so that there is a continuous tip growth, etc. However, as noted above, there are parameter ranges in which the whole sequence of tip formation, annulus formation, bifurcation to the whorl pattern, and resumption of tip growth occur. There is no genetic program that guides the system through its morphogenetic transitions. These are all consequences of the cycle of dynamics generating geometry and then geometry modifying dynamics, the nature of a moving boundary problem. This gives us a "free lunch" view of morphogenesis: just set the process going within the parameter range that allows growth and let it go. The genes define this range, but the organizing principles of the process are embodied in the spatio-temporal properties and behavior of the cytoplasmic-cell wall morphogenetic field. These are described by the field equations, which are influenced, but not defined, by the parameters. The genes, of course, also provide many of the materials out of which the cytoplasm and the cell wall are made (though not calcium and other ions). However, it is clearly not sufficient to know the molecular composition of these structures in order to understand morphogenesis; it is necessary to understand the organizing principles as expressed in the equations. And these are not clear until the model is actually studied, either analytically or by simulation. This is why a genetic program is inevitably inadequate to explain morphogenesis—it simply doesn't show how specific morphologies are generated.

There is another equally important set of influences involved in determining the pattern of morphogenesis in addition to those acting from within the cell. These are the environmental conditions. For instance, changes in the calcium concentration of seawater in which the algae grow can determine whether or not caps, whorls, and tips can be formed, their shape, and the wavelength of the whorl pattern.[12,10,17]

And there is a more active aspect of environmental participation in growth and morphogenesis: electrical currents, due primarily to a flow or chloride ions into the stalk and out of the base, are a necessary accompaniment of growth.[4,26] If these currents are prevented by a current clamp, no growth or morphogenesis occurs. So the dynamics of morphogenesis extends beyond the cell boundaries and includes ionic flux around as well as through the developing organism. Jaffe[18] has argued that this is a widespread accompaniment of growth and morphogenesis.

So we arrive at a causal explanation which is significantly different from the genetic program view of development. Genes are certainly involved. But they define necessary, not sufficient, conditions. For sufficiency, giving us a rational explanation of development, we need to understand the organizing principles of the process as described in the dynamics of the morphogenetic field equations, whose solutions are the morphologies of the developing organism. As in other branches of exact science, the only way of discovering what constitutes a sufficient explanation is to model the process using relevant experimental evidence and to test predictions against further observations.

MULTICELLULAR MORPHOGENESIS—PLANTS

This is all very well for a simple organism like *Acetabularia*, you may be thinking, but what about more complex forms, particularly the vertebrate eye? Let's proceed step by step. First, consider morphogenesis in higher plants. The morphogenerator here is like the growing tip of *Acetabularia*, but now it's multicellular and is called a meristem. This cone-shaped structure grows and generates the leaves in patterns that are highly regular. The process of phyllotaxis, as it is called, produces only three distinct patterns (see Figure 11).

1. Distichous: leaves produced singly on opposite sides of the meristem (180° apart), as in corn and the monocotyledons (grasses) generally.
2. Decussate: leaves produced in groups of 2, 3, 4, or more, depending on the species, with rotation of successive whorls of leaves relative to one another so that one set is located in the spaces above the previous set.
3. Spiral: successive leaves are produced singly at a mean angle of 137.5° from the previous one, producing a spiral.

A close, detailed study of phyllotaxis, both experimental and theoretical, has convinced Paul Green[14,15] that these patterns are basically the only three mechanically stable arrangements of leaf pattern that are possible for a system organized dynamically like the meristem. The generative process involves cell growth and division, resulting in patterns of mechanical strain over the meristem that are reflected

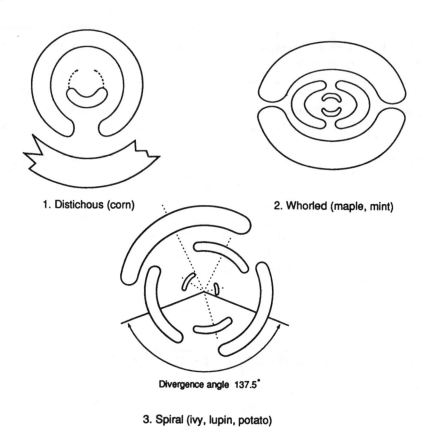

1. Distichous (corn)   2. Whorled (maple, mint)

Divergence angle 137.5°

3. Spiral (ivy, lupin, potato)

FIGURE 11   The three basic patterns of phyllotaxis in higher plants: (1) distichous, (2) whorled, and (3) spiral.

in the orientation of cell wall cellulose microfibrils. These are closely related to the strains in the cytoskeletal elements of cells and calcium concentrations, which together determine the patterns of cell division and localized growth that initiate the leaf primordia. Green observes that there is a self-similarity constraint on the morphogenetic process since the meristem repeatedly generates expanding elements (leaves) that must fit stably into a perpetually self-generating, growing system. Once again it is the spatio-temporal organization of the developmental process, the properties of the morphogenetic field of the multicellular meristem, that appears to provide the explanations of phyllotactic patterns. Different species express different patterns, presumably as a result of specific initial conditions on the growth process and intrinsic wavelengths that are specified by parameters (i.e., genes). And natural selection is again satisfied, as it must be (i.e., tautologically), the different possible arrangements of leaves giving stable growth patterns and providing satisfactory conditions for photosynthesis. However, there is no evidence that one pattern is significantly better functionally than another. More than 80% of plant species have spiral phyllotaxis, but there is no correlation between these and habitat regarding light or shade preference. The explanation for the high frequency of spiral phyllotaxis is, in all probability, simply the greater robustness and stability of this pattern (i.e., a greater domain of attraction of this pattern in the parameter space of the meristem as a developmental generator). However, this conjecture remains to be further studied.

MORPHOGENESIS IN HIGHER ANIMALS

Now, finally, we can return to the initial questions about the vertebrate eye. It is necessary first to look at the basic processes that characterize animal morphogenesis. In plants, we have seen that generation of form is always accompanied by growth, because plant cell walls are relatively rigid and it is only by plastic deformations resulting from new cell wall formation that shape changes occur, whether this is in single cells as in *Acetabularia* or in the meristem of a higher plant where differential growth and oriented cell divisions determine morphogenetic patterns. So plant form always results from outgrowth and the full complexity of the organism's shape is visible to the external observer (including the roots, so long as the plant is grown in liquid or other medium that makes them visible). However, animal embryos can develop complexity by either an outward or an inward deformation of sheets of cells which, lacking walls, are flexible and can change shape with or without growth. This also makes it possible for animal cells to actively migrate over surfaces, either as organized sheets or single cells, or to send out processes such as axons that grow over surfaces in an organized, directed manner. The result is that animals can develop great internal complexity as well as intricate external patterns. But, as we shall now see, this is achieved by basically the same type of cellular organization as that which operates in plants, though now with the freedom of movement that comes from the absence of a cell wall.

The first major morphogenetic movement of vertebrate embryos is gastrulation,[1] the inward movement of cells that is initiated in a particular region of the blastula, the spherical ball of cells that results from cleavage of the fertilized egg (Figure 12). A blastopore forms from the invaginating cells, which spread over the inner surface of the blastula producing a multilayered structure, the gastrula (Figure 13). If the vitelline membrane that normally surrounds the embryo is removed, and the osmotic conditions of the medium in which embryos develop are altered, then instead of invagination followed by migration of cells over the inner surface of the blastula, the blastula buckles out and cells flow over one another to produce an everted sphere, a process called exogastrulation. The lack of the interaction between outer and the inner cell sheets, which normally occurs in the multilayered gastrula and gives rise to further patterns of cell movement and differentiation, results in the failure of the exogastrula to develop normally and a nonfunctional, though coherently differentiated, structure is produced. Under normal conditions, the vitelline

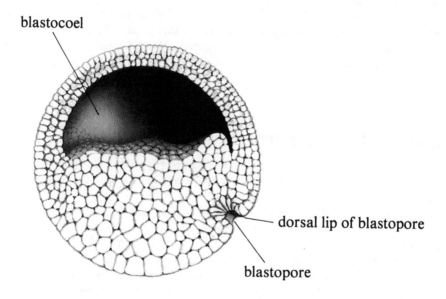

blastocoel

dorsal lip of blastopore

blastopore

FIGURE 12   The initial stage of gastrulation, showing blastopore formation and the structure of the early gastrula.

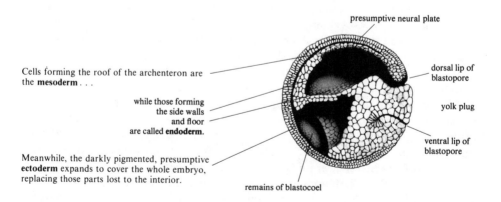

presumptive neural plate

Cells forming the roof of the archenteron are the **mesoderm** . . .

while those forming the side walls and floor are called **endoderm**.

Meanwhile, the darkly pigmented, presumptive **ectoderm** expands to cover the whole embryo, replacing those parts lost to the interior.

dorsal lip of blastopore

yolk plug

ventral lip of blastopore

remains of blastocoel

FIGURE 13   The multilayered late gastrula, showing ectoderm, mesoderm, and endoderm, the yolk plug of cells between dorsal and ventral lips of the blastopore, the region of ectoderm that will become the neural plate, and the diminishing blastocoel.

membrane and osmotic conditions bias the process strongly in favor of invagination and gastrulation. This shows that the mechanical and ionic conditions of the environment are important in determining which of the options (only two in this case) is followed by the developing organism, just as the ionic composition of the medium is important in determining which pattern of growth (local or global deformations) is followed by *Acetabularia*. Organism and environment together define the developmental dynamic and morphogenetic trajectory.

The next major morphogenetic movement of vertebrate embryos is effectively a repeat of the first, but now the infolding of the cell sheet takes place along a line and forms a tube from the neural plate as a result of the influence of the inner cell sheet on the outer and the antero-posterior axis that results from gastrulation. The process is called neurulation and the tube formed is a neural tube, from which the nervous system develops. These deformations of cell sheets to produce the processes of gastrulation and neurulation can be simulated by models that treat cells as excitable media of the same kind as that described for *Acetabularia*: the mechanical state of the cytoskeleton alters in interaction with calcium and cells change their shape, giving rise to propagating waves of cell deformation that result in invagination movements. Figure 14 shows such a simulation by Odell et al.[25] Other aspects of morphogenesis in animal embryos based upon cytoskeleton-calcium dynamics and its consequences with respect to cell shape change, condensation, and migration have been described by Oster and Odell[27] and by Murray and Oster.[21] The

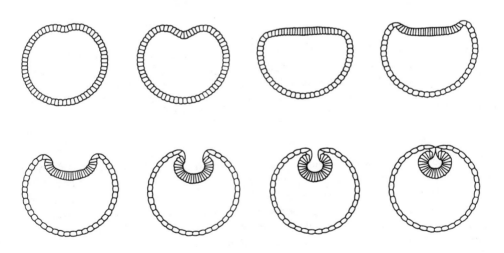

FIGURE 14 A computer simulation of neurulation showing the formation of the flattened neural plate that then folds inward to produce a tube.

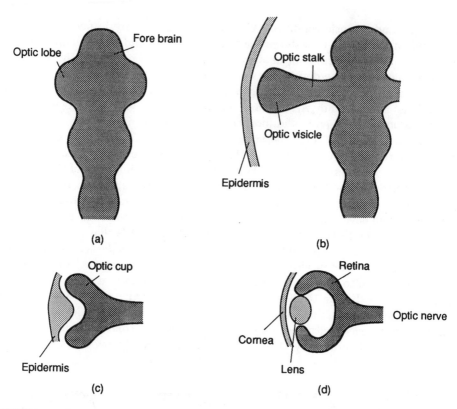

FIGURE 15 Stages in the development of the eye, showing the formation of the optic lobes (a), their outgrowth to form the optic stalk and vesicle (b), interaction between the optic vesicle and the epidermis to form the optic cup and the lens (c), with the final differentiation of the cornea and the retina whose cellular processes form the optic nerve (d).

Acetabularia model was, in fact, inspired by this work. So the case is considerably strengthened for development as a natural process that obeys some basic principles of cytoplasmic dynamics and results in characteristic patterns of morphogenesis that are robust, generic consequences of the organization principles: patterns of cell growth, deformations of cell sheets, spreading of cells and their processes over surfaces, etc. Let us now see how these principles apply to the formation of the eye.

Near the anterior end of the neural tube, tissue begins to bulge out laterally on both sides and grow into bulbous structures, the optic lobes (Figure 15(a)). These continue to grow laterally until they reach the surface layer of the embryo, the epidermis (Figure 15(b)). On contact of the optic lobe with the epidermis, the lobe flattens to form the optic vesicle and then deforms inward in a repeat of the movement that produces the neural tube, forming the optic cup (Figure 15(c)). As this occurs, epidermal cells respond to contact of the optic vesicle by undergoing a transition from squamous (flat) to columnar, resulting in a thickening and inward buckling of the sheet (as in gastrulation and neurulation) which eventually results in the detachment of the thickened cells to form the lens (as did the neural plate in forming the neural tube, but now the geometry is circular rather than cylindrical). The lens becomes transparent, as does the overlying cornea, and the cells of the optic cup differentiate. The outer layer of cells the retina differentiate into neurons whose axons grow over the surface of the retina and down the optic stalk, forming the optic nerve. When they reach the midbrain, they spread out in an organized two-dimensional projection that maps the retina in an ordered manner over the optic tecta of the mesencephelon.

You can now recognize that many of the basic morphogenetic events involved in forming the vertebrate eye are repeats of the basic movements that we have encountered over and over again as the natural state changes of morphogenetic fields organized as described above in terms of calcium-cytoskeleton dynamics, localized cell growth and deformation, bucklings of cell sheets, and directed cell movements over surfaces. These do not account by any means for the full detail of the events that give rise to the highly refined visual systems we see operating in vertebrates. However, that was not our goal, which was the more modest one of suggesting how a primitive but functional system for recording visual images could have arisen independently in many different taxa. It is now clear how simple and natural it is for an embryo to generate a structure of the type shown in Figure 15(c). With a partially transparent epidermis and excitable cells (neurons) in the optic cup, this already functions as a primitive imaging system, a useful visual organ. This is the first necessary step in the evolution of more sophisticated visual systems, which arise by extensions and refinements of basic morphogenetic movements. The processes involved are robust, high-probability spatial transformations of developing tissues, not highly improbable states that depend upon a precise specification of parameter values (a specific genetic program). The latter is described by a fitness landscape with a narrow peak, corresponding to a functional eye, in a large space of possible nonfunctional (low fitness) forms. Such a system is not robust: the fitness peak will tend to melt under random genetic mutation, natural selection being too weak a force to stabilize a genetic program that guides morphogenesis to an improbable functional goal. The alternative is to propose that there is a large attractor (a large range of parameter values) in morphogenetic space that results in a functional visual system—i.e., eyes have arisen independently many times in evolution because they are natural, robust results of morphogenetic processes.

THE EVOLUTION OF GENERIC FORMS

We can now generalize the argument and draw out its evolutionary implications. Let me emphasize again that I am dealing here with a plausible conjecture, not an established result. What we have is a research programme, a direction of enquiry that could make sense of a number of evolutionary phenomena that are extremely puzzling from the neo-Darwinian perspective, in which evolution is to be understood in terms of random genetic variation and natural selection. This view effectively leaves out the basic dynamics of organismic life cycles, the generative principles that underlie the order and regularity of the biological realm. So I shall now suggest that, taking account of these principles and recognizing the likelihood that organismic morphologies are the result of robust morphogenetic processes, evolution is the emergence of the generic forms of the living process. Characterizing these generic forms will then provide us with the principles that can make sense of both the regularity *and* the diversity of the biological realm, just as characterizing the generic modes of any dynamical system explains both its order (regularity) and the variety of expression of this order. There is nothing new about this programme of enquiry in biology. It goes back at least to Goethe's morphological studies in the late eighteenth century, to the work of Geoffroy S. Hilaire in the early nineteenth century, and of William Bateson,[2] of Driesch,[6] Needham,[22] and Waddington.[29,30] The problem is always to discover the generative source of biological order, the structuralist foundations[8,31] that are logically prior to the functionalist and historical preoccupations of neo-Darwinism. So now let's make these issues concrete.

A basic evolutionary problem is how we are to reconcile random variation of genomes, the hereditary lottery, with the systematic morphological order over the biological realm that is revealed by logical classification schemes. If there is a continuum of possible morphologies, all reachable by appropriate genetic variation, while the forms that survive are selected by *external* factors (natural selection), then there is no reason to expect any intrinsic logical order to classification schemes. But the order is there for all to see. So Darwinism invokes a historical principle: because new species derive from old, they share certain characters and differ in others. However, there is no reason for shared characters to persist in lineages adapted to very different habitats, such as frogs and bats in the vertebrate line. Nevertheless, their limbs share a basic structural plan, the tetrapod limb organization. This is believed to persist by inertia, natural selection being unable to get any purchase on variants to produce a better basic design for the flight of the bat than that shared with the jumping frog. So regularity of form is ascribed to historical inertia, an ad hoc, arbitrary assumption since no reason is given why natural selection is successful in some cases and fails in others.

The discontinuities of state between species are regarded as a result of natural selection acting on initially small variations which define a historical trajectory through parameter (genome) space to the fitness peaks of adapted forms, as previously described. The resulting expectation is that fossil remains should provide evidence of the linking trajectories (the missing links between stable species), and the classification scheme should be a tree of progressively bifurcating branches and twigs, all connected to a common ancestor (or a few common ancestors). But the links are still largely missing, despite exhaustive searches; and the tree doesn't fit together as it should. Mayr[19] has gone so far as to say that there is "no clear evidence for any change of a species into a different genus or for the gradual emergence of any evolutionary novelty." Species are discretely distributed; they arise and go extinct with characteristic lifetimes; and, whereas there are hierarchical features of classification schemes, there is no single major cone of taxonomic relationships from the origin of life to contemporary species.[7,13,32]

These difficulties suggest that it is worth considering a different hypothesis. What we see in evolution may be largely a result of the generative dynamics of organismic life cycles, which of course includes inheritance. Random hereditary variation defines a percolation process through parameter space, starting from origins corresponding to morphologically simple organisms. The advancing parameter front hits new attractor domains that generate species with particular morphologies, some of which are stable (the dynamics includes relevant environmental factors, as previously described). The attractors are discretely distributed and vary from small to large, the size of the attractor effectively determining the evolutionary lifetime of a species (i.e., probability of extinction due to random variations). The logical aspects of classification schemes derive from the structure of the generative dynamics. This is intrinsically hierarchical: the sequence of symmetry-breaking bifurcations that generate organismic complexity during development involves progressively shorter wavelengths, as in the example of tip formation, whorl initiation, and the bifurcation of verticils into finer filaments in *Acetabularia* (see Figures 2 and 4). Complexity thus emerges progressively in morphogenesis, finer details developing within the context of previously established larger scale order. As we have seen, the simple spherical structure of the amphibian egg and blastula undergoes a sequence of bifurcations that result in the progressive emergence of more spatial detail, with a repetition of the same generic processes but on smaller wavelengths as in the development of the eye (Figure 15). Such a hierarchical generative dynamic must result in a hierarchical taxonomy.

The basic similarity of eukaryotic morphogenesis, unicellular and multicellular, despite the great variety of molecular materials and their physical properties, which results in the great diversity of biological forms, is what underlies the logical morphological order of the biological realm. This is revealed not only in taxonomy but in other evolutionary phenomena such as:

1. Parallel morphologies evolving in different but related lineages exposed to quite distinct environments, as in the dramatic parallelisms between placental and marsupial mammals.
2. "Mimicry" and "pseudo-mimicry" in butterfly wing patterns, the latter being similar patterns in species evolving in totally different parts of the globe, such as Puerto Rico and Indonesia.

3. Homologous structures persisting in lineages over large evolutionary periods despite great diversity of function, such as tetrapod limbs.
4. The relatively few basic body plans of the phyla (including Burgess Shale species—see Gould[13]) corresponding to the few ways of breaking symmetry in morphogenetic process (see Beloussov[3]).

The basic argument, then, is that the morphological regularity and order that is revealed in evolution arises from the basic generative dynamics of organisms, which is a biological universal. Evolution is the emergence of robust generic forms of the process, driven by a sort of diffusing wave front of parameter variation starting from those values that characterize early life forms. These parameters have been described as specified by genes and aspects of the environment. This is actually too restrictive a description, since the generative dynamics of organisms can change more than parameter values: it can change its dimensionality, the number of levels in the hierarchy, its modularity, its very materials as well as their concentrations. This generative system is complex in the extreme. But this does not imply an absence of robust, generic properties. At the moment we have only the merest hints of how to characterize these. But the remarkable pervasiveness, as well as the subtlety, of biological order strongly suggests that the evolutionary process is an unfolding of the generic states of a complex dynamic system of a distinctive and particularly fascinating type. Understanding biology now means uncovering this order.

REFERENCES

1. Alberts, B., D. Bray, J. Lewis, M. Raff, K. Roberts, and J. D. Watson. *The Molecular Biology of the Cell.* New York/London: Garland, 1983.
2. Bateson, W. *Materials for the Study of Variation.* Cambridge: Cambridge University Press, 1894.
3. Beloussov, L. V. "Generation of Morphological Patterns: The Mechanical Ways to Create Regular Structures in Embryonic Development." This volume.
4. Bowles, E., and N. S. Allen. "A Vibrating Probe Analysis of Light-Dependent Transcellular Currents in *Acetabularia.*" In *Ionic Currents in Development,* edited by R. Nuccitelli, 113–121. New York: Alan R. Liss, 1986.
5. Brière, C., and B. C. Goodwin. "Geometry and Dynamics of Tip Morphogenesis in *Acetabularia.*" *J. Theor. Biol.* **131** (1988): 461–475.
6. Driesch, H. *The Science and Philosophy of the Organism.* 2nd ed. London: Black, 1929.
7. Eldridge, N., and S. J. Gould. "Punctuated Equilibria: An Alternative to Phyletic Gradualism." In *Models in Paleobiology,* edited by T. J. M. Schopf, 82–115. San Francisco: Freeman, Cooper & Co., 1972.
8. Goodwin, B. C. "Structuralism in Biology." *Sci. Progress* (Oxford) **74** (1990): 227–244.
9. Goodwin, B. C., S. A. Kauffman, and J. D. Murray. "Is Morphogenesis an Intrinsically Robust Process?" *J. Theor. Biol.* (1992): submitted.
10. Goodwin, B. C., J. D. Murray, and D. Baldwin. "Calcium: The Elusive Morphogen?" In *Proceedings of the 6th International Symposium on Acetabularia,* edited by S. Bonotto, F. Cinelli, and R. Billian, 101–108. Brussels, Belgium: Belgium Nuclear Center, C.E.N-S.C.K. Mol, 1985.
11. Goodwin, B. C., and L. E. H. Trainor. "Tip and Whorl Morphogenesis in *Acetabularia* by Calcium-Regulated Strain Fields." *J. Theor. Biol.* **117** (1985): 79–106.
12. Goodwin, B. C., J. C. Skelton, and S. M. Kirk-Bell. "Control of Regeneration and Morphogenesis by Divalent Cations in *Acetabularia mediterranea.*" *Planta* **157** (1983): 1–7.
13. Gould, S. J. *Wonderful Life.* New York: Norton, 1989.
14. Green, P. B. "Inheritance of Pattern: Analysis from Phenotype to Gene." *Amer. Zool.* **27** (1987): 657–673.
15. Green, P. B. "Shoot Morphogenesis, Vegetative Through Floral, from a Biophysical Prospective." In *Plant Reproduction: From Floral Induction to Pollination,* edited by E. Lord and G. Bernier, 58–75. Am. Soc. Plant Physical Symp. Series, Vol. 1.
16. Harrison, L. G., K. T. Graham, and B. C. Lakowski. "Calcium Localization During *Acetabularia* Whorl Formation: Evidence Supporting a Two-Stage Hierarchical Mechanism." *Development* **104** (1988): 255–262.

17. Harrison, L. G., and N. A. Hillier. "Quantitative Control of *Acetabularia* Morphogenesis by Extracellular Calcium: A Test of Kinetic Theory." *J. Theor. Biol.* **114** (1985): 177–192.

18. Jaffe, L. F. "The Role of Ionic Currents in Establishing Developmental Pattern." *Phil. Trans. Roy. Soc. B.* **295** (1981): 554–566.

19. Mayr, E. *Toward a New Philosophy of Biology.* Cambridge, MA: Harvard University Press, 1988.

20. Mittenthal, J. E. "Physical Aspects of Organization of Development." In *Lectures in the Sciences of Complexity*, edited by D. Stein, 491–528. Santa Fe Institute Studies in the Sciences of Complexity, Lect. Vol. I. Reading, MA: Addison-Wesley, 1989.

21. Murray, J. D., and G. Oster. "Generation of Biological Pattern and Form." *IMA J. Maths in Med. & Biol.* **1** (1984): 51–75.

22. Needham, J. *Order and Life.* New Haven, CT: Yale University Press, 1932.

23. Newman, S. A. "Generic Physical Mechanisms of Morphogenesis and Pattern Formation as Determinants in the Evolution of Multicellular Organization." In *Principles of Organization of Organisms*, edited by J. Mittenthal and A. Baskin, 241–267. Santa Fe Institute Studies in the Sciences of Complexity, Proc. Vol. XIII. Reading, MA: Addison-Wesley, 1992.

24. Nossal, R. "On the Elasticity of Cytoskeletal Networks." *Biophys. J.* **53** (1988): 349–359.

25. Odell, G., G. F. Oster, B. Burnside, and P. Alberch. "The Mechanical Basis of Morphogenesis." *Devel. Biol.* **85** (1981): 446–462.

26. O'Shea, P., B. Goodwin, and I. Ridge. "A Vibrating Electrode Analysis of Extracellular Ion Currents in *Acetabularia acetabulum*." *J. Cell Sci.* **97** (1990): 505–508.

27. Oster, G. F., and G. M. Odell. "The Mechanochemistry of Cytogels." In *Fronts, Interfaces and Patterns*, edited by A. Bishop. Amsterdam: North Holland, Elsevier Science Division, 1983.

28. Salvini-Plawen, L. V., and E. Mayr. "On the Evolution of Photoreceptors and Eyes." *Evol. Biol.* **10** (1977): 207–263.

29. Waddington, C. H. *The Strategy of the Genes.* London: Allen and Unwin, 1957.

30. Waddington, C. H., ed. *Towards a Theoretical Biology*, Vols. 1–4. Edinburgh: Edinburg University Press, 1968 (1), 1969 (2), 1970 (3), 1972 (4).

31. Webster, G. C., and B. C. Goodwin. "The Origin of Species: A Structuralist Approach." *J. Soc. Biol. Struct.* **5** (1982): 15–47.

32. Wesson, R. *Beyond Natural Selection.* Cambridge: MIT Press, 1991.

L. V. Beloussov
Department of Embryology, Faculty of Biology, Moscow State University, Moscow 119899 Russia

Generation of Morphological Patterns: The Mechanical Ways to Create Regular Structures in Embryonic Development

One of the greatest and as yet unsolved problems of modern biology is morphogenesis, that is, the generation of new shapes and structures during development of organisms. In this paper we associate this problem with the theory of self-organization. Important evidence for the self-organizational properties of morphogenesis is its capacity to reduce the symmetry order (to break symmetry) without necessarily employing any discrete external "dissymmetrizators." Two main kinds of models are competing in present-day science in attempting to interpret these properties, the chemokinetical and mechanical models. We review the explanatory properties of the latter approach. In particular, we develop a simple "internal pressure" model and examine its ability to reconstruct, in a biologically realistic way, the morphogenesis of U-shaped and circular epithelial rudiments. As shown by the modeling, some shapes, which we define as generic, appeared to be produced under much more varied initial conditions and model parameters than the others (nongeneric). Some simple changes in mechanical properties could induce a rapid switch from one generic shape to another. Within this data framework, we discuss the role of natural selection in evolutionary shaping and in the genome-morphogenetic relations.

Thinking About Biology, Eds. W. D. Stein and F. J. Varela, SFI Studies in the Sciences of Complexity, Lect. Note Vol. III, Addison-Wesley, 1993   **149**

1. SYMMETRY BREAKING IN DEVELOPMENT: EXAMPLES AND CONCLUSIONS

Any person who has the courage to open a textbook in embryology, and to study only a few pictures in it, very soon realizes that almost each step of an individual's development is associated with the appearance of certain new structures or, at least, with an increase in complexity of a shape of an observed object—either on a subcellular, or on a cellular-supracellular level. Ooplasmic segregation in early egg development, various kinds of blastomere arrangements during cleavage, creation of a primary gut by invagination of a part of a blastula wall, formation of brain vesicles from the anterior part of a vertebrate neural tube, subdivision of a mass of mesodermal cells into a definite number of somites, or creation of precise radial patterns in Anthozoa polyps and Echinodermata embryos are only a few examples of these many processes, which are usually defined as morphogenetic. Is there something in common among all of them allowing us to look for some general principles of shaping and structure formation in developing organisms?

To answer this question, let us begin from a mere description of morphogenetical processes. For someone familiar with the elements of a symmetry theory, all of the above-mentioned events obey the same regularity: they are associated with the decrease in *symmetry order* (breaking of symmetry, or dissymmetrization). For example, when a spherical blastula with an indefinite number of rotational axes (we assume a spherical blastula for simplicity) transforms into a gastrula with only one rotational axis, a symmetry order reduces from $\infty/\infty \cdot m$ to ∞/m. This means that a body with an infinite multiplicity of rotational axes crossed at arbitrary angles, each one of the axes being also of infinite symmetry order, transforms into a body with one rotational axis only, also of an infinite symmetry order (m defines a mirror symmetry plane, retained in both cases). When an epithelial cylinder of a juvenile Anthozoan sample subdivides into six septae, this is also a dissymmetrization expressed by a transition $\infty \cdot m \rightarrow 6 \cdot m$. Many morphogenetical processes are associated with so-called *translational dissymmetrization*, when a more or less homogeneous row of tissue (a mesenchymal mass, or an epithelial tube) is subdivided longitudinally into a series of parts (segments, vesicles, and so on). On the other hand, only a few developmental processes are associated with an *increase* in symmetry order. These are usually linked to a metamorphosis, for example, the transformation of an asymmetrical (that is, of the first symmetry order) late sea star larva to an adult with a radial symmetry (of order 5).

Dissymmetrization, or symmetry breaking, is a remarkable natural process. As the great French physicist, Pierre Curie, said about a hundred years ago, a dissymmetrization creates an event. He implied by this brief expression that a dissymmetrization gives to a body its characteristic features distinguishing it from the surrounding environment. On the other hand, while discussing the necessary conditions for dissymmetrization, Pierre Curie postulated that this process cannot occur spontaneously: if a "visible" dissymmetrization of a body is taking place, that means that either a body is already internally dissymmetrical, even in some "invisible" manner (nondistinguished by the methods employed), or that it is affected by some "dissymmetrisators" located in the external environment.

This Curie principle should be considered as a particular case of a wide concept of a uniform determinism which since the time of Laplace has moulded the basis of classical science. According to this concept, each natural event should have its own specific cause which should be of the same order of magnitude. This implies that negligibly "small" causes cannot generate noticeable events, and vice versa. One of the main postulates of uniform determinism is also known as the "one cause–one effect" rule. Not only in the nonbiological sciences, but also in biology, including embryology, phenogenetics, and so on, this rule has often been taken for granted and widely used, although mostly in an intuitive and undefined way. But is such an approach really adequate for developmental processes so often associated with symmetry breaking? Let us look at some examples, starting from the first steps of development.

Amphibian eggs acquire their animal-vegetal polarity already in the ovary and soon after hatching may be seen as rotational bodies with a symmetry $\infty \cdot m$, that is, without a preferential plane of symmetry. The next developmental step is in establishing a *saggital symmetry plane*, or, in the other words, the dorso-ventral axis. Does an egg need some external dissymmetrisator to do that? At the first glance, the answer should be positive. As has been shown by observation and experiment, within normal development a spermatozoon acts as such a dissymmetrisator, marking the antero-ventral embryo pole by its entrance point. Further investigation has meanwhile shown that Earth gravity is an even more fundamental dissymmetrisator since the saggital plane always coincides with the plane of the gravitational shift of the heavy endoplasmic particles as related to the egg cortex.[21] Anyway, in both cases the Curie criterion seems to apply. However, neither the spermatozoon nor the gravity force appeared to be the necessary conditions of dissymmetrisation: an egg can acquire its definite saggital plane either by being activated parthenogenetically, without a spermatozoon, or, if a spermatozoon is inserted exactly into the animal pole, by being in this position unable to select any meridial plane for becoming saggital.[16] Similarly, an egg acquires its dorso-ventral polarity also in microgravity conditions.[8] On the other hand, what is ultimately necessary for establishing dorsoventrality is the preservation of certain internal mechanisms which *reinforce* the external, maybe negligibly minor influences. Those appeared to be the arrays of microtubuli on the ecto-endoplasmic border of an egg, obviously promoting the relative sliding of both major egg constituents.[9]

There is even less likelihood of pointing to any definite external dissymmetrisator if one considers such fundamental symmetry breakings of subsequent morphogenesis as are associated with the subdivision of an embryonic neural tube to yield brain vesicles, segmentation of a mesoderm, or subdivision of an Anthozoans body into six or eight septae. And what is probably even more important, quite a similar, although invisible, "spontaneous dissymmerization" takes place during the fundamental process of so-called *embryonic induction*, that is the influence of some embryonic rudiment (inductor) on an adjacent one, stimulating the latter to differentiate and create a new structure. As is generally accepted, the progressive development of at least vertebrate embryos may be considered as a chain of successive inductions.

The most important question for us is whether the spatial structure of an *induced* rudiment is embedded in all its details in the *inductor* itself. (In this case the induction should be considered as a kind of blueprint, implying no dissymmetrisation at all.) On the other hand, the spatial structure of an inductor can be much simpler than that of the induced rudiment so that the induction process, or, rather, a reaction of an induced tissue is associated with a reduction in symmetry order.

As examples we shall consider only two, although quite illustrative, experiments in this field. The first was made by Spemann and Schotte.[20] As is well known, Anuran tadpoles have in their oral region quite a definite pattern of so-called dental rows, lacking in Urodela. These rows can be formed from any piece of the Anurans' embryonic ventral ectoderm, if transplanted onto the oral pole of another embryo of the same species. Therefore, some inductive influence may be exerted by the underlying (mesodermal) oral tissues and one cannot as yet avoid the conclusion that the underlying tissue contains in itself a detailed spatial "prepattern" of the dental rows. The most astonishing thing, however, was that the experimenters found also perfect dental rows when the Anuran's ventral ectoderm was transplanted onto the Urodela oral pole! Since it is clear that the alien Urodelan inductor could not contain in itself a "prepattern" for the Anuran dental rows, we are forced to conclude that only quite general, and spatially nonprelocalized, inductive signals are exerted by an inductor and that these are sufficient for the "competent" embryonic tissue to create quite definite structures. In other words, the inductive reaction involves "delocalization" and a considerable reduction in symmetry order.

Another, even more extreme situation was reproduced by Boris Balinsky in 1924.[1] Transplanting an auditory vesicle into the flank region of Triton embryos, he obtained a supernumerary limb induced by the vesicle from surrounding tissues. It is clear that an auditory vesicle (which can be replaced, with the same result, by a nasal placode or even a piece of celloidin) cannot contain any "structural information" about a limb which is to arise. It simply stimulates the surrounding tissue to "self-organize" itself into a limb.

2. SELF-ORGANIZATION: WHAT DOES IT MEAN AND HOW TO REACH IT (A BRIEF REVIEW)

So, we have pronounced this remarkable word *self-organization*. For contemporary science it is much more than a mere word: it defines a new class of natural processes and systems with largely nonclassical properties. These systems do not obey the classic Curie principle and other maxims of a uniform determinism. They are capable of "spontaneous" dissymmetrisation; negligibly small causes may produce in these systems enormously large effects; one "cause" can generate more than one effect; some strictly localized "causes" may produce delocalized effects, and vice versa. Generally, any kind of distinction into discrete "causes" and "effects" looks inadequate for this class of systems. They should be largely replaced by quite other notions, such as "stability" and "instability" (it is only instability which permits the "spontaneous" dissymmetrisation), parametric and dynamic influences, and positive and/or negative feedback. (For an embryologically oriented discussion of these notions see, Beloussov.[3])

The interdisciplinary branch of science which deals with this kind of system was outlined as a coherent body about two or three decades ago and is usually defined as a theory of self-organization[17] or synergetics.[10] Its creation is one of the most remarkable but as yet underestimated (at least, by the representatives of the biological "mass culture") scientific events of this century. With the appearance of self-organization theory, it became obvious that some of the most fundamental and hardly understood properties of living beings are in fact shared by a large number of nonbiological natural events—chemokinetical, mechanical, hydro- and aerodynamical, and even cosmological. What all of them have in common is that they are far from thermodynamic equilibrium and their elements are linked with strong feedback, described mathematically by nonlinear differential equations.

By attributing the morphogenetic processes within embryonic development to self-organization, we should largely reformulate our research strategy. Instead of trying to discover some unique and specific cause for each developmental process, we should now focus our interest on revealing the fundamental nonlinear feedbacks which underlie the entire flow of the developmental processes and lead, in some of its space-time areas, to loss of stability. Two main research approaches, one empirical and one associated with the construction of models, go hand-in-hand to fulfil this task. Let us review quite briefly some self-organizational models in developmental biology, created within the last two decades.

The first, and probably most popular of these, has been formulated in purely chemokinetical language.[14] The central idea of this class of models was to postulate certain morphogenetically active diffusible chemical substances, activators (A) and inhibitors (H), linked by positive-negative feedback. Particularly, A is assumed to catalyze in a nonlinear way the production of H, whereas H linearly inhibits A production. Organization in space is reached in this model by assuming diffusion of both the reagents, the rate of H diffusion being much greater than that of A diffusion.

We shall not discuss here the numerous pros and cons of this class of model. At all events, it was a milestone in our understanding of developmental phenomena, demonstrating for the first time that self-organizational approaches can really work in this field of biology. But, on the other hand, it is difficult to avoid the impression that the purely chemokinetical models have little in common with real shaping processes, which are mostly associated with mechanical variables, such as deformations, forces, and stresses, rather than with concentrations and free diffusion. And is it not possible that these purely mechanical events, even without the participation of some specifically distributed diffusible substances, are able to self-organize morphological patterns? To reply to this question, let us consider some extremely simple mechanistic examples.

Take a thin plastic bag and blow into it. Almost inevitably you will get a whole series of thin parallel grooves oriented along the main "ribs" of the bag (Figure 1(a)). If, now, an elastic shell acquires the shape of a flattened barrel, its inflation by increase of internal pressure will be associated with the formation of a definite number of the radially oriented folds (depending upon the shell thickness and the barrel's shape). If, meanwhile, the air is sucked out and the internal pressure decreases, the radially oriented folds will be replaced by horizontal ones (Figure 1(b)). In all these cases quite a regular and definite pattern emerges, in the absence of any preexisting chemical heterogeneity (as postulated by chemokinetical models) by the mere action of roughly isotropical forces of internal pressure.

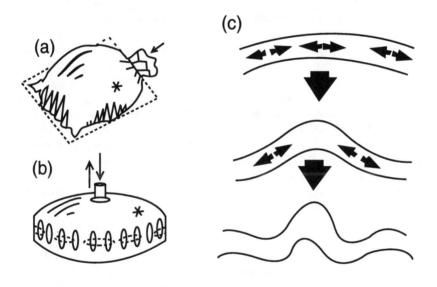

FIGURE 1   How mechanics may create shapes: (a) by blowing into a plastic bag; (b) by increasing (solid arrow) or decreasing (arrow) internal pressure into a flattened elastic barrel (an increase creates a series of vertical folds whereas a decrease create a series of horizontally oriented ones); (c) by curving a slightly bent rod or sheet as increasing internal pressure (diverged arrows). ((a),(b) from L. A. Martynov.[13] "The Role of Macroscopical Processes in Morphogenesis." In *Mathematical Biology of Development*, edited by A. I. Zotin and E. V. Presnov, 135–154. Nauka: Moskva, 1982; (c) original.)

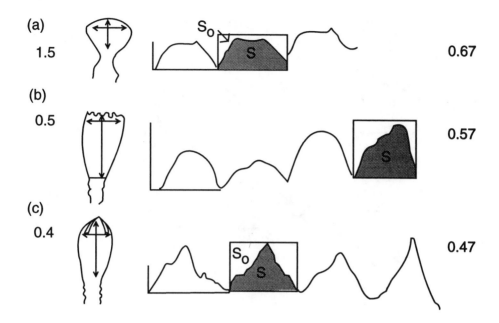

FIGURE 2   Correlations between growth pulsation patterns (at right) and rudiment shapes (at left) in three different species of hydroid polyps, *Dynamena pumila* (a), *Obelia loveni* (b), and *Campanulina lacerata* (c). Numbers at left show the width/length relations of the rudiments and numbers at right S/S_0 relations as being the measure of the "width" of the growth pulsations. Both rows of numbers correlate with each other (from L. V. Beloussov, Ju. A. Labas, and L. A. Badenko.[2] "Growth Pulsations and the Rudimens Shapes in Hydroid Polyps." *Zhurn. Obsch. Biol.* **45** (1984): 796–806 (Russ.)).

Similar results can be achieved by pressure forces acting within shells (or rods) themselves. Consider poorly reinforced iron rails heated by a bright sun. They should buckle due to the increase in internal pressure. When the rails are still straight and not reinforced from any side, their buckling direction is unpredictable (this is a classic Eulerian instability). From the instant of their initial buckling, however, they will continue to buckle to the same side, inevitably creating bending of the opposite sign on the flanks. Those, as pressure continues, will also increase, creating new adjacent bendings and so on, generating, as a result, a wavelike pattern (Figure 1(c)).

Those are some of the most primitive but, nevertheless, real examples of self-organization going in a purely mechanical way. In all of these cases, self-organization means the creation of macroscopical patterns by the action of forces distributed in a much more homogeneous way than the structures that arise. Hence, this kind of transformation also implies a spontaneous breaking of symmetry.

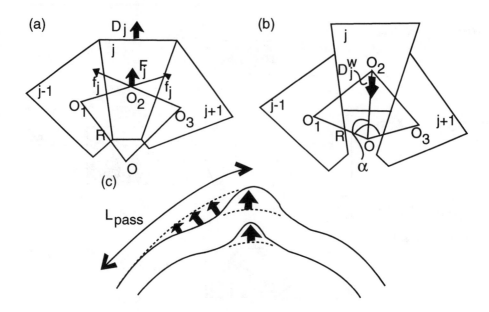

FIGURE 3 The algorithms and parameters employed in the Internal Pressure Model (IPM). In (a) and (b) three adjacent cells $j - 1$, j, and $j + 1$ are shown, with their centers O_1, O_2, and O_3. O is a curvature center. F_j is a resultant of the lateral pressure forces, D_j in (a) symbolizes the centrifugal shift of a cell j; and D_j^w in (b) is a force of elasto-viscous resistance which counteracts F_j. L_{pass} in (c) symbolizes the passive involvement of an adjacent part of a cell layer by an actively shifted cell (dense arrows).

Even *a priori*, mechanical ways of self-organization have some advantages over chemokinetical ones. First, they do not need the refined and vulnerable devices necessary for organizing the synthesis, inhibition, and distribution of diffusible chemicals within the bodies of a complicated geometry. Second, local patterning will depend directly upon the overall shape of a body. Last, but not least, mechanical signals (transmission of the stretching-compression stresses) should spread within the embryonic tissues in a directed manner, at a great (sound wave) rate and over large distances. It would be very strange if organisms, already in early evolution, did not use these almost unavoidable and straightforward mechanisms for regulating their shaping. (For the use of mechanical principles in morphogenesis of bacteria and lower plants, see Harold.[11])

But are pressure forces really present in the developing tissues? For many cases, a positive answer can be given. In some developing organisms at least, the pressure forces are accurately regulated in space and time. For example, in hydroid polyps the cells of the tip region periodically swell and deswell due to osmotically driven water uptake and outflow.[4] During swelling, the corresponding parts of the cell sheets come under increased internal pressure and should buckle like a heated rail. During deswelling, they contract, but a considerable part of the previously achieved deformation is "frozen" by a constantly secreted and rapidly hardened external skeleton, the perisarc. The patterns of extension-contraction periodicity are perfectly regulated, being both stage- and species-specific. Generally, the greater the relative periods of the extension (increase in swelling pressure), the wider are the rudiments which arise, and vice versa (Figure 2). In vertebrate embryonic rudiments, pressure stresses are also present (as can be detected from the extensive overlapping of the tissue edges immediately after their dissection) but are due mostly to some other mechanisms, such as cell proliferation and cell intercalation.[12]

Embryonic tissues are, however, mechanically different from inorganic bodies under internal pressure. The fundamental distinction is that embryos respond to any external influence, including mechanical, *actively*, that is, by generating their own internal mechanical force which tends to modify in some way the external mechanical stress. These are the feedbacks between the externally applied stresses and the internally generated mechanical forces, which seem to be basic for self-organization in embryonic morphogenesis. Revealing these feedbacks is, therefore, one of the central and as yet unsolved morphogenetic problems. Recently Jay Mittenthal (from the University of Illinois) and I suggested that these feedbacks may operate by the internally developed stress tending to reduce the effect of an externally applied one. In doing so, it usually overshoots the initial stress to give a deviation in the opposite direction. (If an initial tension is reduced by an external force, a tissue not only restores, but even exceeds the initial tension. If, instead, the tension is increased, the tissue tends to decrease it, and in doing so can even develop negative tension, that is, internal *pressure* forces.[15]) In this paper I shall not discuss in detail these complicated and disputable aspects of the problem and shall restrict myself to presenting some phenomenologically simple models of morphogenesis implying morphogenetic feedbacks only in quite a rudimentary way.

3. APPLICATION OF THE INTERNAL PRESSURE MODEL (IPM) TO SOME MORPHOGENETIC PROCESSES

3.1 THE MAIN ALGORITHM AND PARAMETERS OF IPM

The main algorithm of IPM is very simple. As one can derive from Figure 3(a), in any area of a nonzero curvature, the pressure forces exerted by cells $j - 1$ and $j + 1$ upon cell j have a radial centrifugal component F_j, tending to shift cell j upwards. We suggest that the distance of cell shift D_j is proportional to F_j (a condition of a viscoelastical deformation). Now it is easy to demonstrate that the cell shift is, approximately, inversely proportional to the radius of local curvature R (see Beloussov and Lakirev[5] for details):

$$D_j = \frac{K}{R}. \tag{1}$$

Here K is a coefficient of proportionality which reflects the *activity* of a j cell's response to the external pressure. It is obvious that the greater is coefficient K, the larger is the distance of cell migration in the direction of relaxation of the externally apposed pressure stress.

Another universal model parameter W expresses the viscoelastic linkage between cells, resisting their centrifugal shifts and smoothing the deformed surface. According to the model's equations,[5] W varies from 0 to 1. The mode of its action is illustrated by Figure 3(b) by the symbol D_j^w.

3.2. IPM FOR U-SHAPED RUDIMENTS.

Many epithelial rudiments (buds of hydroid polyps, primary guts in Echinodermata, and Hemichordata embryos, vertebrate neural tubes as seen in frontal projections) at the very beginning of their development may be geometrically approximated as U-shaped tubes (Figure 4(a)–(c)). We used IPM for reconstructing their further morphogenesis and compared it with the real pictures. Since, as earlier mentioned, the shape of hydroid buds is largely dependent upon pulsatorial patterns (Figure 2), we imitated these patterns by employing different sets of periodically changing K values (see for details Beloussov and Lakirev[5]). Particularly, we imitated a "constant" (nonpulsatorial) regime (K values remaining unchanged within modeling time) as well as regimes of "wide," "intermediate," and "narrow" pulsations (corresponding to the Figure 2(a)–(c)).

The results of modeling initiated from the shape shown in Figure 4(a) are presented in Figure 5(a)–(h). One can easily see that the shapes obtained are affected in the expected way by the pulsatorial parameters (the "wider" the pulsation, the larger is the transversal rudiment's diameter, as seen in comparing the figures from left to right) and also by W values (the greater this parameter, the smoother is the rudiments surface). In spite of all the variety, most of the resulting shapes reproduce the typical differentiation of the rudiment tips into three parts, a central and two lateral lobes. The same "trefoiled archetype" was obtained in most cases if the modeling was started from the other modifications of U-shape (Figure 4(b)–(c) and Figure 6(a)–(d)).

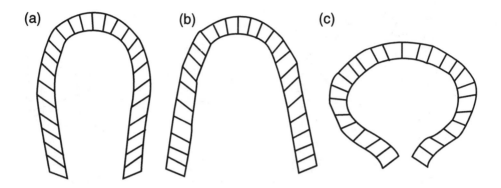

FIGURE 4   (a)–(c) The initial configurations of U-shaped rudiments used in subsequent modeling.

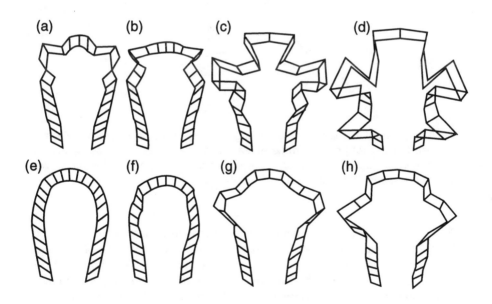

FIGURE 5   Results of shape modeling starting from the configuration. (a) For all the frames $K = 18$, for (b),(c) $W = 0.1$, for (d),(e) $W = 0.5$. (b),(d) – show "narrow" pulsations, (c),(e) – show "wide" pulsations.

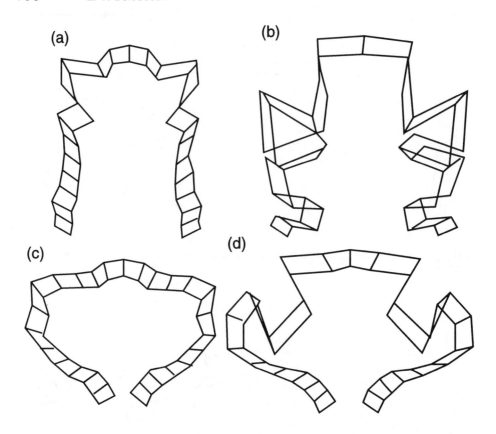

FIGURE 6   Results of shape modeling starting from configuration Figure 4(b) [(a) and (b) here] and Figure 4(c) [(c) and (d) here]. For all the frames $K = 18$, $W = 0.1$. (a), (c) – show "narrow" pulsations, (b), (d) – show constant regimes. Note the qualitative similarity of the shapes obtained with those in Figure 5.

Consequently, we were able to reproduce a morphogenesis which combines great variability in its local features with the *structural stability* of an entire (in this case, "trefoiled") archetype. This situation looks biologically realistic. The trefoiled differentiation of the tips of U-shaped tubes can be observed in a great number of rudiments having nothing in common. Consider, for example, hydroid polyps (the lateral lobes forming either tentacles or hydranths), archenterons of Echinodermata and Hemichordata embryos (here the lateral lobes create coelomic sacs and in Hemichordata often two pairs of them are simultaneously formed, as in Figure 5(d)), and anterior parts of the vertebrate neural tubes, the lateral lobes being now represented by eye vesicles.

On the other hand, the model contains an oversimplified and insufficiently general assumption. Thus, it suggests that, within each step of the modeling, the

curvature changes take place *simultaneously* in all the points of the rudiments surface. Such an assumption is not *robust*; in the other words, it is nonstable to even
small perturbations of the temporal pattern. However, one of the postulates of a
self-organization theory is that the natural processes are as a rule robust. In the
given case, this means that curvature increase processes are more likely to proceed
heterochronously (for example, in a wavelike manner) rather than synchronously.
Actually, the wavelike manner of the curvature increase has been confirmed in
many cases, for example, in hydroid polyps[4] and in brain rudiments of amphibian
embryos,[18] and there are good indications that this is a general phenomenon. Accordingly, in the elaborated IPM version, we imitated these waves by successive,
propagating impulses of curvature increase (each of them obeying the algorithm
(1)) travelling around a rudiment perimeter and involving *passively* some part of
an adjacent tissue in the centrifugal shift (Figure 3(c), L_{pass}). Formally, this was
expressed by introducing a wave-length parameter Ta which is directly correlated
with the amount of passive involvement, as described by the formula

$$D_n = D_j \left[\frac{1 - (n_j)}{Ta} \right] \qquad [Ta \geq (n - j)] \qquad (2)$$

where D is an active shift of a j cell [obeying algorithm (a)], D_n is a passive shift
of the nth cell, and $n - j$ is the distance between nth and jth cells measured in cell
numbers. As one can see from Eq. (2), the length of the passive involvement wave,
as measured in cell diameters, cannot exceed Ta.

For the symmetrical U-shaped rudiments described above, the introduction
of this wave did not qualitatively affect the shapes obtained, with one particular
exception, where two symmetrical upward waves of the curvature increase, meeting
together at the upper pole of the rudiment. In this case, the ingression was always
formed at the pole. However, as was shown by direct observations, the waves never
really met together at the apical pole of hydroid buds, although they probably
did so in some other tubular rudiments. On the other hand, in some other kinds
of rudiments which we shall now consider the introduction of a wavelike pattern
exerts prime-order effects upon morphogenesis.

3.3 A "WAVE" MODIFICATION OF IPM FOR EMBRYONIC BRAIN

As mentioned above, Saveliev[18] observed steep waves of curvature increase (each
of them involving only few adjacent cells) travelling anteriorly along the surface of
a neural tube of amphibian embryos every 15 minutes. That permitted us to em-
ploy the "wave" modification of IPM for modeling the morphogenesis of embryonic
brain, as seen in the saggital section. We created two model constructions differing
from each other by the border conditions. In the first, we considered, solely, an
anterior part of a neural tube (Figure 7(a)) whereas in the second, which is closer
to reality, we introduced in addition an anterior notochord extremity (Figure 7(b),

nch), completely immobilizing the shape changes of the overlaid part of a neural tube.

Black contours in Figure 7(a)–(b) represent the initial shape, while the light contours show the resulting shape, of neural tubes. One can see that a typical subdivision of initially smooth neural tube into three distinct primary brain vesicles is perfectly reproduced in both cases, while the addition of notochord makes the picture even more realistic, smoothing a considerable part of the neural tube surface and preventing closure between the third and second vesicles (cf. Figure 7(a), arrow). It is of interest that the presence of a notochord has effects that spread beyond the borders of its direct contact area with a brain tube. Note also that in this (as well as in the next) construction, the diameters of the protuberances which arise have nothing in common with the length of the passive involvement wave (*Ta*) which is much shorter, not exceeding two cell diameters.

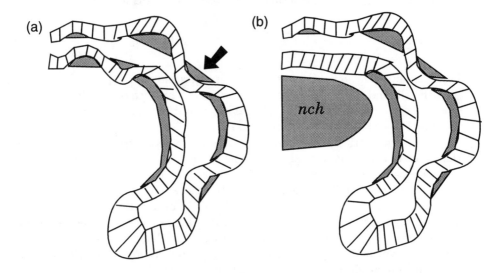

FIGURE 7 Modeling of early brain morphogenesis in the anterior part of an amphibian embryo neural tube. A dark contour shows an initial shape and the light contours the resulting shape. In this and the next figures, crosses indicate starting points of the successive waves of curvature increase which move in a clockwise direction around the entire rudiment. In (b), the notochord (*nch*) is introduced which is assumed to immobilize the above laid cells of a neural tube. $K = 80$, $W = 0.6$, and $Ta = 2$.

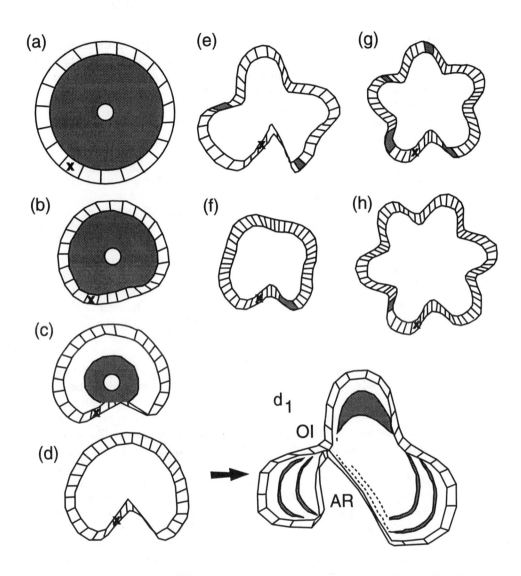

FIGURE 8 Modeling the morphogenesis of circular shapes, as dependent upon cell number. In frames (a)–(k) cell numbers (N) are changed in the succession 20, 26, 30, 36, 46, 50, 66, 80, 86, and 106. For all the frames $Ta = 3$. Dark contours in (a)–(d_1) indicate successive changes in dimensions and shape. If a cell number does not exceed 20, an initial circular shape remains stationary (a). (b)–(f) show the different variants of an resulting gastrula shape. (d) is a dipleurulalike shape produced from d shape by a mere continuation of the modeling. AR - "archenteron", OI - "oral invagination". (g)–(j) demonstrate the generation of successive symmetrical 5- to 8-fold shapes.

3.4 A WAVE MODIFICATION OF IPM FOR INITIALLY CIRCULAR RUDIMENTS

In all of the above cases, we started modeling from certain shapes, obtaining nonzero curvature gradients. Is it possible also to generate some regular and biologically realistical shapes starting from those without curvature gradients, that is, from ideally circular ones? While it is intuitively clear that the "simultaneous" model cannot do so (in this case a circle will merely expand without deviating from its initial shape), the capacities of the wave-model version deserve to be examined. Here is a summary of the results obtained in this class of models (for more details see Beloussov and Lakirev[6]).

1. The circular shapes, while being subjected to successive rotational waves of curvature increase, initiating from the same point, produced quite regular shapes which were mostly dependent upon K values (Eq. (1)) and on cell number N which is taken as constant for the whole duration of modeling. For the smallest N values, a circular shape appeared to be stable (Figure 8(a)), whereas with increase in N, the gastrulalike shapes were produced (Figure 8(c)–(f)); some of them, by a mere continuation of modeling, created shapes astonishingly reminiscent of the dipleurula larvae of Echinodermata (Figure 8(d)–(d$_1$)). On even further increase in N values, a series of radially symmetrical shapes with increased fold numbers was created (Figure 8(g)–(k)). On the other hand, a gradual increase in K values led to a reduction in the radial symmetry order with formation of gastrulalike shapes.[6]

2. The shapes obtained could be divided into two distinct classes: the highly robust (structurally stable) ones, which could be generated under a large variety of parameter values (including K, W, N, and Ta), and those much less robust, which required for their creation quite specific combinations of a narrow range of the parameters. To the first class belonged both gastrulalike and radially symmetrical shapes from 5-fold onward, whereas to the second belonged the stationary circular ("blastulalike") and the radially symmetrical 3- and 4-fold shapes. For example, in order to create a perfectly symmetrical 3-fold star, the parameter values need to be held in the immediate vicinity of $K = 350$, $Ta = 7$, whereas a 5-fold symmetrical star is compatible with K value variations from 150 up to 300 and Ta variations from 2 to 4. There is nothing mystical in the superior stability of 5- and higher order stars over that of 3- and 4-fold stars. Simply, the parameter values required to construct the stars with small fold numbers are too close to those generating gastrulalike shapes with a single dominating invagination. It is this invagination which prevents the 3- and 4-fold stars from becoming radially symmetrical (as is clear from Figure 8(e)–(f)). The 5-fold stars appear to be the first ones to avoid such domination.

3. The introduction of some quite simple border conditions may create a sudden transition from one structurally stable shape to another. For example, while the parameter values remain those to be required for generating a 6-fold star (Figure 9(a)), a non-extensible shell removed from the initial circular surface

to distance of half a cell height may be introduced (such a shell permitting only small centrifugal cell shifts). In this case, instead of the star, a perfect gastrula appears (Figure 9(b))! The shell may imitate either a vitelline membrane closely adhering to the embryo surface, a barely extensible hyaline layer of early Echinodermata embryos, or an increased elasticity of the embryonic surface proper.

4. EVOLUTIONARY CONCLUSIONS: GENERIC SHAPES, NATURAL SELECTION, AND HOW GENES REGULATE MORPHOGENESIS

The above models are certainly quite far from covering the whole realm of morphogenetic processes. Nevertheless, we believe that they lead us to some conclusions which are important for elaborating more adequate views on how morphogenesis and its evolution can be organized.

The main idea argued here is as follows: Each single morphogenesis, as well as its evolution, is guided by some simple, universal, and robust (structurally stable)

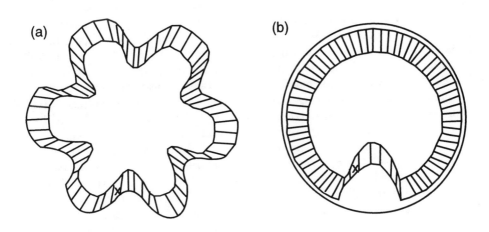

FIGURE 9 A radial (in this particular case, 6-fold) shape (a) and a gastrulalike shape (b) which are generated under the same parameter values ($K = 160, Ta = 2, N = 70$). In (b) a nonextensible external membrane is introduced, removed at a distance of half a cell height from the cells' outer surfaces. In (a) the cell sheet is freely extensible.

laws of self-organization. So far as morphogenesis is inevitably linked with deformations, stresses, and so on, an essential part of these laws should imply mechanical components. While considering here some phenomenologically quite simple mechanical (or, better to say, cytomechanical) models of morphogenesis, we could generate a wide variety of biologically realistic shapes and demonstrate that some of them appear much more often than others. To the first belong, as we saw, the gastrulalike, the radially symmetrical (beginning from 5-fold), and also, proceeding from U-formed tubes, the trefoiled shapes. The others, for example, stationary circular, 3- and 4-fold radially symmetrical, and, say, 2- and 4-foiled, require much more specific conditions (parameters values) to be created. We can define the first class of shapes as *generic*, and the second as *nongeneric*. Obviously, the generic shapes should often appear within the course of evolution even if they have no adaptive values and are not supported by natural selection. On the other hand, the creation of the nongeneric shapes may be expected only if they have some real adaptive value allowing the selection of the specific combinations of parameters, or if they had these values in some past epoch, and then continued to live on in the conditions of a strict breeding isolation.

Another conclusion is that the generic shapes can rapidly switch from one to another, without any intermediate steps. One example is the transformation of a starlike shape to a gastrulalike one, or vice versa (Figure 9). Such a process may well be guided by natural selection, although for quite another property having nothing in common with the shaping process.[7] Assume, for example, that it is of adaptive value to acquire a barely extensible outer shell. In this situation, almost automatically, a gastrula shape will emerge out of a radially symmetrical one. Is it not possibly that the earlier Deuterostomia developed perfect gastrulation as a side result of selection directed to the acquisition of a tightly adhered vitelline membrane or a nonextensible hyaline layer, both lacking in Protostomia?

Until now, we have not mentioned genes. Does it mean that we do not ascribe to them any role in regulating morphogenesis? Certainly not. The fact is that so many vague and poorly defined statements have been made in this field that we need to be quite precise now in formulating and discussing the problem.

First of all, it is usually ignored that we are dealing here with a transcendent case of transmission of *nonspatial* genetical "information" into some regular *spatial* (or, more precisely, spacio-temporal) structures. How is this possible? How can a factor which is, initially at least, homogeneously distributed throughout embryonic space (what we mean is a textbook rule that practically all the embryonic cells have the same equivalent genomic sets) produce heterogeneous structures implying refined regionally different cell behavior?

Again, we refer here to self-organization theory as the only one which is capable of giving a general and theoretically complete answer to this question. The theory argues that this phenomenon is possible if the homogeneously distributed factors play the role of *parameters* of the nonlinear equations describing structure formation. Returning to biology, we should conclude that genetic factors affect the space-time constant developmental parameters. In our model framework these are

K, W, and N. It is quite natural to accept that these values, that is, the measure of internal cell mechanochemical activity (K), elasto-viscous resistance (W), and total cell number (N) are directly determined by the cell genome. As seen from the above constructions, different values of these parameters produce regular morphological differences, sometimes of quite a local character (a different number of protuberances) in spite of the equality of the parameter values in all the cells of a given modeled rudiment.

Certainly, one may object that we artificially simplify the situation by ignoring the fact that, in spite of the initial genetic equivalency, sooner or later within development regular heterogeneous patterns of gene expression arise. These have been extensively studied, for example, in Drosophila embryos.[19]

However, from the broadest point of view, this objection is irrelevant. Any kind of heterogeneity increase (= symmetry breaking), be it manifested on the gene expression or cell sheets deformation level, cannot be interpreted without implying the same self-organizational principles. On the other hand, from the biological point of view, it surely makes a great difference whether these rules are working within the genetic level considered as a closed entity or whether they involve more or less important and numerous feedbacks arising at the morphogenetic level. Or, formulating the question more freely: is a developing organism no more than a container for the genes which operate as sole masters by their own rules and which change from time to time the container's shape, or is it an integrated multileveled system, each level possessing its own self-organizational properties and being able to affect the others?

In my view (I hope, shared by many others), it is the second possibility which corresponds to normal development, whereas the first one may lead, in ultimate cases at least, only to malformations (carcinogenesis). However, only few attempts have been made, until now, to investigate the higher level effects upon molecular processes in living beings, including gene expression. This will be an inspiring task for the coming generation of investigators.

REFERENCES

1. Balinsky, B. I. *Introduction to Embryology.* Pennsylvania, PA: Saunders, 1975.
2. Beloussov, L. V., Ju. A. Labas, and L. A. Badenko. "Growth Pulsations and the Rudimens Shapes in Hydroid Polyps." *Zhurn. Obsch. Biol.* **45** (1984): 796–806 (Russ.).
3. Beloussov, L. V. "Ontogenesis and Self-Organization Theory." *Rivista di Biologia (Biology Forum)* **81(2)** (1988): 159–183.

4. Beloussov, L. V., Ju. A. Labas, N. I. Kazakova, and A. G. Zaraisky. "Cytophysiology of Growth Pulsations in Hydroid Polyps." *J. Exp. Zool.* **249** (1989): 258–270.

5. Beloussov, L. V., and A. V. Lakirev. "Generative Rules for the Morphogenesis of Epithelial Tubes." *J. Theor. Biol.* **152** (1991): 455–468.

6. Beloussov, L. V., and A. V. Lakirev. "Why Five?" *Rivista di Biologia (Biology Forum)* **85(1)** (1992): 93–102.

7. Cherdantsev, V. G. "Induction and Differentiation of Axial Structures in Lamprey. On the Evolution of Primary Embryonic Induction." *Zhurn. Obsch. Biol.* **52** (1991): 89–103 (Russ.).

8. Dorfman, J. G., and V. G. Cherdantzev. "A Structure of Morphogenetical Movements of Gastrulation in Anuran Embryos. I. A Destabilization of Ooplasmic Segregation and Cleavage as a Result of Clinostating." *Ontogenez (Sov. J. Devel. Biol.)* **8** (1977): 238–250.

9. Elinson, R. P., and B. Rowning. "A Transient Array of Parallel Microtubules in Frog Eggs: Potential Tracks for a Cytoplasmic Rotation that Specifies the Dorso-Ventral Axis." *Devel. Biol.* **128** (1988): 185–197.

10. Haken, H. *Synergetics.* Berlin: Springer-Verlag, 1978.

11. Harold, F. M. "To Shape a Cell: An Inquiry Into the Causes of Morphogenesis of Microorganisms." *Microbiol. Rev.* **54** (1990): 381–431.

12. Keller, R. "Cell Rearrangements in Morphogenesis." *Zool. Sci.* **4** (1987): 763–779.

13. Martynov, L. A. "The Role of Macroscopical Processes in Morphogenesis." In *Mathematical Biology of Development,* edited by A. I. Zotin and E. V. Presnov, 135–154. Nauka: Moskva, 1982.

14. Meinhardt, H. *Models of Biological Pattern Formation.* New York: Academic Press, 1982.

15. Mittenthal, J., and L. V. Beloussov. "Rheological Rules of Cell Behavior that Generate Gastrulation and Neurulation of *Xenopus laevis.*" In *Lectures in Theoretical Biology,* Vol. 2, edited by K. Kull and T. Tijvel. Valgus: Tartu (Estonia), in press.

16. Nieuwkoop. Personal communication, 1982.

17. Nicolis, G., and I. Prigogine. *Self-Organization in Non-Equilibrium Systems.* New York: Wiley & Sons, 1977.

18. Saveliev. Personal communication, 1990.

19. Scott, M. P., and P. O'Farrell. "Spatial Programming of Gene Expression in Early *Drosophila* Embryogenesis." *Ann. Rev. Cell Biol.* **2** (1986): 49–80.

20. Spemann, H. *Experimentelle Beitrage zu einer Theorie der Entwickelung.* Jena: Fisher Verlag, 1936.

21. Vincent, J. P., G. F. Oster, and J. C. Gerhart. "Kinematics of Grey Crescent Formation in *Xenopus* Eggs. The Displacement of Subcortical Cytoplasm Relative to the Egg Surface." *Devel. Biol.* **113** (1986): 484–500.

Lewis Wolpert
Department of Anatomy and Developmental Biology, University College, and Middlesex School of Medicine, London

Gastrulation and the Evolution of Development

Reprinted, with modifications, by permission from "Gastrulation" *Development* (1992).

INTRODUCTION

By the evolution of development, I mean the evolution of development itself: the way the various developmental mechanisms and embryonic stages evolved and the selective pressures that were acting. Thus one wants to understand the origin of the key process in development like patterning, morphogenesis and cell differentiation, the origin of embryonic stages like gastrulation, and how developmental novelty, for example, the neural crest, could have arisen. One also needs to understand variants in developmental processes, like different modes of gastrulation and patterns of cleavage, radial, and spiral. In considering selective pressures it is necessary, too, to try and understand redundancy. We may never know the true answers, but that is no reason not to attempt to try and answer such questions, for they lie at the heart of the whole of evolution. In a sense, all evolution of multicellular animals is the brilliant result of altering developmental programmes.

If we consider the three basic processes in development—differentiation, spatial patterning, and change in form—these are already, it can be argued, well developed in the eukaryotic cell.[24] One may assume that the original eukaryotic cells had all the basic structures that characterize all cells—membranes, nucleus, mitochondria and so on, and, in fact, it is remarkable how similar the basic components of all cells are. Given these characteristics, little more was required to generate multicellular animals, and so development.

The cell cycle can be taken as a paradigm for several of the key processes in development. Every cell cycle involves a temporal program of gene activity, a spatial differentiation that assigns the results of growth to the daughter cells, and cell motility which ensures that the chromosomes are distributed to daughter cells, and in animal cells, brings about cell cleavage by active constriction. The evidence for a temporal program of gene activity is well documented[1] and it does not seem unreasonable to consider G_1, S, and G_2, as being homologous with different differentiated cell states.

Two further features of the cell cycle merit attention. The first involves the decision whether or not to enter the cycle following mitosis. This decision is closely linked to growth and the nuclear/cytoplasmic ratio. It also involves some sort of threshold event. In more general terms it involves intracellular signalling and a switch of a kind that is very similar to many developmental processes.

The other feature meriting attention is mitosis and cleavage. Here there is highly organized spatial patterning within the cell, which is also linked to motility. The development of the plane of cleavage at right angles to the spindle foreshadows the orthogonality that is fundamental to embryonic development. In addition, mitosis provides the opportunity for unequal distribution of components to daughter cells that could be the basis for the origin of divergence in cell fate and differentiation. This autonomous generation of differences is clearly present in the division of yeast cells in relation to mating type.[22]

It requires little imagination to derive other properties from the primitive cell for development. Cell adhesion and cell-to-cell signalling may be novel but require little modification of a system in which there is flow of material to the membrane with an external coat. Even the provision of an extracellular matrix would seem to present little difficulty given an endoplasmic reticulum and Golgi. There is, perhaps, one cellular process that may require a novel evolution and that is cellular memory. The inheritance of the differentiated state through a cell cycle might require new methods for controlling gene action.

The single-cell protozoa show complex patterning with very precise spatial location of organelles.[10] Many of the mechanisms required for the evolution of metazoan development are present in the protozoa,[11] their presence suggesting that they represent fundamental biological processes possibly linked to the cytoskeleton that arise, as it were, easily and naturally in evolution. Protozoa clearly show polarity in form. The rows of cilia suggest that a mechanism for generating spacing patterns—stripes—is well developed. In the ciliated protozoa, fission involves considerable reorganization of existing structures. For example, in *Tetrahymena* a new oral apparatus, which is normally located at one end of the animal, begins to develop near the zone of fission—in fact, two new cell surface organizations are formed within what was originally one, the two now being in tandem alignment along the *anteroposterior* axis. This reorganization is essentially complete before fission occurs.

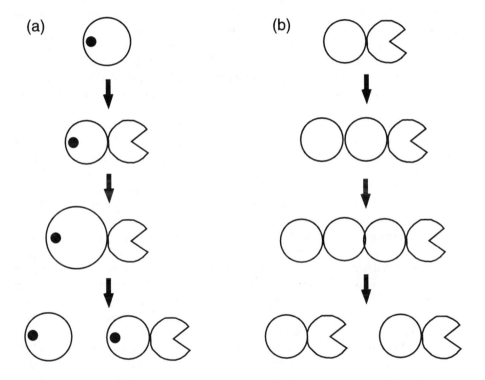

FIGURE 1 Possible modes of development and reproduction in primitive metazoa, made up of just two cell types, one of which can be thought of being specialized for feeding. In (a) development and reproduction is based on lineage and a stem cell. One cell retains some determinant that maintains this nonfeeding cell as a stem cell. In (b) development and reproduction is based on cell-to-cell interactions. One of the cells divides and the organism then reproduces by fission, a new feeding cell developing in just one of the two cells that have split off. This requires both cell signalling and polarity.

Thus it is clear that single-cell organisms have many of the mechanisms required for multicellular development. But these mechanisms must now be used for multicellular patterning. With a mechanism for spacing patterns and positional fields, many patterns are possible.[23]

THE EVOLUTION OF EARLY DIFFERENTIATION AND PATTERNING

The simplest conceivable multicellular organism would consist of two cells that differ from one another.[24] This unlikely ancestor would have manifested two key processes of multicellular development—diversity in cell state, or cell differentiation—and spatial heterogeneity, for it was essential that the two cells be different. From the beginning there needed to be a mechanism to ensure this difference and either cytoplasmic differences or a signal could be the basis (Figure 1). Both became firmly established. Almost all eggs—mammals providing a major exception—have a well-defined polarity due to cytoplasmic differences, but in most embryos at least some later differences arise not because of cytoplasmic localization, but because of cell-to-cell interactions.

It is a major problem in the evolution of development to understand why some animals have very well defined cell lineages, the asymmetric behavior of daughter cells being a key feature of their development, whereas others rely largely on cell-to-cell interactions. What does this diversity reflect? One possible scenario for thinking about the difference is to relate it to how cells are specified. In the former, cell fate is specified on a cell-by-cell basis, whereas in the latter, groups of cells are specified and thus may be related to asexual reproduction.

Again, what underlies the differences between spiral and radial cleavage? Metazoans arose from protozoans which were almost certainly ciliated. All known protozoa are asymmetrical with no known examples of bilateral symmetry. This may provide the clue to spiral cleavage. For the asymmetric structure of protozoa may have lead inevitably to spiral cleavage, the essence of which is cytoplasmic asymmetry. It also seems reasonable to think of some primitive organisms relying on this asymmetry to generate differences between cells by asymmetric distribution of some component. Once, however, cellular interactions are involved, then asymmetrical divisions are no longer required. This provides a plausible explanation for the rather good correlation between spiralean development (and cytoplasmic localization) and its loss in systems based on cell interactions. While this in no way implies that in spirally cleaving embryos there are no cellular interactions, it is quite impressive that in vertebrates and insects, which are not spirally cleaving, there are no good examples of autonomous asymmetric cell divisions.

GASTRULATION: CONSERVATION AND DIMENSIONALITY

In the evolution of gastrulation and its conservation in different animals, there are at least two separate processes that need to be considered—the setting-up of the main body plan and the morphogenetic movements. Curiously it seems that the latter as distinct from the former has been most widely conserved. There is nothing equivalent to the action of the organizer of vertebrates or the micromeres of sea urchins in spirally cleaving embryos. By contrast, in a very wide variety of animals, as pointed out a long time ago by Haeckel, the formation of the endomesoderm involves invagination or at least a related process.[21] As far as is known, there is no case in which a metazoan develops its three germ layers without undergoing a substantial rearrangement of the cells, usually of the type characterized by invagination. Indeed, the similarity of invagination in *Drosophila* and sea urchins is remarkable; a picture of one could easily be mistaken for the other (compare Leptin and Grunewald[18] with Gustafson and Wolpert[12]).

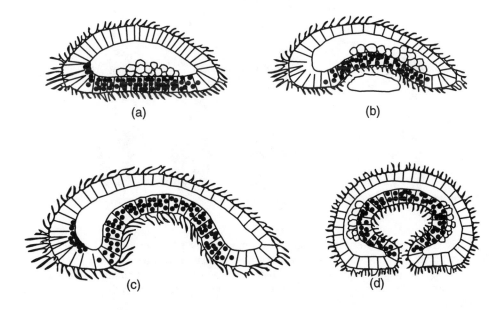

(a)　　　　　　　　　　　　(b)

(c)　　　　　　　　　　　　(d)

FIGURE 2 A possible scenario for the development of the gastrula, based on Jaegerstern.[17] In (a) the "Blastea" may have sexual cells inside. Phagocytosis may have become limited to the ventral side, and this could have lead to a small invagination to assist feeding and decomposing larger food particles. (b) Further development leads to the development of a primitive gut with ciliary activity aiding the movement of food particles (c,d).

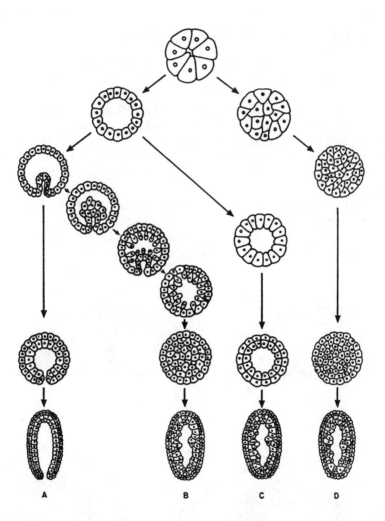

FIGURE 3 Some different modes of development leading to the formation of the two-layered planula in Cnidaria. This may occur by gastrulation of the cells as a sheet (a) or by multipolar ingresssion (b). A quite different mode is by delamination (c). A yet further variant starts from a solid, rather than a hollow blastula, which then develops a cavity (d).

A possible explanation for the remarkable conservation of gastrulation movements throughout the animal kingdom is that a fundamental requirement in specifying pattern in the early embryo is that this occurs in cell layers, often only one cell thick. Given this requirement, then, gastrulation movements are the inevitable result. And, in a way, the "layer" requirement makes biological sense, for it reduces the problem of specifying pattern in three dimensions to specification in two dimensions, which is a considerable simplification. In this way the third dimension is acquired or specified by cell movement.

The origin of the mesoderm and endoderm in a very wide variety of animals is rather similar, namely the movement of cells from an outer sheet into the hollow interior. Variants often reflect the yolkiness of the egg as in birds. The timing of such events—that is, whether the endoderm or mesoderm moves in first, and when they enter—accounts for much variation, but the same basic mechanism is involved. However, a major variant is seen when the inner layers form by delamination; that is, tangential planes of cleavage give rise to inner cells from the outer layer. Many of these variants can be seen in Cnidarian gastrulation (see below).

There have been attempts to classify animals into two major *superphyla* on the basis of gastrulation—the Protostomea and the Deuterostomea. In protostomes, the blastopore is supposed to become the mouth, whereas in deuterostomes the mouth is supposed to form secondarily, at some distance from the blastopore which may become the anus or close up. However, classification on this basis has been questioned, since in almost every phylum there are animals that form their blastopores into anus when their classification requires they should there be forming mouths.[21] It seems rather that the relationship between mouth and anus to the blastopore should be seen as a continuum.

THE ORIGIN OF THE GASTRULA

The origin of gastrulation is linked to the origin of the metazoa. Haeckel's views on metazoan origins have had a dominant influence: he supposed that some protozoan became colonial and gave rise to the "Blastea," a hollow sphere of cells, the ancestor of all metazoa. Further development of the Blastea gave rise to the "Gastrea" which was a two-layered structure, the inner layer entering through invagination at the posterior pole. A somewhat different explanation was offered by Metschnikoff in 1886 (see Jaegerstern[17]) and has been promoted by Hyman.[16] Invagination is seen as a secondary process, multiple ingression of cells being primary; in essence the idea is that phagocytosis would be carried out by outer cells and digestion by the inner cells, while in Haeckel's "Gastrea" the invagination provides a primitive gut with a mouth. As Jaegerstern points out, there is much to support the view of the origin of gastrulation as being linked to the evolution of an animal—the gastrea—whose form corresponds to a very large number of animals at early stages of development. In support of the Gastrea theory, Jaegerstern has provided quite a detailed scenario for the transformation of the Blastea into a Gastrea based on its settling on the ocean floor and ingesting food particles from the bottom (Figure 2). In his view the primitive intestine retained the cilia and these may have assisted feeding. He also proposes that this bottom dweller had dorso ventrality and was bilaterally symmetrical.

The Cnidaria provide a valuable phylum in which to try to understand the evolution of early stages of development.[6] The phylum presents some of the simplest animal structures which are made up of only two germ layers but which develop into a variety of forms characteristic of other metazoa phyla. There are clear signs of segmental and spacing patterns and moderately complex spatial differentiation. Lacking a mesoblast they are less complex internally than other phyla. Most Cnidaria give rise to a ciliated larva known as a planula. This is essentially a two-layer hollow cylinder. An extraordinary feature of Cnidarian development is the apparent diversity of pathways from fertilization to this rather simple structure (Figure 3). Cleavage, which is somewhat variable depending often on the yolkiness of the egg, is usually radial to begin with, the first few cleavages being at right angles to one another. In some cases, by contrast, cleavage is anarchic. Cleavage leads to two common blastula types—the stereoblastula which is a solid mass of cells, and the coeloblastula in which there is a sheet of cells surrounding a hollow interior. There is, at first sight, no obvious correlation between these two kinds of blastulae and either the yolkiness or size of the egg (see Campbell,[6] Table I), but the correlation may well be with relative cell size (see below).

Gastrulation, the formation of a two-layer structure, is no less diverse. Even within one class there is great diversity. Gastrulation may occur by invagination of the sheet from one pole, by ingression of cells from one pole, by ingression from multiple sites, or by delamination by tangential cleavage. This last mechanism seems to be quite different from the others, being based on a specific plane of cleavage such that one cell remains at the embryo's surface and the other is effectively placed in the interior.

The other modes of gastrulation might be thought of as having a common theme. Studies on sea urchin gastrulation have shown that cell movement and change in cell adhesion are key processes.[13,14] Initially the mesenchyme cells enter at the vegetal pole and this is followed by invagination of the gut—ingression and invagination are rather similar. Since most of the different forms of gastrulation in Cnidaria involve movement of cells from the surface layer towards the center, the differences between them may merely reflect spatial differences as to where this occurs and differences in the mutual adhesion of the cells. Multipolar ingression is not hard to understand, though there is the problem of specifying which cells will go in; unipolar ingression localizes the process that involves motility and loss of adhesion. Invagination will occur if loss of adhesion in unipolar ingression is delayed, a feature that can be directly observed in sea urchin morphogenesis when, in some embryos, the entry of the primary mesenchyme cells is delayed and a small invagination occurs. It is but a small step, the loss of adhesion, that might lead to invagination proper. A quite different theory about the origin of the gastrula has been proposed by Buss[5] and critically discussed by Wolpert.[24]

So many aspects of the variability can be seen as variations on a basic theme. But the problem, then, is why should there be this variability? What, if any, is the selective pressure? And why should delamination sometimes occur?

Berill[4] has analyzed development of scyphomedusae and claims a correlation of egg size and type of gastrulation. In general, he claims, the smaller eggs gastrulate by unipolar ingression and the largest by invagination alone, the intermediate sizes combining both methods. He rightly draws attention to the relationship between blastula wall thickness and blastula size. The most relevant correlation may lie here—polar ingression and invagination correlate with a thin wall and many cells, and multipolar invagination with thick wall and few cells. Thus differences in gastrulation may relate to the mechanics of gastrulation, this being determined by the number of cleavages prior to gastrulation. This, I would suggest, might be related to selection for yolk/nucleus (see Elinson,[8] for vertebrates).

From his consideration of the embryology of the annelids and arthropods, Anderson[2] has concluded that "The blastula, or its equivalent stages of embryonic development, has a greater stability of functional configuration than any stage that precedes or follows it, no doubt because this is the stage at which the fundamental framework of bodily organization is established." Many of the differences may again be accounted for by changes in the amount of yolk and the manner in which it is used.

Nevertheless, it is far from clear that the divergent modes of gastrulation give a functional significance. It simply may not matter to the embryo which way it gastrulates, provided the end result is appropriate. There may thus be no selection for "improving" gastrulation, providing that it is reliable. Variants may reflect unimportant differences in developmental terms.

SELECTION ON DEVELOPMENTAL PROCESSES: THE PRIVILEGED EMBRYO

Selection on development will act so as to ensure reliable generation of reproductive organisms. Reliability will be at a premium. But that selective process is not so much on development as on the organism itself. For the evolution of development, the more interesting question is whether there is selection on the developmental processes themselves. Is economy of energy of one developmental process over another to be preferred? Will irrelevant development pathways be lost? Is there a tendency to shorten a developmental process? These are not only questions of interest in their own right, but they relate directly to the variable pathways embryos use to achieve similar results. Are all these variants merely drift or are they the result of selection for something else, like yolkiness or rapid development, with a knock-on effect on development?[7,19]

It is possible that the embryo is evolutionarily privileged; that is, the main selective pressure is for reliable development. When one compares an embryo to an adult, this privileged status is evident: the embryo need not seek food, avoid predators, mate, reproduce, all functions for which there is such strong selective pressure. There is not even any evidence that conservation of energy is important, and cell death, which is clearly wasteful, accompanies many developmental processes. If this is the case, it may facilitate the evolution of novelty in development.

EVOLUTION OF DEVELOPMENTAL NOVELTY: THE NEURAL CREST

A feature of developmental systems is that there are examples of apparently different pathways leading to a similar structure. Gastrulation in Cnidaria has already been described; other examples are the different ways of forming the neural tube in vertebrates—folding and cavitation—and the different ways of forming somites in Amphiocus. It is possible that there is no selective advantage of one mechanism over the other and there are merely variants that have arisen by chance. Indeed, examination of the timing of the stages of gastrulation in sea urchins shows a considerable variation within embryos from the same species. Such variations may be neutral since the embryo is privileged.

If morphological variants that do not affect the later stages in development are not subject to selection, then they will persist and can provide the basis for the evolution of novelty. Over long periods, new variants will arise and this greatly increases the possibility that some of these will alter later development in a positive way. The variants, because of their neutrality, offer the opportunity for further diversity. They provide, in a sense, an opportunity to explore new developmental modes, some of which will be deleterious and so selected against, but others which will be advantageous. These ideas can be illustrated with respect to the origin of the neural crest.

The origin of the neural crest is an important and unexplained event in the evolution of vertebrates. I propose the following scenario that naturally links it to the folding of the neural tube. The development of the neural tube involves two main processes. In the first, the cells in an epithelial sheet change shape so that a sharp curvature results and the two ridges come together in the midline. In the second process, the cells in the sheet fuse across the midline and the outer sheet, the ectoderm, separates from the inner sheet, the neural tube. This second process involves rearrangement of cell contacts. It does not seem unreasonable that during this process, in the primitive condition at least, there were errors in the fusion process and some cells were left in the space between the ectoderm and neural tube. These cells would have provided for the origin of the neural crest.

These primitive neural crest cells would not have been subject to selection since they would be neutral to the development of the animal. They would have persisted for many generations and during this period new variants could have arisen to give rise to Schwann, pigment, or sensory cells. The key point is that the primitive population persisted long enough to, as it were, explore new possibilities.

This view of the evolution of development makes it much easier to understand how novelties could have arisen in development. Mutants, for example, that caused local invagination or ingression of cells may have been treated as neutral, further variants being generated until one provided the basis for a new structure and became subject to positive selection.

Selection on developmental processes acts primarily on reliability and this requires consideration of buffering and redundancy in developmental processes.

BUFFERING, REDUNDANCY, AND PRECISION

The idea that redundancy may be quite common in cell developmental biology probably has its modern origin in Spemann's idea[20] of double assurance, a term taken from engineering: "The cautious engineer makes a construction so strong that it will be able to stand a load which, in practice, it will never have to bear." Spemann's observations on lens induction led him to this view. In *Rana esculenta* a lens will form without being induced by the eyecup even thought the eyecup can induce a lens. Two different processes "are working together, either of which would be sufficient to do the same alone." Thus redundancy could provide a mechanism for ensuring reliability and precision.

More recently there have appeared an increased number of reports of redundancy in both cells and developmental systems.[26] To everyone's surprise the complete loss of certain specific genes—like some actins—has had remarkably little, if any, effect on the phenotype of the cell or embryo. Another example in development is provided by the specification of the spatial pattern of expression of the gap gene *Krüppel* in early *Drosophila* development. This pattern is controlled both by *bicoid* and *hunchback*.[15]

A number of different types of redundancy can be imagined, an example would be two very similar genes coding for two very similar proteins which fulfill the same function. Here I will focus on the case where there are two different genes controlling the same process but in different ways, as in the case of *Krüppel* or the three cyclin genes in budding yeast, any two of which seem to be dispensable. We use buffering to mean the control of some process in the face of variations in ambient temperature, pH, and other molecules. If the process is the production of a defined concentration of some biochemical species, then negative feedback is the control mechanism that is almost invariably used. In such systems there is no evidence for redundancy. And it is hard to see how a redundant system would be better and so have a selective advantage.

By contrast, in those many systems in cells and embryos where a particular molecular species is constrained to vary in space and time to give a defined spatio-temporal profile, negative feedback cannot be the mechanism for buffering, precisely because things are not kept constant. Moreover, such spatio-temporal profiles are usually themselves the result and cause of on/off switches, like gene activity. Such switching mechanisms for gene activity in eukaryotes, at least, almost never involve negative feedback, but positive feedback is often present to keep a gene switched on by autocatalysis. In many systems the amount of protein made is directly related to the number of gene copies as in the amplification of chorion genes or when additional copies have been involved.

How can buffering be achieved in a system with spatio-temporal switching? An answer would seem to lie in providing multiple parallel mechanisms with differing— oppositely directed—responses to environmental variations. In this way buffering and, hence, precision could be achieved. Such a mechanism has been referred to as canalization when applied to developmental systems.

As an example, consider the specification of the activity of the *Krüppel* in the early *Drosophila* development. *Krüppel*, a gap gene, has a broad band of expression about half way along the early embryo. Both activation and repression of its pattern of expression is controlled by proteins which have graded concentration. Activation is by low levels of both *bicoid* and *hunchback* and repression by high levels of these same two genes, and there are, indeed, separate binding sites adjacent to the *Krüppel* gene, both of which specify its expression in the same region. Two further genes, *tailless* and *giant*, control its final expression. There is apparent redundancy since the pattern of expression of *Krüppel* is very similar in the absence of either *bicoid* or *hunchback*. So why should specification require this apparent redundancy? We propose that this is required for precision in the expression of *Krüppel*. In general terms, precision will be improved if there are multiple parallel mechanisms for the specification, and this will be particularly so if the effect of environmental perturbation, like temperature, move the region of activation in opposite directions. We have, however, no evidence for this. On the other hand, it seems very unlikely that two or more behaviorally identical parallel processes, such as would be provided by two identically redundant genes, would improve buffering or precision.

The same arguments would apply to other systems where it is claimed that redundancy is present. We assert that in all such cases the redundancy is only apparent and the presence of more than one system in parallel has a selective advantage ensuring the precise buffering of the process. Where absence of a gene has no clear phenotype effect, we believe this merely reflects that the system has not been tested in the appropriate environment.

We do, however, have to recognize that some cases of redundancy may reflect evolutionary relics, since genes coding for processes which no longer have a selective advantage, but for which there is no selective disadvantage, may persist for millions of years.

Precision in specifying spatio-temporal patterns is a major problem for the developing embryo. Waddington's theory of canalization was proposed in order to account for this.[9] He suggested that selection would operate against the causes of deviation from optimal shapes, such as that of the insect wing. Apparent redundancy thus provides the mechanism for such a process.

CONCLUSION

Gastrulation can be thought of as one of the most important processes in early development. Its evolution probably did not require any significant mechanisms absent from the eukaryotic cell. The striking conservation of the formation of the endomesoderm by ingression is that it enables patterning to occur in cell sheets—that is, two dimensions—and the three-dimensional structure to be created by gastrulation movements.

Because of its importance, it is necessary for the end result to be reliable. However, the precise pathway by which gastrulation takes place may be quite variable. Reliability and precision probably requires apparently redundant processes.

Gastrulation may represent a very primitive metazoan whose development has been elaborated and extended in evolution. It is one of the few examples, possibly, of the discredited theory of recapitulation.

REFERENCES

1. Alberts , B., D. Bray, J. Lewis, K. Roberts, and D. Watson. *Molecular Biology of the Cell.* New York: Gardland, 1989.
2. Anderson, D. T. *Embryology and Phylogeny in Annelids and Arthropods.* Oxford: Pergamon, 1973.
3. Balinsky, B. I. *An Introduction to Embryology.* Philadelphia: Saunders, 1965.
4. Berill, N. J. "Developmental Analysis of Scyophomedusae." *Biol. Rev.* **24** (1949): 393–410.
5. Buss, L. W. *The Evolution of Individuality.* Princeton: Princeton University, 1987.
6. Campbell, R. D. "Cnidaria." In *Reproduction of Marine Invertebrates*, vol. 1, 133–199. New York: Academic Press, 1974.
7. Dickinson, W. J. "On the Architecture of Regulatory Systems: Evolutionary Insights and Implications." *Bio. Essays* **8** (1988): 204–208.
8. Elinson, R. P. "Change in Developmental Patterns: Embryos of Amphibians with Large Egg." In *Development as an Evolutionary Process*, edited by R. A. Raff and E. C. Raff, 1–121. New York: Liss, 1987.
9. Falconer, D. S. *Introduction to Quantitative Genetics.* Edinburgh: Oliver & Boyd, 1964.
10. Frankel, J. "The Patterning of Ciliates." *J. Protozool.* **38** (1991): 519–525.
11. Goodwin, B. C. "Unicellular Morphogenesis." In *Cell Shape: Determination, Regulation and Regulatory Role*, edited by W. D. Stein and F. Bronner. London: Academic Press, 1989.

12. Gustafson, T., and L. Wolpert. "The Cellular Basis of Sea Urchin Morphogenesis." *Intl. Rev. Cytology* **15** (1963): 139–213.
13. Gustafson, T., and L. Wolpert. "Cellular Movement and Contact in Sea Urchin Morphogenesis." *Biol. Rev.* **42** (1967): 442–498.
14. Hardin, J. "Contact-Sensitive Cell Behavior During Gastrulation." *Sem. Devel. Biol.* **1** (1990): 335–345.
15. Hulskamp, M., and D. Tautz. "Gap Genes and Gradients—The Logic Behind the Gaps." *Bio. Essays* **13** (1991): 61–268
16. Hyman, L . *The Invertebrates*, vol . II. New York: McGraw-Hill, 1942.
17. Jaegerstern, G. "The Early Phylogeny of the Metazoa. The Bilaterogastrea Theory. " *Zool. Bidrag. (Uppsala)* **30** (1956): 321–354.
18. Leptin, M., and B. Grunewald. "Cell Shape Change During Gastrulation in *Drosophila*." *Development* **110** (1990): 73–84.
19. Sander, K. "The Evolution of Patterning Mechanisms: Gleanings from Insect Embryogenesis and Spermatogenesis." In *Development and Evolution*, edited by B. C. Goodwin, N. Holder, and C. C. Wylie, 137–159. Cambridge: Cambridge University Press, 1983.
20. Spemann, H. *Embryonic Development and Induction*. New Haven: Yale University Press, 1938.
21. Willmer, P. *Invertebrate Relationships*. Cambridge: Cambridge University Press, 1990.
22. Wolpert, L. "Stem Cells: A Problem in Asymmetry." *J. Cell Sci. Suppl.* **10** (1989): 1–9.
23. Wolpert, L. "Positional Information Revisited." *Development* (1989): 3–12
24. Wolpert, L. "The Evolution of Development." *Biol. J. Linn. Soc.* **39** (1990): 109–124.
25. Wolpert, L., and W. D. Stein. "Positional Information and Pattern Formation." In *Pattern Formation*, edited by G. M. Malacinski and S. V. Bryant, 2–21. New York: Macmillan, 1984.
26. Wolpert, L., and W. D. Stein. "Redundancy and Buffering in Cells and Embryos." In preparation (1993).

Mae-Wan Ho and Fritz-Albert Popp
Developmental Dynamics Research Group, Open University, Walton Hall, Milton Keynes, MK7 6AA, U.K., and International Institute of Biophysics, Technology Centre, Opelstrasse 10, 6750 Kaiserslautern 25, F.R.G.

Biological Organization, Coherence, and Light Emission from Living Organisms

TO BE OR NOT TO BE
WHAT IS LIFE?

The question "what is life?" has been posed in one form or another since the beginning of modern science. Is living matter basically the same as nonliving, only more complicated, or is something else required? Descartes placed living matter firmly within the ken of the laws of physics, or more specifically, mechanics. Since then, generations of vitalists, including the embryologist Hans Driesch, the philosopher Henri Bergson, and the physiologist J. S. Haldane, have found it necessary to react against the mechanical conception of life by positing an additional *entelechy*, or *élan vital*, which is outside the laws of physics and chemistry (see Needham[40]).

The vitalists were right not to lose sight of the fundamental phenomenon of life that the mechanists were unable to acknowledge or to explain. But we no longer live in the age of mechanical determinism. Contemporary physics grew out of the breakdown of Newtonian mechanics at the beginning of the present century, both at the submolecular quantum domain and in the universe at large. The full implications for biology have yet to be worked out; although some major thinkers like Whitehead[67] already saw the need to explain physics in terms of a general

Thinking About Biology, Eds. W. D. Stein and F. J. Varela, SFI Studies in the Sciences of Complexity, Lect. Note Vol. III, Addison-Wesley, 1993   **183** theory of the organism, thus turning the hierarchy of explanation upside down. Whitehead's view is not accepted by everyone, but, at least, it indicates that the traditional boundaries between scientific disciplines can no longer be upheld, if one is to really understand nature. Today, physics has made further inroads into the "organic" domain, in its emphasis on nonlinear phenomena far from equilibrium, on coherence and cooperativity which are some of the hallmarks of living systems. The vitalist/mechanist opposition is of mere historical interest, for it is the very boundary between living and nonliving that is the object of our enquiry, and so we can have no preconceived notion as to where it ought to be placed.

As a first tentative answer to the question of "what is life," we propose that *life is a process of being an organizing whole.* By "whole," we do not mean an isolated, monadic entity. Instead, it is an open system that structures or organizes itself by simultaneously "enfolding" the external environment and spontaneously "unfolding" its potential into highly reproducible or dynamically stable forms. Our task is to discover the precise nature of this organizing whole with the help of contemporary physics and chemistry. This is distinct from working out the logic of living organization—an approach undertaken by Varela,[63] for example.

A SENSITIVE VIBRANT WHOLE

Biology textbooks often state that the most important characteristic of organisms is the ability to reproduce, and then proceed to an account of DNA replication and protein synthesis as though that were the solution to the fundamental problem of life. The ability to reproduce is only one of the properties of living organisms, and it could be argued, not even the most distinguishing one. For there are a number of other characteristics which leave us in no doubt that they are alive: their extreme sensitivity to specific cues from the environment, their extraordinary efficiency and rapidity of energy transduction, their dynamic long-range order and coordination, and, ultimately, their wholeness and individuality.[25]

For example, the human eye is an exquisitely sensitive organ as it can detect a single quantum of light, or photon, falling on the retina. The photon is absorbed by a molecule of rhodopsin, the visual pigment situated in special membrane stacks in the outer segment of a rod-shaped cell. This results in a nervous impulse ensuing at the opposite end of the cell, the energy of which is at least a million times that contained in the original photon. The amplification of specific incoming signal is in part well understood as a molecular cascade of reactions: the specific receptor protein, in this case, rhodopsin, activates many molecules of a second protein, transducin, each of which then activates a molecule of the enzyme phosphodiesterase to split many molecules of cyclic GMP. The cGMP keeps sodium ion channels open, but the split noncyclic GMP is ineffective. The result is that the sodium channels close, giving rise to a hyperpolarization of the cell membrane—from about $-40\,\text{mV}$ to $-70\,\text{mV}$—which initiates the nerve impulse.[59]

There are notable caveats in the account. For one thing, the component steps have time constants that seem to be too large to account for the rapidity of visual perception in the central nervous system, which is of the order of 10^{-2} s. Thus, it takes 10^{-2} s just to activate *one* molecule of phosphodiesterase after photon absorption. Furthermore, much of the amplification is actually in the initial step, where the single photon-excited rhodopsin passes on the excitation to at least 500 molecules of transducin within one millisecond. How that is achieved is still a mystery, except that as rhodopsin and transducin molecules are bound to a common membrane, the membrane must play a crucial role in both the amplification and long-range transfer of excitation.

Another instructive example is muscle contraction (see Alberts et al.[2]). Skeletal muscle consists of long, thin, muscle fibres, several centimeters in length, each of which is a giant cell formed by the fusion of many separate cells. A single fibre, some 50μ in diameter, consists of a bundle of *myofibrils*, each 1 to 2μ in diameter. A myofibril has regular, 2.5μ repeating units or *sarcomeres* along its length. Each sarcomere consists of alternating thin and thick filaments made up respectively of polymers of actin, complexed with other proteins, and myosin. The thin filaments are attached to an end-plate, the Z disc (see Figure 1). Contraction occurs as the alternating myosin and actin fibres slide past each other by cyclical molecular treadmilling between myosin head groups and serial binding sites on the actin fibre. The sarcomere shortens proportionately as the muscle contracts. Thus, when a myofibril containing a chain of 20,000 sarcomeres contracts from 5 to 4 cm, the length of each sarcomere decreases correspondingly from 2.5 to $2\,\mu$. The energy for contraction comes from the hydrolysis of the high-energy intermediate, ATP, into ADP and inorganic phosphate.

The sequence of events leading up to contraction begins with the firing of the nerve supplying the muscle, which triggers an action potential in the muscle-cell plasma membrane, the electrical excitation spreads rapidly into a series of membranous folds, the *transverse tubules*, that extend inwards from the plasma membrane to surround each myofibril at the region of the Z disc. Here, the electrical signal is somehow transferred to the *sarcoplasmic reticulum*, a specialized endoplasmic reticulum separate from the plasma membrane, and wrapped intimately around each myofibril. The sarcoplasmic reticulum then releases into the cytosol large amounts of the Ca^{2+} stored in its lumen. The resultant sudden rise in free Ca^{2+} initiates contraction simultaneously in the *entire cell* within milliseconds. This means that the numerous cycles of attachment and release of all the individual myosin heads to and from the binding sites on the actin filaments are precisely coordinated over the whole cell, each molecular event requiring the transfer of energy contained in one molecule of ATP. Muscle contraction is known to be close to 100% efficient in that the free energy contained in ATP is completely converted into mechanical work.[21,33] (This high efficiency of energy transduction is characteristic of all the major bioenergetic reactions. For example, in the mitochondrial inner membrane, the coupling of electron flow to ATP synthesis is estimated to be 95% efficient [see Slater[56]].) Careful measurements of the Ca^{2+} release in muscle cells shows that it begins almost immediately after electrical excitation; in other words, it takes hardly any time at all for the signal to traverse the entire cell (see Rios and Pizarro[50]). This long-range coordination over macroscopic distances and the rapidity and

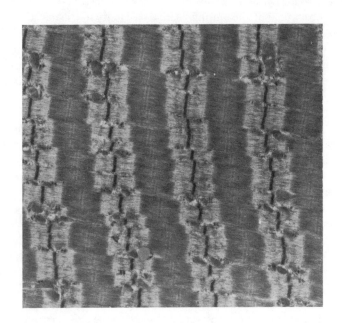

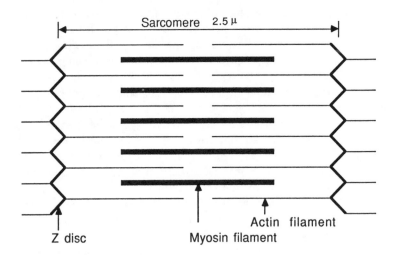

FIGURE 1   The ultrastructure of skeletal muscle. (a) Electron micrograph of rabbit skeletal muscle. (b) Diagram of a sarcomere.

efficiency of energy transduction in living organisms cannot be explained in terms of mechanisms and molecules in bits alone.

So what constitutes this sensitive, vibrant whole that is the organism? An organism that, furthermore, develops from a relatively featureless fertilized egg or seed to a complicated shapely creature that is nonetheless the same essential whole?

TO EQUILIBRATE OR NOT TO EQUILIBRATE
THE SPACE-TIME CATENATION OF LIVING PROCESSES

One cannot fully appreciate the problem of biological organization without taking into account the organism's space-time structure, which calls into question the applicability of traditional equilibrium thermodynamics to living systems.

To begin with, the organism qualifies as a dissipative structure[47] in the sense that it is maintained in a steady state by a flow of energy and chemicals. As soon as that flow is interrupted, disintegration and death begins. However, that steady state is not a static bulk phase in a rigid unvarying container. Even a single cell has its characteristic shape and anatomy, all parts of which are in constant motion; its electrical potentials and mechanical properties similarly are subject to cyclic and noncyclic changes as it responds to and counteracts environmental fluctuations. Spatially, the cell is partitioned into numerous compartments, each with its own steady states of processes that can respond directly to external stimuli and relay signals to other compartments of the cell. Within each compartment, microdomains[67] can be separately energized to give local circuits, and complexes of two or more molecules can function autonomously as "molecular machines" (see below). In other words, the steady "state" is not a state at all but a conglomeration of processes which are spatio-temporally organized; i.e., it has a deep space-time structure and cannot be represented as an instantaneous state or even a configuration of states (see Ho[24]). Relaxation times of processes range from $< 10^{-14}$s for resonant energy transfer between molecules to 10^7s for circannual rhythms. The spatial extent of processes, similarly, span at least ten orders of magnitude from 10^{-10} m for intermolecular interactions to meters for nerve conduction and the general coordination of movements in larger animals.

The processes are catenated in both time and space: the extremely rapid transient flows (very short lived pulses of chemicals or of energy) are propagated to longer and longer time domains of minutes, hours, days, and so on via interlocking processes which ultimately straddle generations. These processes include the by now familiar enzyme activation cascades (see previous section), some of which culminate in the expression of different genes and in morphological differentiation as seen in the response of many cells to changes in their immediate environment. The processes, rather than constituting the system's memory as usually conceived, are, in effect, projections into the future. They determine how the system develops and responds in times to come. Concomitantly, the locus of change propagates spatially to the rest of the cell or organism, stimulating other processes to take place which feed back on the earlier process to dampen or amplify its effects. It may result in the organism moving, or altering its environment.

The dynamism of the living system is such that each single cell is simultaneously crisscrossed by many circuits of flow, each with its own time domain and direction, specified by local pumping, gating, and chemical transformation. Thus, classical equilibrium constants are quite irrelevant[69] for each "constant" is in reality a cont'nuous function of variables including the flow rates, the electrical and mechanical field strengths, and so on. Furthermore, since the reaction products change the chemical potentials of all the components by altering the variables, the equilibrium "constants" will also be functions of time. How can we describe such a space-time structure? Does the thermodynamics of equilibrium processes have any relevance for the living system? Williams[69] advocates a shift from a conventional thermodynamic approach to a dynamic approach. (This has already begun to some extent, with the work of Hill[22,23] who introduces linear approximations of nonequilibrium thermodynamical concepts into biochemistry.)

THE THERMODYNAMICS OF LIVING SYSTEMS

It has been remarked by many scientists that, whereas the physical world runs down according to the second law of thermodynamics such that useful energy continually degrades into heat or random molecular motion (technically referred to as *entropy*), the biological world seems capable of doing just the opposite in increasing organization by a flow of energy and matter. Physicists and chemists feel that, as all biological processes require either chemical energy or light energy and involve real chemical reactions, the second law, as much as the first law of thermodynamics (the conservation of energy) ought to apply to living systems. One explanation in the right direction is that, because living systems are open, they can create a local decrease in entropy at the expense of the rest of the universe, so that the entropy of living systems plus the universe always increases in all real processes[54] and there is no violation of the second law. But there may be a more fundamental reason.

The second law of thermodynamics, as usually formulated, is a statistical law applied to a system consisting of a large number of particles, i.e., a bulk phase system with no space-time structure. As is clear from the previous description, living systems consist of compartments and microdomains each with its own steady state; complexes of a few molecules can act as efficient cyclic molecular machines with no immediate reference to the steady state of the surroundings. This implies that, if thermodynamics were to apply to living systems, it must apply to individual molecules. Such is McClare's[37] contention.

In order to formulate the second law of thermodynamics so that it applies to single molecules, McClare introduces the important notion of a characteristic time interval, τ, within which a system reaches equilibrium at temperature θ. The energies contained in the system can be partitioned into stored energies versus thermal energies. Thermal energies are those that exchange with each other and reach equilibrium in a time less than τ (so, technically, they give the so-called Boltzmann distribution characterized by the temperature θ). Stored energies are those that remain in a nonequilibrium distribution for a time greater than τ, either as characterized by a higher temperature, or such that states of higher energy are more populated than states of lower energy. So, stored energy is any form which does not thermalize or degrade into heat in the interval τ.

McClare goes on to restate the second law: useful work is only done by a molecular system when one form of stored energy is converted into another. In other words, thermalized energy is unavailable for work and it is impossible to convert thermalized energy into stored energy. But this is unnecessarily restrictive, and possibly untrue, for thermal energy *can* be harvested to do useful work in a cooperative system (see below). It is only necessary to recognize that useful work can be done by a molecular system via a transfer of stored energy in a time less than τ.

The major consequence of McClare's formulation arises from the explicit introduction of time. For there are now two quite distinct ways of doing useful work, not only slowly according to conventional thermodynamic theory, but also quickly—both of which are reversible and at maximum efficiency as no entropy is generated. (This is implicit in the classical formulation $\delta S \geq 0$, for which the limiting case is $\partial S = 0$.) Let us take the slow process first. A slow process is one that occurs at or near equilibrium. By taking explicit account of characteristic time, a reversible thermodynamic process merely needs to be slow enough for all thermally exchanging energies to equilibrate, i.e., slower than τ, which can be, in reality, a very short period of time. So high efficiencies of energy conversion can still be attained in thermodynamic processes that occur quite rapidly, provided that equilibration is fast enough. This may be where spatial partitioning and the establishment of microdomains is crucial for restricting the volume within which equilibration occurs, thus reducing the equilibration time. This means that local equilibrium may be achieved at least for some biochemical reactions in the living system.

At the other extreme, there can also be a process occurring so quickly that it, too, is reversible. In other words, provided the exchanging energies are not thermal energies in the first place, but remain stored, then the process is limited only by the speed of light. Resonant energy transfer between molecules is an example of a fast process. As is well known, chemical bonds, when excited, will vibrate at characteristic frequencies, and any two or more bonds which have the same intrinsic frequency of vibration will resonate with one another. More importantly, the energy of vibration can be transferred through large distances (theoretically infinite, if the energy is radiated, as electromagnetic radiations travel through space at the speed of light; in practice, it may be limited by nonspecific absorption in the intervening medium). Resonant energy transfer occurs typically in 10^{-14} s, whereas the vibrations themselves die down, or thermalize, in 10^{-9} s to 10^1 s. It is 100% efficient and highly specific, being determined by the frequency of the vibration itself, and resonating molecules (like people) can attract one another. By contrast, conventional chemical reactions depend on energy transfer that occurs only at collision; it is inefficient because a lot of the energy is dissipated as heat, and specificity is low, for nonreactive species could collide with each other as often as reactive species.

Does resonant energy transfer occur in the living system? McClare[38] suggests it occurs in muscle contraction, where it has already been shown that the energy released in the hydrolysis of ATP is almost completely converted into mechanical energy in a molecular machine which can cycle autonomously without equilibration with its environment (see previous section). Similar cyclic molecular machines are involved in other major energy transduction processes: in the coupled electron transport and ATP synthesis in oxidative phosphorylation (see Slater[56]), photophosphorylation (see Tien[62]), and in the Na^+/K^+ ATPase (see Stein[57]). Ultrafast, possibly resonant, energy transfer processes are implicated in photosynthesis where the first step of the charge separation in the chlorophyll molecules of the reaction center is a readily reversible reaction that takes place in less than 10^{-13} s.[11]

Thus, the living system may use both means of efficient energy transfer: slow and quick reactions, always with respect to the relaxation time, which is itself a variable according to the processes and the spatial extents involved. This insight is offered by taking into account the space-time structure of living systems explicitly.

Another important insight is the fundamental quantum nature of biological processes. McClare[37] defines a molecular energy machine as one in which the energy stored in single molecules is released in a specific molecular form and then converted into another specific form so quickly that it never has time to become heat. It is also a quantum machine because it sums the effects produced by single molecules. Muscle contraction is the most obvious example, as described above. Even in conventional enzyme kinetics, more and more quantum mechanical effects are recognized. Electron tunnelling is already well known to be involved in charge separation and electron transport across membranes (see Tien[62]) as well as across proteins such as cytochrome c.[5] Hydrogen transfer reactions may also involve tunnelling across energy barriers via an overlap of quantum mechanical wave functions between substrates and products.[32] It may be that very few reactions occurring in organisms involve thermalization of stored molecular energy. This does not preclude thermal excitation where the activation energy barrier is sufficiently low, as in the hydrogen bonding involved in the maintenance of protein conformations. Indeed, such conformational fluctuations in the nanosecond range have been found in a large number of globular proteins.[9] But, in order to do useful work, these fluctuations have to be coordinated. Otherwise, there will be equal probability for the reaction to go forwards as backwards.

There seems to be no escape from the fundamental problem of biological organization: how individual quantum molecular machines can function in collective modes extending over macroscopic distances. Just as bulk phase thermodynamics is inapplicable to the living system, so perforce, some new principle is required for the coordination of quantum molecular machines. This principle is coherence, perhaps even quantum coherence.

COHERENCE AND BIOLOGICAL ORGANIZATION

Nobel laureate biochemist Szent-Györgyi[60] was one of the first to suggest that we can only begin to understand the characteristics of living systems if we take into account the collective properties of the molecular aggregates, such as those observed under special conditions in solid state physics, or condensed matter physics as it is called nowadays.

Thus, the molecules in most physical matter have a high degree of uncoordinated or random thermal motion; but when the temperature is lowered to below a critical level, all the molecules may condense into a collective state and exhibit the unusual properties of superfluidity and superconductivity. In other words, all the molecules of the system move as one, and conduct electricity with zero resistance (by a coordinated arrangement of conducting electrons). Liquid helium, at temperatures close to absolute zero, was the first superfluid substances known. Various pure metals and alloys are superconducting at liquid helium temperatures. Today, technology has progressed to superconducting materials which can work at $125°$ K[4].

The solid state physicist Fröhlich[13,14] points out that as living organisms are made up predominantly of dielectric molecules rather densely packed together, they may indeed be considered as special solid state systems where electric and viscoelastic forces constantly interact. Under such conditions, condensation into collective modes of activity can occur, analogous to superconductors working at physiological temperatures. Specifically, metabolic energy, instead of being lost as heat, is stored in the form of collective modes of electromechanical and electromagnetic vibrations that extend over macroscopic distances within the organisms. He calls these collective modes *coherent excitations*. They can vary from a stable or metastable highly polarized state (resulting from mode softening of interacting frequencies towards a collective frequency of 0), to limit cycle oscillations, to much higher frequencies when the energy supply exceeds a certain threshold. Each collective "mode" can be a band of frequencies, with varying spatial extents, as consistent with the spatiotemporal structure of the living system. Nevertheless, the frequencies are coupled together so that energy fed into any specific frequency can be readily communicated to other frequencies. Conceptually, Fröhlich achieves energy exchange via a "heat bath," a quasi-equilibrium approximation; though, as we shall see later, this energy coupling may have a more fundamental origin.

THE CELL AS A SOLID STATE SYSTEM

The question as to whether an organism can be regarded as a solid state system has been debated for many years. Let us take a look at a cell as depicted in many textbooks. The cell membrane is supported by and attached to the membrane skeleton composed of a basketwork of contractile filamentous proteins lying immediately underneath it. The membrane skeleton in turn connects with the three-dimensional network of the cytoskeleton linking up the inside of the cell like a system of tele-graph wires terminating onto the membrane of the nucleus. The nuclear membrane and the cell membrane are also in communication via concentric membrane stacks, the Golgi apparatus and the endoplasmic reticulum, occupying a large proportion of the cell volume. The remaining volume of the cytoplasm is the cytosol (or "soluble" cytoplasm), in which are found organelles such as the mitochrondria and ribosomes. In the nucleus, the chromosomes (organized complexes of DNA and proteins) are anchored directly to the inside of the nuclear membrane. There is still something missing in this account.

For many years now, Clegg[11] has championed the idea that the cell is, in effect, a solid state system (although many biologists still find this hard to imagine, as few of us have the requisite biophysical background to appreciate the implications). He suggests that even the cytosol is filled with a dense network of actinlike protein strands, or "microtrabecular lattice," connecting with almost all cytoplasmic ultra-structure and to which practically all the enzymes and much of the substrates are attached. (Direct binding of many glycolytic pathway enzymes to actomyosin com-plexes has been observed in bovine muscle, and the binding is greatly enhanced on electrical stimulation.[10]) The surface area of the lattice is estimated to be about 50 times that of the cell, and its volume about 50% that of the cytosol. If this is so, then very little of the macromolecular constituents of the cytosol is dissolved in the cell water. On the contrary, 60–100% of the cell water itself might be bound, or struc-tured by the microtrabecular lattice, intracellular membranes, and the cytoskeleton. Structured water transmits dipole interactions or oscillations, and proton currents could flow in the layer of water molecules next to membranes, as suggested by observations on both oxidative and photosynthetic phosphorylation.[31,68] The flow of protons along membranes has been experimentally demonstrated in artificial phospholipid monolayers.[51]

In spite of, or perhaps because of, its highly condensed and connected nature, the whole cell is extremely dynamic; the connections between the parts as well as the configurations of the cytoskeleton, the membranes, the chromosomes, and so on, can all be remodeled within minutes subject to appropriate signals from the environment: for example, the presence of food and light, hormones, growth factors, or mechanical or electrical stimulation (see, for example, Sato and Rodan[52]). The entire cell acts as a coherent whole so that information or disturbance to one part propagates promptly to all other parts.

Thus, many of the basic principles of condensed matter physics may apply to the living system. The difference is that the living system has a much more complex and dynamic organization which empowers it (by electrical currents, proton flows and ions fluxes, and so on) to metabolize, grow, differentiate, and maintain its individuality. Still it is instructive to explore what condensed matter physics may have to tell us about the living system.

For example, it has been noted that the primary sites of energy transduction, the cell membranes, are closely analogous to the pn junction, a semiconductor device[61] which facilitates charge separation and is capable of generating an electric current when excited by energy in the form of heat or light. (It is the basis of the solar cell.) In common with these semiconductor devices, various biological membranes and artificially constituted phospholipid membranes also exhibit thermoelectric, photoelectric, and piezoelectric effects due, respectively, to heat, light, and mechanical pressure. In addition, many semiconductor devices are luminescent, producing light as the result of heating, electrical pumping (the basis of electroluminescent devices that are used in producing laser light), and also stimulation by light.[61] We shall see later that organisms do indeed exhibit luminescence, which can be stimulated in a highly nonlinear way with heat and light.

COHERENT EXCITATIONS AND LONG-RANGE ORDER

Coherent excitations can give rise to long-range order in the living system, as well as rapid and efficient energy transfer. This is because all the molecules involved in the exchange are vibrating in phase, so they can respond simultaneously to the same excitation signal. Fröhlich[14] proposes that biological membranes, by virtue of their dipolar structure and the existence of large transmembrane potentials of some 10^7 V/m, are particularly prone to such collective vibrational modes. This could explain how the absorption of a single photon, or the binding of a single ligand by a receptor protein could excite hundreds of membrane-bound molecules simultaneously as a first step in the amplification of an external signal arriving on the cell membrane. It is also possible that oscillations in the lipid network could induce simultaneous conformational changes of various proteins anchored in the membrane.[31,65] Proteins are giant dipoles which can undergo coherent dipole excitations over the entire molecule. And in an array of densely packed giant dipoles such as muscle and the cytoskeleton, the excitation could be coherent throughout the array, accounting for the kind of long-range coordination of molecular machines that is required in biological functioning. Similarly, RNA and especially DNA are also enormous dielectric molecules that can sustain coherent excited modes which may have important biological functions,[36] say, in determining which genes are transcribed or translated.

COHERENT EXCITATIONS AND SENSITIVITY TO EXTERNAL ELECTRO-MAGNETIC FIELDS

Coherent excitations render the organism very sensitive to external electromagnetic fields, for weak signals can be greatly amplified and then affect biological functioning. Indeed, the biological effects of electromagnetic fields have increasingly become the focus of public attention.[56]

There have been many observations suggesting that diverse organisms are sensitive to electromagnetic fields of extremely low intensities—of magnitudes that are similar to those occurring in nature.[46] These natural electromagnetic sources, such as the earth's magnetic field, provide information for navigation and growth in a wide variety of organisms, while major biological rhythms are closely attuned to the natural electromagnetic rhythms of the earth, which are in turn tied to periodic variations in solar and lunar activities. In many cases (described below), the sensitivity of the organisms to electromagnetic fields is such that they detect signals below the level of thermal noise. This points to the existence of amplifying mechanisms in the organisms receiving the information (and acting on it). Specifically, the living system itself must also be organized by intrinsic electrodynamical fields, capable of receiving, amplifying, and possibly transmitting electromagnetic information in a wide range of frequencies—rather like an extraordinarily efficient and sensitive, and extremely broadband radio receiver and transmitter, much as Fröhlich has suggested. In our laboratory, we have just completed a study showing that brief exposures of early *Drosophila* embryos to weak static magnetic fields (0.5 to 9 mT) result in a high proportion of characteristic body pattern abnormalities in the larva hatching 24 hours later (see Figure 2).[26] As the energies involved are below thermal threshold, there can be no significant effect unless there is a high degree of cooperativity or coherence in the pattern determination processes reacting to the external field. This sensitivity may be mediated via the cell membrane, which may also be responsible for the changes in Ca^{2+} efflux in cells and tissues exposed to low-frequency alternating electromagnetic fields.[1] Liburdy and Tenforde[36] have shown that static magnetic fields of 10 mT and greater can induce the release of drugs from artificial phospholipid liposomes due to the summation of diamagnetic anisotropies of molecular aggregates near phase transition. How that might interact with the dynamic process of pattern determination in *Drosophila* to generate the specific abnormalities—which are often global in nature—is not yet known. The biological effects of weak electromagnetic fields since the 1970s are reviewed by Becker[6] and Adey.[1]

QUANTUM COHERENCE

A key notion in our new perspective of living organization is coherence. Coherence in ordinary language means correlation, a sticking together, or connectedness; also, a consistency in the system. So we refer to people's speech or thought as coherent if the parts fit together well, and incoherent if they are uttering meaningless nonsense or presenting ideas that don't make sense as a whole. Thus, coherence always refers to wholeness. However, in order to appreciate its full meaning, it is necessary to make incursions into its quantum physical description, which gives us some insights that are otherwise not accessible.

We begin with Young's classic two-slit experiment, in which a source of monochromatic light is placed behind a screen with two narrow slits. When only one

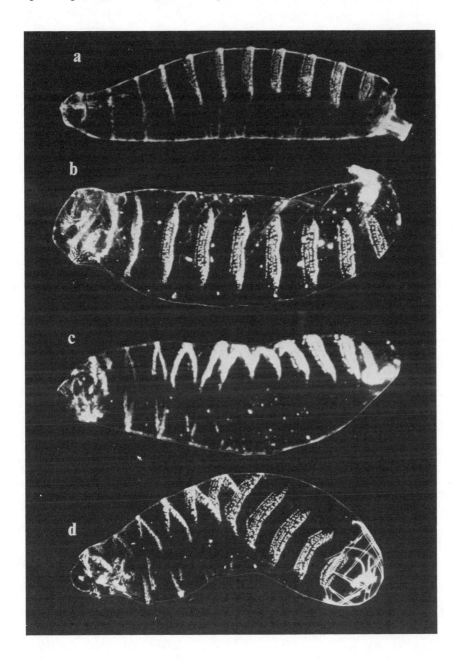

FIGURE 2   Some disturbances in global segmentation pattern in *Drosophila* larvae that have been exposed to weak, static magnetic fields during early development. (From Ho, M. W., T. A. Stone, I. Jerman, J. Bolton, H. Bolton, (continued)

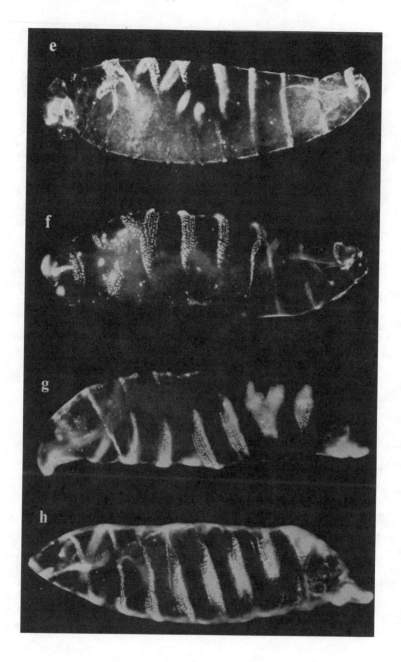

FIGURE 2 (cont'd.) B. C. Goodwin, P. T. Saunders, and F. Robertson.[26] "Brief Exposure to Weak Static Magnetic Fields During Early Embryogenesis Cause Cuticular Pattern Abnormalities in Drosophila Larvae." *Physics in Med. & Biol.* **37** (1992): 1171–1179.)

of the slits is opened, light behaves as particles, which, having passed through the slit, form a collection of discrete spots on a photographic plate, adding up to an image of the single slit. When both slits are opened, however, even single photons—generated one at a time at very low light intensities—behave as waves in that they pass through both slits at once, and falling upon the photographic plate, form a characteristic interference pattern. The intensity or brightness of the pattern at each point depends on a "probability" that light falls on that point.

The "probability" is placed between quotation marks because it is not probability in the ordinary sense. One way of representing these special probabilities is as correlation functions consisting of the product of two complex amplitudes. Light arriving at the point p on the photographic plate (Figure 3) has taken different paths, tp and bp. The intensity at p is then given as the sum of four such correlation functions:

$$I = G(t, t) + G(b, b) + G(t, b) + G(b, t)$$

where $G(t, t)$ is the intensity with only the top slit opened, $G(b, b)$ the intensity with only the bottom slit opened, and $G(t, b) + G(b, t) = 2G(t, b)$ is the additional intensity (which take on both positive and negative values) when both slits are opened. At different points on the photographic plate, the intensity is

$$I = G(t, t) + G(b, b) + 2G(t, b) \cos \theta$$

where θ is the angle of the phase difference between the two light waves. The fringe contrast in the interference pattern depends on the magnitude of $G(t, b)$. If this correlation function vanishes, it means that the light beams coming out of t and b are uncorrelated; if there is no correlation, we say that the light at t and b are incoherent. On the other hand, increase in coherence results in an increase in fringe contrast, i.e., the brightness of the bands. Since $\cos \theta$ is never greater than one (i.e., when the two beams are perfectly in phase), then the fringe contrast is maximized by making $G(t, b)$ as large as possible and this signifies maximum coherence. But there is an upper bound to how large $G(t, b)$ can be. It is given by the Schwarz inequality:

$$G(t, t,) \times G(b, b) \geq |G(t, b)|^2.$$

The maximum of $G(t, b)$ is obviously obtained when the two sides are equal:

$$G(t, t) \times G(b, b) = |G(t, b)|^2.$$

Now, it is this equation that gives us a description of coherence. A field is coherent at two space-time points, say, t and b, if the above equation is true. Furthermore, we have a coherent field if this equality holds for all space-time points, X_1 and X_2. This coherence is referred to as first-order coherence because its refers to the correlation between two space-time points, and we write it more generally as

$$G_{(1)}(X_1, X_1) \times G_{(1)}(X_2, X_2) = |G_{(1)}(X_1, X_2)|^2.$$

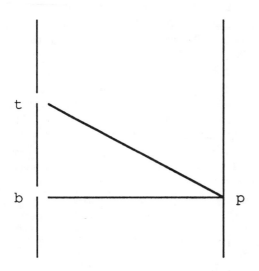

FIGURE 3   Young's two-slit experiment.

This equation tells us that, paradoxically, the correlation between two space-time points in a coherent field factorizes, or decomposes, neatly into the self-correlations at the two points separately, and that this decomposability or factorizability is a sufficient condition for coherence. What this means is that any two points in a coherent field behave statistically independently of each other. If we put two photon detectors in this field, they will register photons independently of each other.

Coherence can be generalized to arbitrarily higher orders, say, to n approaching ∞, in which case, we shall be talking about a fully coherent field. If nth-order coherence holds, then all of the correlation functions which represent joint counting rates for m-fold coincidence experiments (where $m < n$) factorize as the product of the self-correlations at the individual space-time points. In other words, if we put n different counters in the field, they will each record photons in a way which is statistically independent of all the others with no special tendency towards coincidences, or correlations (see Glauber[16]). Coherence can therefore exist to different orders or degrees.

A coherent state thus maximizes both global cohesion and also local freedom! Nature presents us a deep riddle that compels us to accommodate seemingly polar opposites. What she is telling us is that coherence does not mean uniformity: where everybody must be doing the same thing all the time. An intuitive way to think about it is in terms of a symphony orchestra or a grand ballet, or better yet, a jazz band where every individual is doing his or her own thing, but is yet in tune or in step with the whole. This is precisely the biochemical picture we now have of the living state: microcompartments and microdomains, right down to molecular machines, all functioning autonomously, doing very different things at different rates, yet all in step with the whole organism.

Many of the most paradoxical properties of the living system follow from coherence defined in this more rigorous sense. For example, factorizability optimizes communication by providing an uncorrelated network of space-time points which can be modulated instantaneously by specific signals. Furthermore, it provides the highest possible fringe contrast (or visibility) for pattern recognition, which may account for the great specificities in response of organisms to diverse stimuli. The factorizability of coherent fields may also underly the efficiency of bioenergetic processes in two respects. First, it concentrates the highest amount of energy of the field in localized zones by constructive interference, as well as creating effectively field-free zones within the field by destructive interference. Second, since the higher order correlations are the lowest in a completely coherent field, the smallest possible amount of energy is subject to correlated transfer between an arbitrarily large number of space-time points in the field with minimum loss. A coherent field is also fluctuationless or noiseless, the sort that any communications engineer working at radio frequencies, for example, would say is coherent.

One way to be coherent is to occupy a single mode, precisely the sort of thing that happens in superconductivity and superfluidity. That "mode" does not have to be a single frequency; it is only necessary that it represents one degree of freedom. Hence, there can be a broad band of frequencies coupled together or intercommunicating, so that energy fed into any frequency is propagated to all other frequencies.

An important consequence of coherence has to do with energy storage. Coherence is associated with a time and a volume over which phase correlation is maintained. The coherence time for a quantum molecular process is just the characteristic time interval τ over which energy remains stored in McClare's[38] formulation of the second law. So, in conformity with the second law of thermodynamics, the longer the coherence time, τ, the more extended is the time scale over which efficient energy transfer processes can take place, provided that the relaxation times are either much less than τ, or, in the quasi-equilibrium approximation, if they take place slowly with respect to τ. In other words, efficient energy transfer processes can, in principle, occur over a wide range of time scales, depending on the coherence times in the system (see Ho,[24] for more detailed discussion on this point).

We may now offer an answer to the question which was posed at the beginning of this chapter: What is it that constitutes a whole or an individual? It is *a domain of coherent, autonomous activity*. Defined thus, it opens the way to envisaging individuals which are aggregates of individuals, such as a population engaging in coherent activities (see Ho[24]). But we shall not pursue this further here.

HOW COHERENT IS THE ORGANISM?

There is at present no indication as to what degree of coherence may exist in living organisms, assuming that coherence does, indeed, exist. There are signs of it from many areas of biological research.

Frequency coupling is well known in biological rhythms, which often show harmonic relationships with one another,[8] such as the relationship between respiratory rhythm and heartbeat frequency. Similarly, the phenomenon of sub-harmonic resonance has been observed in metabolic oscillations, where entrainment is obtained to driving frequencies which are approximately integer multiples of the fundamental.[21] Recently, a gene has been isolated in *Drosophila* mutations which alters the circadian period.[34] Remarkably, the wing beat frequency of its love song is correspondingly speeded up or slowed down according to whether the circadian period is shortened or lengthened. This correlation spans six to seven orders of magnitude, linking the circadian period of about 10^5 s with the period of the love song, which is about 10^{-2} s.

A high degree of coordination exists in muscle contraction, as already pointed out. Insect flight muscle oscillates synchronously with great rapidity, supporting wing beat periods of milliseconds (see McClare[37]). Many organisms, tissues, and cells show spontaneous oscillatory contractile activities that are coherent over large spatial domains with periods ranging from 10^{-1} s to minutes.[7] Similarly, spontaneous oscillations in membrane potentials can occur in a variety of "nonexcitable" cells as well as in cells traditionally regarded as excitable, the neurons, and these oscillations range in period from 10^{-3} s to minutes, again involving entire cells or tissues (such as the heart, stomach, and the intestine). Finally, recent applications of supersensitive SQUID magnetometers to monitor electrical activities of the brain has revealed an astonishing repertoire of rapid coherent changes (in milliseconds) which sweep over large areas of the brain.[49] These, and the observations on synchronous firing patterns (40 to 60 hz) in widely separated areas of the brain recorded by conventional electrodes, are compelling neurobiologists to consider mechanisms that can account for such long-range coherence.[17] It has been suggested that the synchronization of the oscillatory response in spatially separate regions may serve as "a mechanism for the extraction and representation of global and coherent features of a pattern," and for "establishing cell assemblies that are characterized by the phase and frequency of their coherent oscillations."

It has recently been proven mathematically that synchrony is the rule in any population of oscillators where each oscillator interacts with every other via the absorption of the energy of oscillation, thus resulting in phase locking (see review by Stewart[58]). As the coupling between oscillators is fully symmetric, once they have locked together, they cannot be unlocked. Examples of such phase-locked synchronously oscillating systems may include the flashing of fireflies in various parts of Southeast Asia, the chirping of crickets in unison, as well as the pacemaker cells of the heart, the networks of neurons in the circadian pacemaker and hippocampus, the insulin-secreting cells of the pancreas, and the simultaneously contracting smooth muscle cells in the intestine and the stomach. The coupling energies (or signals) could be visual or auditory in the case of populations of whole organisms, though more subtle electromagnetic interactions cannot be ruled out, especially in the case of cells in tissues.

At the molecular level, muscle contraction is known to occur in definite quantal steps that are synchronous over entire muscle fibres, and measurements with high-speed ultrasensitive instrumentation suggest that the contraction is essentially fluctuationless (as characteristic of a coherent quantum field, see above).[20] Similarly, the beating of cilia in mussels and other organisms also occurs in synchronized quantal steps with little or no fluctuations.[3]

Finally, we proceed to describe some novel observations on light emission from living organisms that present deep challenges to our understanding of coherence in the context of biological systems.

THE LIGHT THAT THROUGH THE GREEN FUSE
BIOPHOTONS AND COHERENCE IN LIVING SYSTEMS

Practically all organisms emit light at a steady rate from a few photons per cell per day to several hundred photons per organism per second.[43,44] The emission of biophotons, as they are called, is distinct from well-known cases of bioluminescence in fireflies and luminescent bacteria, for example. It is universal to living organisms, occurring at an intensity generally many orders of magnitude below that of bioluminescence and, in contrast to the latter, is not associated with specific organelles. Nevertheless, biophoton emission is strongly correlated with functional states of the organisms, and responds to many external stimuli or stresses (see below). The response to temperature is highly nonlinear: an abrupt increase in emission rate as temperature rises, reaching saturation in a plateau, and characteristic hyteresis as temperature drops. Spectral analyses of the emitted light reveals that it typically covers a wide range of optical frequencies, with an approximately equal distribution of photons throughout the range, deviating markedly from the equilibrium Boltzmann distribution.

Biophotons can also be studied as rescattered emission after a brief exposure to light of different spectral compositions. It has been found, without exception, that the rescattered emission decays, not according to an exponential function characteristic of noncoherent light, but rather to a hyperbolic function (see Figure 4) which can be proven to be a sufficient condition for a coherent light field.[43,44] The hyperbolic function takes the general form,

$$x = A(t + t_0)^{-1/\delta}$$

where x is the light intensity, A is a constant, and t is time after light exposure. The light intensity is inversely proportional to time, which points to the existence of memory in the system and, hence, of time measure.

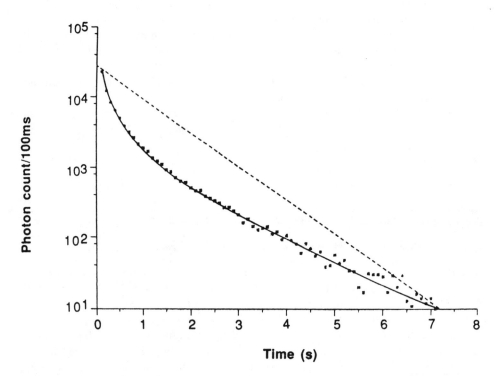

FIGURE 4   Hyperbolic decay of rescattered emission in *Drosophila* embryos after 1 minute of stimulation by white light. (From Ho, M. W., F. A. Popp, X. Xu, and S. Ross.[28] "Light Emission and Rescattering in Synchronously Developing Populations of Early *Drosophila* Embryos—Evidence for Coherence of the Embryonic Field and Long-Range Cooperativity." In *Advances in Biophotons Research*, edited by F. A. Popp, K. H. Li, and Q. Gu. Singapore: World Scientific, 1992.)

This phenomenon can be intuitively understood as follows. In a system consisting of noninteracting molecules emitting at random, the energy of the emitted photons are lost completely to the outside, or converted into heat, which is the ultimate noncoherent energy. If the molecules are emitting coherently, however, the energy of the emitted photons is not completely lost. Instead, part of it is coherently coupled back or reabsorbed by the system (rather like the interaction between coupled oscillators described above). The consequence is that the decay is delayed, and follows, typically, a hyperbolic curve with a long tail. Other nonlinear forms of delayed decay kinetics, such as oscillations, are predicted for a coherent field, and are also often observed.

The typical hyperbolic and nonlinear decay kinetics is uniform throughout the visible spectrum as evidenced both by the rescattering of monochromatic light or of light of restricted spectral compositions and by the spectral analysis of the rescattered emission.[39,43] The rescattered emission also covers a broad spectrum, and retains its spectral distribution even when the system is perturbed to such an extent that the emission intensity changes over several orders of magnitude. These observations are consistent with the idea that the living system is one coherent photon field far from equilibrium, with coherence simultaneously in the whole range of frequencies that are nonetheless coupled together to give a single degree of freedom *as a statistical average.*

As is made clear in the description of quantum coherence above, coherence does not mean uniformity, or that every part of the organism must be doing the same thing or vibrating with the same frequencies. There can indeed be domains of local autonomy such as those that we know exist in the organism. Furthermore, as organisms have a space-time structure, any measurement of the degree of freedom performed within a finite time interval will deviate from the ideal, which is that of a fully coupled system with little or no space-time structure. Another source of variation may arise because some parts of the system are temporarily decoupled from the whole, and the degree of coherence will reflect changes in the functional states of the system. Such a variation in the degree of coherence appears to be associated with the development of malignancy in cells.

LONG-RANGE COMMUNICATION BETWEEN CELLS AND ORGANISMS

In considering the possibility that cells and organisms may communicate at long range by means of electromagnetic signals, Presman[46] points to some of the perennial mysteries of the living world: how do birds in a flock, or fish in a shoal, move so effortlessly in unison? During emotional mobilization, the speed and strength of action of the organism are much greater than the normal working level. The motor nerve to the muscle conducts at 100 times the speed of the vegetative nerves that are responsible for activating processes leading to the enhancement of the contractile activity of the muscles required in a crisis: adrenalin release, dilatation of muscular vessels, and increase in the heart rate. Thus, it appears that the muscle receives the signals for enhanced coordinated action long before the signals arrive at the organs responsible for the enhancement of muscle activity! This suggests that there may be a system of communication that sends emergency messages simultaneously to all organs, including those perhaps not directly connected with the nerve network. The speed with which this system operates seems to rule out all conventional mechanisms. (This may also be true for visual perception and muscle contraction described at the beginning of this chapter.) Presman proposes that electromagnetic signals are involved, which is consistent with the sensitivity of animals to electromagnetic fields. Electromagnetic signals of various frequencies have been recorded in the vicinity of isolated organs and cells, as well as close to entire organisms.[41]

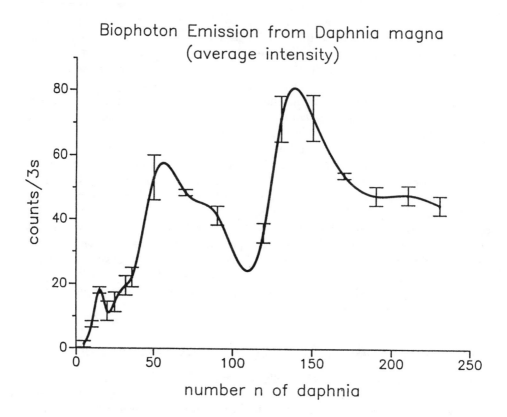

FIGURE 5 Self-emission rate in *Daphnia* as a function of population density. The maxima and minima turn out to be harmonics of the body size. (Redrawn from Galle et al.[15])

In our laboratory, we have recorded profuse electrical signals from fruit fly embryos (1hz to 30hz) during the earliest stages of development.[27] Whether these signals are involved in communication is not yet known.

Schamhart and van Wijk[53] investigated the photon emission characteristics of normal and malignant cells. They found that while normal cells exhibit decreasing light reemission with increasing cell density, malignant cells show a highly nonlinear increase with increasing cell density, suggesting long-range interactions between the cells as being responsible for their differing social behavior: the tendency of disaggregation in the malignant tumor cells as opposed to attractive long-range forces between normal cells. The difference between cancer cells and normal cells may lie in their communicative capability, which in turn depends on their degree

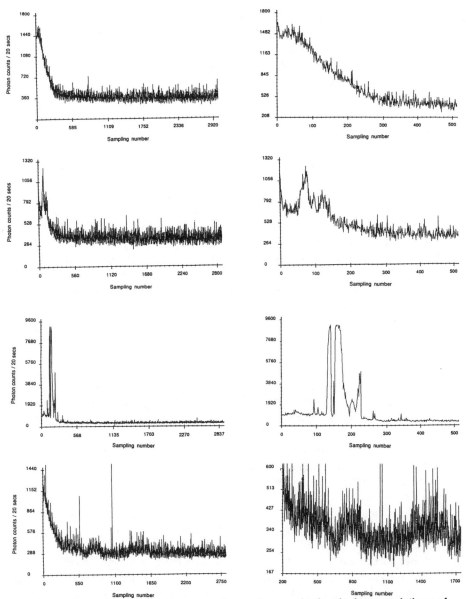

FIGURE 6 Superdelayed luminescence in synchronously developing populations of
Drosophila embryos. Top to bottom: control batch of 289 embryos not exposed to
white light stimulation, exhibiting typical pattern of self-emission; batch of 107 embryos
exposed to white light for one minute at 10 min. of development; batch of 232 embryos
exposed to white light at 5 min. of development; batch of 172 embryos exposed to
white light at 5 min. of development. The traces on the right are expanded versions
of those on the left. (From Ho, M. W., F. A. Popp, X. Xu, (continued)

FIGURE 7   (cont'd.) and S. Ross.[28] "Light Emission and Rescattering in Synchronously Developing Populations of Early *Drosophila* Embryos—Evidence for Coherence of the Embryonic Field and Long-Range Cooperativity." In *Advances in Biophotons Research*, edited by F. A. Popp, K. H. Li, and Q. Gu. Singapore: World Scientific, 1992.)

of coherence. The parameter $1/\delta$ in the hyperbolic decay function (see above) can be taken as a measure of incoherence, as it is directly correlated with the inability of the system to reabsorb emitted energy coherently. This parameter was shown to increase with increasing cell density in the malignant cells, whereas it decreased in normal cells.

Similar long-range interactions between organisms have been demonstrated in *Daphnia* where the light emission rate varies periodically with cell number in such a way as to suggest a relationship to average separation distances which are harmonics of the body size (see Figure 5).[15]

Finally, in synchronously developing populations of early *Drosophila* embryos, we have recently discovered the remarkable phenomenon of superdelayed lumines-cence in which intense, often prolonged, multiple flashes of light are reemitted with delay times of 20 minutes to 8 hours after a single brief light exposure (Figure 6).[28] The phenomenon depends, among other factors, on the existence of synchrony in the population. Although the timing of the light exposure must fall within the first 40 minutes of development in order to obtain superdelayed luminescene, the occur-rence of the flashes themselves do not obviously correlate with specific embryonic events. This suggests that the flashes give information concerning the physical state of the embryos at the time of light stimulation—such as the existence of a high de-gree of coherence—rather than at the time during which the flashes themselves occur. For the multiple prolonged flashes, the total luminescence produced shows significant first- and second-order dependence on the number of embryos present. These observations suggest that superdelayed luminescence results from coopera-tive interactions among embryos within the entire population, such that all the embryos reemit in synchrony.

The phenomenon bears some resemblance to various nonlinear quantum electro-dynamic effects (see next section). As yet, we do now know whether any functional significance could be attached to it. *Drosophila* females typically lay eggs just be-fore sunrise, so the external light source could be used as an initial synchronizing signal or *Zeitgeber*, which maintains the circadian and other biological rhythms. The superdelayed reemission could then be a means of maintaining communication and synchrony among individuals in the population. On the other hand, the flashes may simply be an indicator of the embryos' globally coherent state at the time when light stimulation is applied. This enables the embryos to interact nonlinearly to generate light emission that is coherent over the entire population, and orders of magnitude higher than the self-emission rate.

WHERE DO BIOPHOTONS COME FROM?

In this last section, we consider some possibilities regarding the source of biophotons. Preliminary attempts at imaging single *Drosophila* embryos by means of their superdelayed luminescence suggests that the latter is emitted over the entire embryo.[28] This is consistent with the broad spectrum of the emitted light, and also with *a priori* expectations as explained below.

Light is generally emitted from an excited atom or molecule, when an electron in the outermost orbital, having absorbed a quantum of energy, is promoted to a higher energy level. The excited atom or molecule can then either undergo a chemical reaction, or the electron can relax back to the ground state radiatively, by emitting a photon, or nonradiatively, by a transfer of phonons (sound waves) or by giving off the energy as heat. The energy of the emitted photon will be equal to the difference between the energy levels of the excited and the ground state, which determines the frequency of the emitted photon,

$$E_e = E_1 - E_o = h\nu$$

where E_e, E_1, and E_o are, respectively, the energies of the emitted photon, the excited level, and the ground level; h is Planck's constant; and ν is the frequency of the emitted photon.

In a solid state system, such as a semiconductor device, the atomic or molecular orbital electrons are no longer isolated as in the gaseous or liquid state. Instead, as the result of interactions in the crystal lattice, they coalesce into continuous bands of frequencies separated in energy by regions where there are no energy levels at all. The lowest energy band involved in bonding—the valency band—is filled with electrons. The next higher band is the conduction band to which electrons can be promoted by absorbing energy such as heat or light. Electrons in this band are the mobile charge carriers. Between the highest valency level and the lowest conduction level lies the *band gap*, corresponding to the threshold of energy which must be absorbed to promote a valency electron to a conducting one. Thus, excitation always involves a separation of charges. If the excited electron is not conducted away, it may relax back to the ground state (to recombine with the positively charged "hole" left behind) by radiating a photon or nonradiatively as described above. Another source of photons in a solid state system is from "excitons," i.e., an excited electron-hole pair which can propagate over long distances before giving up the energy by emitting a photon.

As we have seen, charge separation underlies the primary bioenergetic transduction processes associated with biological membranes, and the formation of excitons and their propagation is believed to be widely involved in energy transduction and in biocommunication. Rattemeyer and Popp[48] suggested that the DNA molecule is an excited duplex, or exciplex, in which photons are stored and, hence, can be a source of biophotons. Exciplex formation in DNA has been shown to predominate even at room temperature.[64] So it is not surprising that living systems could emit light from processes taking place all over the cell. It is more difficult to imagine how coherence would manifest itself in light emission, or, put the other way round, to work out what the characteristics of light emission tell us about coherence in the living system.

Self-emission is generally too weak to be analyzed in detail. There is no possibility of observing interference effects directly, as the light is not monochromatic. More opportunity presents itself in the analysis of rescattered light. As mentioned earlier, the hyperbolic decay kinetics is a sufficient condition for a coherent field. This is consistent with the observation that the hyperbolic decay kinetics is uniform over the entire optical spectrum, implying that all the frequencies are coupled in a single mode as required for coherence.

In the case of superdelayed luminescence in *Drosophila* embryos, we have to explain both the extraordinarily long delay in reemission and the greatly enhanced reemission rate. There are two kinds of novel electrodynamical phenomena in physics which bear certain formal similarities to superdelayed luminescence in our biological system. (The similarity is formal only because the long time constants involved in the biological systems have as yet no equal in the physical systems.) One of these, cavity quantum electrodynamics, is the control of spontaneous radiation from excited atoms in resonant cavities[18,19]; the other is the localization of light in semiconducting material containing dielectric microstructures.[30] Both physical phenomena emphasize delocalization of energy and the collective interactions on which biological organization depend.

Cavity quantum electrodynamics is associated with experiments demonstrating that spontaneous radiation from excited atoms can be greatly suppressed or enhanced by placing them in a special reflecting cavity which restricts the modes that the atom can radiate into. Strong coupling between the atoms and the radiation field within the cavity can then lead either to a suppression of spontaneous emission of excited atoms or its great enhancement, corresponding to the subradiant and superradiant modes, respectively. The early embryo can be regarded as a cavity resonator in which the spontaneous emission of atoms coherently excited by light exposure has become suppressed for various periods of time before they become reemitted in a greatly enhanced rate possibly as the result of some natural change of state as development proceeds subsequent to light stimulation. (The light exposure has no deleterious effects on development, and qualifies as a noninvasive probe.) The situation is further complicated by our not dealing with one single cavity resonator, but rather with a population of nearly identical resonators which can further interact to emit, cooperatively and simultaneously, over the entire population.

Localization of light occurs as the result of coherent scattering and interference effects in semiconducting materials containing dielectric microstructures which are close to the dimensions of the wavelength of light that is being scattered, i.e., $a \approx \lambda/2\pi$. Under these conditions, there are no propagating modes in any direction for a band of frequencies. Any impurity atom with a transition frequency in this band gap will not exhibit spontaneous emission of light. Instead, the emitted photon will form a bound state to the atom. In other words, the photon will be trapped indefinitely in the material. When a collection of impurity atoms is present in the dielectric, a single excited atom can transfer its bound photon to neighboring atoms by resonant dipole-dipole interactions. The distance for such photon tunneling is approximately ten times the dimension of the microstructures. Thus, a photon-hopping conduction results, involving a circulation of photons among a collective, which greatly increases the likelihood of coherent stimulated emission or laser action.

The conditions for light localization and subsequent coherent stimulated emission may well be present in the early embryo. There are many candidates for dielectric microstructures of the dimensions of the wavelengths of light. Globular proteins range from 5 to 10 nm; protein complexes and ribosomes, 20 to 30 nm. These can contribute to the localization of various frequency bands of photons. As the embryo develops, however, conditions favorable for stimulated coherent emission of the trapped photons could become realized. In connection with the trapping or storage of photons, it has been observed that death in organisms is invariably initiated by an intense light emission (at least three orders of magnitude above the self-emission rate) which can go on for more than 48 hours.[35,41]

Light emission offers a noninvasive probe for coherence in living organisms. The challenge is to develop a theory which can predict the rich repertoire of behavior that can be exhibited by living systems, with the help of recent advances in solid state physics. The key to living organization, wholeness, and individuality lies therein.

This chapter began with an enquiry into living organization—a sensitive, vibrant wholeness characterized by long-range dynamic order, rapid and efficient energy transfer and transformation, and a complex space-time structure. We show how all these properties can be understood in terms of coherence, in particular, quantum coherence. The latter provides a framework for interpreting and investigating many novel phenomena, including light emission from living organisms. These phenomena connect biology with the rapidly advancing fields of nonlinear quantum optics and quantum electrodynamics. The tantalizing riddle of what is life beckons once more, enticing us to move on in our endless quest.

REFERENCES

1. Adey, W. R. "Collective Properties of Cell Membranes." In *Resonance and Other Interactions of Electromagnetic Fields with Living Systems*. Royal Swedish Academy of Sciences Symposium, May 22, 1989.

2. Alberts, B., D. Bray, J. Lewis, M. Raff, K. Roberts, and J. D. Watson. *Molecular Biology of The Cell*. New York: Garland, 1983.

3. Baba, S. A. "Regular Steps in Bending Cilia During the Effective Stroke." *Nature* **282** (1979): 717–772.

4. Batlogg, B. "Physical Properties of High-Tc Superconductors." *Physics Today* **June** (1991): 44–50.
5. Bechtold, R., C. Kuehn, C. Lepre, and S. S. Isied. "Directional Electron Transfer in Ruthenium-Modified Horse Heart Cytochrome c." *Nature* **322** (1986): 286–288.
6. Becker, R. O. *Cross Currents: The Promise of Electromedicine, The Perils of Electropollution.* Los Angeles: Jeremy P. Tacher, 1990.
7. Berridge, M. J., P. E. Rapp, and J. E. Treherne, eds. *Cellular Oscillators*, vol. 81. Cambridge: Cambridge University Press, 1979.
8. Breithaupt, H. "Biological Rhythms and Communications." In *Electromagnetic Bio-Information*, edited by F. A. Popp, R. Warnke, H. L. Konig, and W. Peschka, 2nd ed., 18–41. Muchen: Urban & Schwarzenberg, 1989.
9. Careri, G., P. Fasella, and E. Gratton. "Enzyme Dynamics: The Statistical Physics Approach." *Ann. Rev. Biophys. Bioeng.* **8** (1979): 69–97.
10. Clarke, F. M., F. D. Shaw, and D. J. Morton. "Effect of Electrical Stimulation post mortem of Bovine Muscle on the Binding of Glycolytic enzymes. Function and Structural Implications." *Biochem. J.* **186** (1980): 105–109.
11. Clegg, J. S. "Properties and Metabolism of the Aqueous Cytoplasm and Its Boundaries." *Am J. Physiol.* **246** (1984): R133–R151.
12. Fleming, G. R., J. L. Martin, and J. Berton. "Rates of Primary Electron Transfer in Photosynthetic Reaction Centres and Their Mechanistic Implications." *Nature* **333** (1988): 190–192.
13. Fröhlich, H. "Long-Range Coherence and Energy Storage in Biological Systems." *Intl. J. Quantum Chem.* **2** (1968): 641–649.
14. Fröhlich, H. "The Biological Effects of Microwaves and Related Questions." *Adv. Electronics & Electron Phys.* **53** (1980): 85–152.
15. Galle, M., R. Neurohr, G. Altman, and W. Nagl. "Biophoton Emission from Daphnia Magna: A Possible Factor in the Self-Regulation of Swarming." *Experientia* **47** (1991): 457–460.
16. Glauber, R. J. "Coherence and Quantum Detection." In *Quantum Optics*, edited by R. J. Glauber. New York: Academic Press, 1969.
17. Gray, C. M., P. Konig, A. K. Engel, and W. Singer. "Oscillatory Responses in Cat Visual Cortex Exhibit Inter-Columnar Synchronization Which Reflects Global Stimulus Properties." *Nature* **338** (1989): 334–337.
18. Gross, M., and S. Haroche. "Superradiance: An Essay on the Theory of Collective Spontaneous Emission." *Phys. Rep.* **93(3)** (1982): 301–396.
19. Haroche, S., and D. Kleppner. "Cavity Quantum Electrodynamics." *Physics Today* **42** (1989): 24–30.
20. Hess, G. "The Glycolytic Oscillator." *J. Exp. Biol.* **81** (1979): 7–14.
21. Hibbard, M. G., J. A. Dantzig, D. R. Trentham, and V. E. Goldman. "Phosphate Release and Force Generation in Skeletal Muscle Fibres." *Science* **228** (1985): 1317–1319.
22. Hill, T. L. *Free Energy Transduction in Biology.* New York: Academic Press, 1977.

23. Hill, T. L. *Free Energy Transduction and Biochemical Cycle Kinetics.* New York: Springer-Verlag, 1989.
24. Ho, M. W. "Towards an Indigenous Western Science—Causality in the Universe of Coherent Space-Time Structures." In *Reassessing the Metaphysical Foundations of Science,* edited by W. Harman. [city?]: Noetics Sciences Institute, 1992.
25. Ho, M. W. "Coherent Excitations and the Physical Foundations of Life." In *Theoretical Biology. Epigenetic and Evolutionary Order from Complex Systems,* edited by B. Goodwin and P. Saunders, 162–176. Edinburgh: Edinburgh University Press, 1989.
26. Ho, M. W., T. A. Stone, I. Jerman, J. Bolton, H. Bolton, B. C. Goodwin, P. T. Saunders, and F. Robertson. "Brief Exposure to Weak Static Magnetic Fields During Early Embryogenesis Cause Cuticular Pattern Abnormalities in Drosophila Larvae." *Physics in Med. & Biol.* **37** (1992): 1171–1179.
27. Ho, M. W., S. Ross, H. Bolton, F. A. Popp, and X. Xu. "Electrodynamic Activities and Their Role in the Organization of Body Pattern." *J. Sci. Explor.* **6** (1992): 59–77.
28. Ho, M. W., F. A. Popp, X. Xu, and S. Ross. "Light Emission and Rescattering in Synchronously Developing Populations of Early *Drosophila* Embryos—Evidence for Coherence of the Embryonic Field and Long-Range Cooperativity." In *Advances in Biophotons Research,* edited by F. A. Popp, K. H. Li, and Q. Gu. Singapore: World Scientific, 1992.
29. Iwazumi, T. "High-Speed Ultrasensitive Instrumentation for Myofibril Mechanics Measurements." *Am. J. Physiol.* **252** (1987): 253–262.
30. John, S. "Localization of Light." *Physics Today* **44** (1991): 32–40.
31. Kell, D. B., and G. D. Hitchens. "Coherent Properties of the Membranous Systems of Electron Transport Phosphorylation." In *Coherent Excitations in Biological Systems,* edited by H. Fröhlich and F. Kremer, 178–198. Berlin: Springer-Verlag, 1983.
32. Klinman, J. Y. "Quantum Mechanical Effects in Enzyme-Catalysed Hydrogen Transfer Reactions." *TIBS* **14** (1989): 368–373.
33. Kushmerick, M. J., R. E. Larson, and R. E. Davies. "The Chemical Energetics of Muscle Contraction. I. Activation Heat, Heat of Shortening and ATP Utilization for Activation-Relaxation Processes." *Proc. Roy. Soc. Lond. B* **174** (1969): 293–313.
34. Kyriacou, C. B. "The Molecular Ethology of the Period Gene." *Behav. Genetics* **20** (1990): 191–212.
35. Li, K. H., F. A. Popp, W. Nagl, and H. Klima. "Indications of Optical Coherence in Biological Systems and Its Possible Significance." In *Coherent Excitations in Biological Systems,* edited by H. Fröhlich and F. Kremer, 117–122. Berlin: Springer-Verlag, 1983.
36. Liburdy, R. P., and T. S. Tenforde. "Magnetic Field-Induced Drug Permeability in Liposome Vesicles." *Radiation Res.* **108** (1986): 102–111.

37. McClare, C. W. F. "Chemical Machines, Maxwell's Demon and Living Organisms." *J. Theor. Biol.* **30** (1970): 1–34.

38. McClare, C. W. F. "A 'Molecular Energy' Muscle Model." *J. Theor. Biol.* **35** (1972): 569–575.

39. Musumeci, F., M. Godlevski, F. A. Popp, and M. W. Ho. "Time Behaviour of Delayed Luminescence in *Acetabularia acetabulum.*" In *Advances in Biophoton Research*, edited by F. A. Popp, K. H. Li, and Q. Gu. Singapore: World Scientific, in press.

40. Needham, J. *Order and Life.* New Haven, CT: Yale University Press, 1935.

41. Neurohr, R. Unpublished observation, 1989.

42. Pohl, H. A. "Natural Oscillating Fields of Cells. Coherent Properties of the Membranous Systems of Electron Transport Phosphorylation." In *Coherent Excitations in Biological Systems*, edited by H. Fröhlich and F. Kremer, 199–210. Berlin: Springer-Verlag, 1983.

43. Popp, F.-A. "On the Coherence of Ultraweak Photoemission from Living Tissues." In *Disequilibrium and Self-Organization*, edited by C. W. Kilmister, 207–230. Dordrecht: Reidel, 1986.

44. Popp, F.-A., and K. H. Li. "Hyperbolic Relaxation as a Sufficient Condition of a Fully Coherent Ergodic Field." In *Advances in Biophoton Research*, edited by F. A. Popp, K. H. Li, and Q. Gu. Singapore: World Scientific, in press.

45. Popp, F. A., B. Ruth, W. Bahr, J. Bohm, P. Grass, G. Grohlig, M. Rattemeyer, H. G. Schmidt, and P. Wulle. "Emission of Visible and Ultraviolet Radiation by Active Biological Systems." *Collective Phenomena* **3** (1981): 187–214.

46. Presman, A. S. *Electromagnetic Fields and Life.* New York: Plenum Press, 1970.

47. Prigogine, I. *Introduction to Thermodynamics of Irreversible Processes.* New York: Wiley, 1967.

48. Rattemeyer, M., and F. A. Popp. "Evidence of Photon Emission from DNA in Living Systems." *Naturwissenschaften* **68** (1981): S572–S573.

49. Ribary, U., A. A. Ioannides, K. D. Singh, R. Hasson, J. P. R. Bolton, F. Lado, A. Mogilner, and R. Llinas. "Magnetic Field Tomography (MFT) of Coherent Thalamocortical 40hz Oscillations in Humans." *Proc. Natl. Acad. Sci.* **88** (1991): 11037–11039.

50. Rios, E., and G. Pizarro. "Voltage Sensor of Excitation-Contraction Coupling in Skeletal Muscle." *Physiol. Rev.* **71(3)** (1991): 849–908.

51. Sakurai, I., and Y. Kawamura. "Lateral Electrical Conduction Along a Phosphatidylcholine Monolayer." *Biochim. Biophys. Acta* **904** (1987): 405–409.

52. Sato, M., and G. A. Rodan. "Bone Cell Shape and Function." In *Cell Shape: Determinants, Regulation and Regulatory Role*, edited by W. D. Stein and F. Bronner, 330–362. London: Academic Press, 1989.

53. Schamhart, S., and R. van Wijk. "Photon Emission and Degree of Differentiation." In *Photon Emission from Biological Systems*, edited by B. Jezowska-Trzebiatowski, B. Kochel, J. Slawinski, and W. Strek, 137–50. Singapore: World Scientific, 1986.

54. Schrodinger, E. *What is Life?* Cambridge: Cambridge University Press, 1944.

55. Shulman, S. "Cancer Risks Seen in Electro-Magnetic Fields." *Nature* **345** (1990): 46.

56. Slater, E. C. "Mechanism of Oxidative Phosphorylation." *Ann. Rev. Biochem.* **46** (1977): 1015–1026.

57. Stein, W. D. "Energetics and the Design Principles of the Na/K-ATPase." *J. Theor. Biol.* **147** (1990): 145–159.

58. Stewart, I. "All Together Now...." *Nature* (News and Views) **350** (1991): 557.

59. Stryer, L. "The Molecules of Visual Excitation." *Sci. Am.* **257** (1987): 42–50.

60. Szent-Gyorgi, A. *Introduction to a Submolecular Biology.* New York: Academic Press, 1960.

61. Thornton, P. R. *The Physics of Electroluminescent Devices.* London: Spon, 1967.

62. Tien, H. T. "Membrane Photogeophysics and Photochemistry." *Prog. Surf. Sci.* **30(1/2)** (1989): 1–199.

63. Varela, F. "What is the Immune Network For?" This volume.

64. Vigny, P., and M. Duquesne. "On the Fluorescence Properties of Nucleotides and Polynucleotides at Room Temperature." In *Excited States of Biological Molecules*, edited by J. B. Birks, 167–77. London: Wiley, 1976.

65. Wayne, R. P. *Principles and Applications of Photochemistry.* Oxford: Oxford Science Publications, 1988.

66. Welch, G. R., B. Somogyi, and S. Damjanovich. "The Role of Protein Fluctuations in Enzyme Action: A Review." *Prog. Biophys. Mol. Biol.* **39** (1982): 109–146.

67. Whitehead, A. N. *Science and the Modern World.* Harmondsworth: Penguin, 1925.

68. Williams, R. J. P. "On First Looking into Nature's Chemistry. Part I. The Role of Small Molecules and Ions: The Transport of the Elements." *Chem. Soc. Rev.* **9(3)** (1980): 281–324.

69. Williams, R. J. P. "On First Looking into Nature's Chemistry. Part II. The Role of Large Molecules, Especially Proteins." *Chem. Soc. Rev.* **9(3)** (1980): 325–64.

Francisco J. Varela,† **Antonio Coutinho,**‡ **and John Stewart**‡
†CREA, Ecole Polytechnique, 1 rue Descartes, 75005 Paris and ‡Unité d'Immunobiologie,
Institut Pasteur, 25 rue du Dr.Roux, 75015 Paris

What is the Immune Network For?

SECOND GENERATION IMMUNE NETWORKS: BEYOND DEFENSE

In recent years, immunology has undergone an important change by admitting that immune components might operate as a network. Initially the concept was applied restrictedly to a web of variable regions (V-regions) in immunoglobulin molecules (Ig), and had little significance other than some form of regulation of immune responses. More recently, this view of the network has been fleshed out to include not only antibodies that link to other antibodies (i.e., anti-idiotypic antibodies), but also V-regions expressed on the surface of B and T lymphocytes at various development stages, as well as components of the somatic self (i.e., markers on cell surfaces and soluble macromolecules circulating in the body fluids.) The initial ideas on immune networks (IN) were incomplete because they concentrated on the regulation of clonal immune *responses*, which are a manifestation of the system's capacity to defend the body from infections, rather than on properties of the immune system (IS) that emerge from its network organization, such as natural tolerance and memory. We have called *second generation* immune networks this wave of research that includes theoretical advances, observations in unimmunized mice and humans, and novel therapeutics in autoimmune diseases, these generating a new burst of interest on IN (reviewed in Varela and Coutinho[4,19,20,21]). Our own views have been extensively presented elsewhere.

The main point of the present chapter is to consider the *next* step; one that follows naturally from assuming the second generation stance, i.e., that INs *are* a biological reality. In this perspective, the focus of interests change quite drastically from the previous paradigm. Classically, immune responses represent the bulk of immunological lore. In the new perspective, immune responses are relegated to a peripheral role since infections are not always present and, when they are, the corresponding specific responses are mounted by an array of normally inactive, disconnected B and T cells. These stand in high contrast to the naturally or internally activated, highly connected lymphocytes, the core of the IN. We speak, therefore, of a *peripheral* immune system, which is concerned with "conventional" immune responses to microbial antigens, accountable by the clonal selection theory. This contrast with the *central* immune system concerned with internally activated cells, tightly arranged in an interacting network.

Now, everybody knows what immune responses are *for*, and thus how to understand the peripheral IS in functional and evolutionary terms. The question that follows immediately is: What is the central IS *for*? If we maintain, as we do, that the central IS is the most interesting and, possibly, the more ancient and primordial dimension of "immunity," then how are we to understand its function and significance? If not concerned with defense and surveillance functions, then with what is it concerned?

A CONSERVATIVE VIEW

A minimalistic answer of the question "what is the central IS for?" is given by the traditional position in the general problem of self-nonself discrimination. This view starts from two notions: (1) the circulating antibodies, i.e., the V-region repertoires of binding sites are "complete": they can recognize virtually any molecule that is presented to them to same degree; thus recognition of autologous tissues cannot be avoided; (2) all effector activities of lymphocytes (carrying and producing the repertoire) are destructive of the target structures.

It follows that evolution of clonal immune responses with great amplification potential would represent a time bomb for the organism, were it not for mechanisms that would control potentially autoaggressive clones. On the other hand, since lymphocytes respond to a whole degenerate diversity of molecular shapes, and not just to a highly specific one, it would be impossible to inactivate self-reactive cells without activating many others. In essence, therefore, it would have been impossible to evolve a competent peripheral IS and avoid self-destruction, without a central IS where self-reactivities could be "controlled." In other words, this minimal view considers central IS constitution and operation as the "price to pay" for avoiding autoimmune destruction, thus giving predominance to peripheral IS functions and evolutionary significance.

Obviously, this conservative view would ascribe no other particular function to the central IS, which is essentially useless without a peripheral IS. It is, nevertheless, more interesting than the conventional views that base "self-nonself discrimination" upon the elimination of all self-recognizing clones.

SELF-ASSERTION IN A HISTORICAL NETWORK

To explore an alternative to this classical or conservative view, it is necessary to set it in contrast with some key ideas that need to be recapitulated here. We have maintained that the main characteristic of the central IS is its *autonomy*, that is, it develops and operates close to normality by its own self-driven dynamic, even in the absence of antigenic challenge foreign to the normal body components. Thus the IN is concerned with what we have so far called *self-assertion* of a molecular and cellular identity.[4,20,23] This identity is the self-constructed establishment of an immune repertoire which constitutes a fingerprint for the individual's molecular identity. In classical clonal immunology, self is always conceived as a negative: that to which the system fails to respond. In our view the self is taken as positively defined. The presence in low quantities of natural antibodies to all known molecular profiles[1] is perhaps one of the clearest manifestations of the fact that the individual repertoire at any time is the reflection of the self-antigen-directed positive selection and activation of the corresponding lymphocytes (although clonal deletion of lymphocytes with above-threshold levels of ligand binding certainly contributes to the composition of V-region repertoires).[1] Natural "tolerance" to self-antigens is nothing but the manifestation of that process of selection. By speaking of self-assertion, then, we wish to underscore the essential idea that *the molecular self should be conceived as an emergent property of both the immune network's global regulation and the history of the individual's somatic components.* It is the converse of a negative self; it is an autonomously self-asserted one.

This also makes clear that there is no contradiction between the peripheral IS and central IS, since they are not isolated from one another. Immune responses to infectious agents certainly do take place, and they are likely to alter significantly the configuration of the circulating antibody populations, which, in turn, are an integral of the central IS dynamics, as are self molecular components. This necessary integration, between peripheral IS and central IS, implies that network and clonal selection mechanisms are *not in opposition to each other*, but are complementary processes, addressing different capacities of the immune system.

In our previous studies of the IN (and in much of the other research on second generation networks), attention has been mostly directed at understanding the dynamical properties of the network itself, not an easy task. We have arrived at a modicum of understanding as to how such a network behaves in various dynamical regimes, including relative stability, cyclic fluctuations, and even near-chaotic flows, a phenomenon accessible both to empirical[11,22] and theoretical studies.[6,17]

Further, we have also considered how, by explicit rules, the system undergoes metadynamical (or plastic) changes by recruitment of new lymphocyte populations.[5,8,18] These recruitment rules are sufficiently well understood that they can even be applied to the design of artificial devices so as to endow them with adaptability.[2]

DEFINING A RESEARCH STRATEGY

One straightforward approach to answering the question "what for" is simply to remove the central IS and see what happens. The results of this approach, although they must be taken into account, are not immediately helpful because, apparently, there is not much that we can see happening. Invertebrates don't have an IN, and yet from octopus to insects, they seem able to develop quite sophisticated forms of animal life. In mammals, several forms of immunodeficient mice, with an abnormal central IS or none at all are known ("nude," SCID, RAG-1, and RAG-2 deficient mutants): if protected from infections, they seem to exist quite happily.

This rather negative result, however, is not conclusive, because biological systems have been put together by evolutionary tinkering and, very typically, contain all sorts of redundancies and fail-safe mechanisms. This is precisely the reason why a variety of recent "knock out" experiments, that inactivate genes believed to be of utmost importance in, say, development, have led to no blatantly atypical phenotype. For instance, "knock out" mice lacking MHC genes or for several interleukins, seem to do very well, apparently as well as normals in some cases. This organizational feature of living organisms is so important that it is worth furnishing another example. Ecosystems are constituted by a network of interspecific relations; they can certainly collapse, but, more often than not, the removal of any one species can be compensated for. To take an example that concerns us human beings: if an irate God were to remove us all, as a species, tomorrow, the planetary biosphere would get on perfectly well without us. So, what are the human beings for?

Nevertheless, although this line of argument is valid as far as it goes, it will reduce to mere hand waving and empty special pleading if we are not able, at some point, to identify some experimentally measurable differences between animals with, and without, an IN. To keep ourselves honest, we must bear in mind that if no such differences exist, the conclusion would indeed be inescapable that, even if the IN exists, it does not do much of anything. In this case, it could well be argued that the conservative view is, after all, correct.

At this point, the difficulty is to know where to look. Without some sort of indication, the problem cannot be solved. This is the point where computer simulations of the IN come in. Our strategy is thus the following: (i) we postulate that an IN does indeed exist (in conformity with a certain amount of experimental evidence, as seen above); (ii) we use computer simulations to gain ideas as to how such a system would behave; (iii) on this basis, we extend our insights to the sorts of things an IN could conceivably do for an animal; and (iv) we translate the conclusions of (iii) into actual feasible experiments.

THE METAPHOR OF THE ECO-SOMA

In order to make our main point clearer, let us invoke here a pictorial metaphor. Imagine that we can arrange in a circle all the molecular profiles present in an individual's body. Further, imagine that we place the entire repertoire of free and bound V-regions involved in self-assertion (i.e., the central IS) inside this perimeter. From all we know, any given idiotype will bind significantly not to one, but to several points around this perimeter. Let us now draw a link between any two points of the somatic circle that have an affinity to the same idiotype of the repertoire, and let this line be thicker, the higher the affinity of the interactions, the concentration of the free antibody, and the size of the corresponding B-cell clone. The resulting image would be something like Figure 1: a web of passages drawn in a dense manner with anisotropic distribution.

Let this image now acquire a dynamic. In fact, since members of the repertoire fluctuate in time, they will touch points in the perimeter also in a fluctuating manner. Graphically this means that the traces will have varying thicknesses. Conversely, somatic components will also change in time (e.g., during hormonal cycles, the resulting modulations in genetic expression, or cell proliferation), and will in turn affect the entire network through their points of attachment.

On the whole, the image we wish to evoke is one that makes evident the interdependence between the IN and its somatic environment, much like the interdependence between a terrestrial niche and the living species that inhabit it. Like in the ecological case, the immune events are a *two-way* affair: the environment is modified while modifying the network of links that interact with it at specific points. Tugging this ecosystem (or eco-soma) at one point produces avalanche effects in a distributed mode mediated by the network dynamics. The main point, then, is that the somatic perimeter and the central network are *codefined* or mutually determined. This kind of mutual definition, reminiscent of the ecological scenario, is the picture that the reader should keep in mind for what follows.

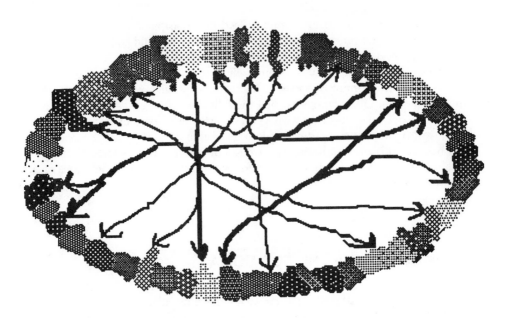

FIGURE 1   A figurative image of the somatic environment, represented here in a circle, linked by two-way interactions mediated by the immune network. The main hypothesis is that the immune network provides the glue which makes a given set of cellular and molecular profiles a coherent ensemble.

SIMULATIONS: WHAT BEHAVIOR DOES A NETWORK DISPLAY?

The simulations of the IN that we and others have carried out to date comprise two major aspects, which we have called dynamics and metadynamics.[20] By dynamics we mean the temporal patterns followed by the interacting V-regions, be they free immunoglobulins or the corresponding B-cell clones. By metadynamics, in contrast, we designate the continuous processes of removal of certain clones from the population, and the *recruitment* of new clones from the pool of lymphocytes, freshly produced by the bone marrow. Such recruitment is highly significant, reaching up to 20% per day in mice.[9] It is our contention that these processes (which take place on a time scale of days to weeks and months) give rise to emergent phenomena that are qualitatively different from those that arise from the dynamics of the system which occur at a faster time scale (seconds to hours); both together constitute the most characteristic and biologically important dimensions of the IN.

By itself, and in the context of the metadynamics, the dynamics of the system are, of course, important. They can be described using the classical tools of differential equations and have been shown to include various regimes: relative stability, cyclic fluctuations, or near-chaotic flows (see above). For what follows, bear in mind that the dynamics itself subdivides into two components: (a) molecular events (circulating immunoglobulin (Ig) secretion and removal) which occur on a rapid time scale of minutes to hours; and (b) cellular events (B-cell maturation and proliferation) which occur on a slower time scale of days, and which probably overlap with the time scale of metadynamics.

In presenting the main results of the simulations, we start by describing the metadynamics because this is the context within which the dynamics takes on significance. For a description of the metadynamics, the vital conceptual tool is that of *molecular shape space*,[14,15] that the universe of the stereochemical shapes that determine intermolecular affinities can be represented as points in a low-dimensional space. It is heuristically useful to consider that the coordinates of this shape space may correspond to physically identifiable properties such as ionic charge, hydrophobicity, geometric configurations (concavity or convexity), etc. We shall assume that any molecular shape occupies a point in shape space, and use the simplified version of the shape space concept[19] according to which each point in shape space correspond to a *pair* of complementary shapes (conventionally, "black" and "white"). To render the results visually perceptible, we use a shape space of two dimensions.

Qualitatively, the essential results of the metadynamical simulations are the following. First, the IN has the capacity to constitute itself, by appropriate recruitment and adjustment of the clones concerned, so that it comes to form coherent, quasistable configurations in shape space: given the conditions of representation for the model, these appear as parallel chains of "black" and "white" clones (Figure 2). This renders more precise our long-held intuition that the IN is characterized by a capacity for self-organization, and that it does not need to be driven by external antigenic contacts for its full operation.[5,21,23] Secondly, when this "morphogenesis in shape space" takes place in the presence of self-antigens (represented, in this preliminary study, by points in shape space whose presence is unconditional, i.e., not affected in return by the IN), the emergent configurations of the IN are such as to *integrate* the self-antigens. In the simulations this appears as the self-antigens being part of a chain of the same color (see Figure 3).

To use this result to answer our question "What is the IN for?", we need to turn the point of view around, and instead of asking how the somatic environment affects the development of the IN, we ask what is the effect of an IN on the somatic environment. In terms of the shape space concept, the somatic molecular environment can be assumed to have a definite morphology of its own. One of the great difficulties here is that the nature of this morphology is at present unknown. One could certainly leave aside extremes such as a small bounded domain (which would go against the diversity of known components), or a complete filling of the entire space (which would go against the finiteness and individuality of genes and proteins). However, whether this molecular "homunculus" is regular or fragmented, or whether it has some deep topological or geometrical properties, remains a great challenge to immunology. In our state of ignorance, the most reasonable hypothesis is, perhaps, that the shapes of somatic molecules bear little intrinsic relationship to each other, with one major exception. This is that we can anticipate many pairs of complementary shapes: any biologically active molecule (enzymes, hormones, neurotransmitters, interleukins, even toxins) will have a corresponding receptor with a complementary shape. In terms of our "black-white" shape space representation, these will appear as pairs of black and white points close to each other. By and large, however, there will be few cross-reactions (for rather obvious functional reasons), so that to a first approximation we may suppose that these pairs are also randomly scattered over shape space, perhaps clustered in families of homology, but with no significant overall relationships among them.

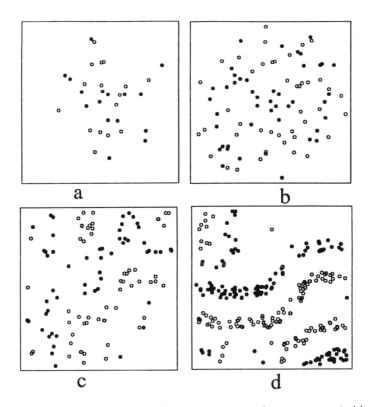

FIGURE 2   Antibodies from a developing immune network are represented in a simplified two-dimensional shape space, where each molecular shape has a perfect complement of the opposite color (black and white dots). The emergent morphogenetic patterns shown in succesive times of this simulations are due to rules of recruitment of new lymphocytes. The simulations occur on a toroid shape space. (a) 90 time steps (corresponding to 1.5 days in the life of a mouse). (b) 180 time steps (3 days), (c) 360 time steps (6 days), and (d) 3000 time steps (50 days).

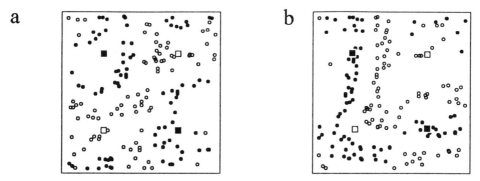

FIGURE 3 Similar to Figure 2, but in this case the morphogenesis occurs in the presence of self-molecules (indicated by squares) kept at a fixed amounts. (a) 480 time steps (8 days), and (b) 600 time steps (10 days).[18]

What then will be the effect of adding an IN? We propose that *the IN gives rise to chains in shape space which link together many of these previously unrelated somatic molecules to each other.*

Consideration of dynamics and time scales leads us to further specify this hypothesis. (i) The constitution of the linking chains, including the determination of which somatic molecules become linked to which, is a metadynamic event on a time scale of weeks to months. (ii) Fluctuations in the strengths of these connections, due to cell proliferation or death, is also relatively slow (days to weeks). (iii) For a given set of chains, changes in molecular concentration at one point will influence Ig concentration and the sensitivity (or "field") at another point; such influence will run along the lines of *paired* or complementary chains. The time scale of such influence, which is primarily due to molecular interactions, will be rapid (seconds to hours).

ON MOTION IN SHAPE SPACE

The above discussion gives actual immunological substance to the impressionistic arrows drawn in Figure 1. Let us pursue this point further. We can surely say that both the somatic and the idiotypic components *must* be strongly coupled and complementary, as follows from the evidence of the existence of natural antibodies and autoreactivities to all autologous macromolecules. To the extent that the set of autologous molecules and the idiotypic repertoire have a morphology or shape in shape space, the morphologies of each must be in "concert" in shape space; i.e., they must be mutually definitory.

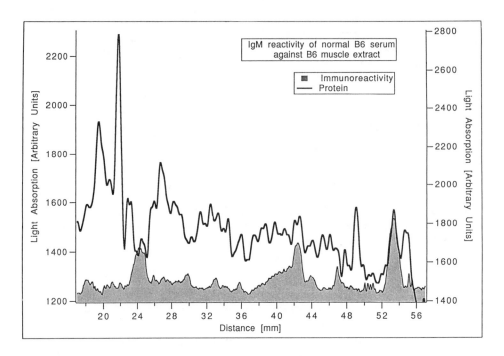

FIGURE 4 Protein peaks separated by size in gel electrophoresis from mouse muscle extract (solid line). Below (shaded) the amount of immunoreactivity of total serum from the same animal. As is clear from the mismatch between the two plots, of the many proteins (and their respective antigenic determinants) available for reaction with the antibody reactivities, only a few are actually targets for recognition, and the strength of the reactions bear little or no correlation with the concentration of each somatic protein.[13]

We can also assert that this complementarity is not a simple isomorphic covering: the somatic environment's view of the IN is a slanted perspective based on the way the web of V-regions touches the somatic self in highly nonuniform manner. It follows that the somatic self is defined by the central IS in a much distorted and individual manner, in relation to the concentrations of its components. Such an "immununculus"[3] has recently been revealed in experiments conducted by Nobrega et al.,[13] which were designed precisely to detect natural antibody reactivities on somatic self-proteins present in extracts of various tissues. A typical example is shown in Figure 4, using extracts from rat muscle as the universe of self-proteins. Of the many proteins and respective antigenic determinants available for reaction with the antibody reactivities in the serum of this animal (continuous lines show the protein peaks separated by size in gel electrophoresis), only a few are actually targets for recognition as seen in the important noncoincidence between the protein position and the reactivities in the serum (shaded area shows the level of immunoreactivity).

Thus, the strength of the immune affinities bear little or no correlation with the concentration of each somatic protein. Like in our metaphoric eco-soma in Figure 1, the links are not evenly distributed in all directions, but are highly nonuniform and historically determined.

On this basis we now reformulate our hypothesis about the function of the IN in somewhat more suggestive terms, as follows: the *IN provides a somatic morphology with motion in shape space*. That is, it provides the somatic morphology with the capacity to move around within shape space as a coherent codetermined unity. Without the mediating links, such coherent motion of a distributed morphology would be, of course, impossible. The IN provides the elastic links that make it possible to consistently shift the entire ensemble in a coherent fashion. Let us hasten to remind the reader that motion here is not literal displacement in space/time, but displacement in shape space. In fact, the morphology of self will change by addition/deletion of components and by variations in their concentrations, along the life of an individual, and throughout its ontogeny. This is motion in its most literal sense, for there is motion in the relative amounts or concentrations of components even when shape is stable. Further, if the self-morphology were color coded, one would see it changing colors constantly, constituting a more subtle kind of motion.

To illustrate this point, think of the nervous system. Certainly a nervous system is not required in order for an organism to have sensors and effector organs. For instance, coelenterates have pressure receptors on their external surface and secreting glands for digesting their pray. What is new in certain coelenterates like *Hydra* is that these sensors and effectors enter into *contact by dynamic links,* provided by the long axonic profiles of primitive neurons. The nervous system provides the link between sensory and motor surfaces far distant in the body, allowing the animal to move in physical space as a coherent perceptuo-motor unit. Thus the nervous system does not invent sentient behavior, but it does introduce true behavior by linking sensorium and motorium, a key evolutionary step in the animal lineage (unlike in plants which, correspondingly, lack motion).[12]

In our eyes, the IN provides a strikingly similar revolutionary dimension to a multicellular organism, with the key difference that motion happens not in physical space, but in shape space. The advantages of this added capacity are as clear as the advantages of having a nervous system: global regulation, memory and recognition capacities in the context of an adaptable coherent unity.

THE EFFECTOR MECHANISMS OF THE NETWORK

At this point, we need to raise the question of the relationship of Igs to effector functions. Even in classical immunology, with quantitatively massive immune responses, Igs *per se* have little effect: bacteria can live quite happily and undisturbed by a thick coat of Ig. By and large, Igs function in the elimination of their ligands only by triggering distinct effector mechanisms (e.g., the complement cascade, macrophage activity, inflammation, etc.) Exceptions arise when there are direct functional consequences of molecular interaction and antibody binding. This is the case whenever antibody binds directly to the active site of a molecule, or in any other way (allosterically or by mere hindrance) inhibits the biologically relevant interactions of the target molecule with the respective ligands (e.g., receptors). One can then say that the antibody "neutralizes" the biological activity of the target molecule. Although in conventional immunology this type of effector activity is limited to antitoxin antibodies and to the neutralization of viral particles (possibly through binding to the viral proteins necessary for the recognition of cell surface viral receptors on the target tissues), these exceptions acquire particular importance in our discussion, for they concern antireceptor antibodies, as well as all autoantibodies. Thus, as seen above, no molecular profile in the body makes sense without a complementary structure; that is, all somatic molecules to be functional require rather specific interactions with other body components, and the active sites of both interacting partners can obviously be targets for autoantibody binding.

So what could be the effector mechanisms for the IN? As is so often the case in biology, we may turn to human clinical pathology for some clues; in particular, to *autoimmune* disease which we would interpret here as a consequence of IN dysfunction. By and large, autoimmune pathologies fall into the following categories: (i) haematopoietic disorders; (ii) rheumatisms and kidney disease; (iii) endocrine disorders; and (iv) nervous system disorders. Of these, haematopoietic disorders may be related to the genesis of the IN itself; since causes and effects may be intertwined, the situation is difficult to interpret and thus not clearly informative. Rheumatisms and nephritis, on the other hand, may be grouped with allergies and inappropriate inflammation, and it may be that simply for mechanical reasons joints and glomeruli are particularly sensitive to a bit of "dust in the works," setting up a vicious cycle where the cure (inflammation) is worse than the disease.

We are left, then, with autoimmune disorders of two systems: the endocrine and nervous system. What these systems have notably in common is that neither of them represents a direct effector function: they are both *communication* systems whose function is to link perception and action in a cycle.

As for hormones, they do seem to be crucial for the development of higher forms of multicellular life, the evidence being that they have evolved independently in many different branches, not only in vertebrates, but also in invertebrates and even in plants. In general, the integration involved is not that between an organism and its external environment, but rather the internal coordination of the various organs and cells.

Again for this to be more than mere metaphor, we have to ask ourselves: how does motion in shape space translate into molecular and cellular mechanisms? At this point, however, the answer is quite straight forward, for all we have to focus on are the the kinds of interactions that are *known* to exist between free and bound Igs, the various cell types in the body, and their impact on organismic coherency.

Let us consider as an example which is provided by one very active field in autoimmunity, namely that of antireceptor antibodies and diseases, the two-way exchange between Ig's binding onto cell surface determinants. Antibodies to a hormone are generally inhibitory of the respective hormonal activity, whereas antibodies against cell surface hormone receptors may be inhibitory or stimulatory (that is, antagonist or agonist of hormonal actions). Furthermore, it is often found that anti-idiotypic antibodies to antihormone antibodies may function as surrogates of the hormone itself (this has actually been used as an experimental strategy to isolate the respective receptors). Thus, it is well established that natural antibodies to hormones and/or their receptors do have functional, biologically relevant activities. In turn, the presence of the respective somatic structure (autoantigen) has been shown in a few instances ("obese" chickens with hypothyroidism due to antithyroid antibodies, certain rheumatoid factors, etc.) to be strictly necessary to the development of the autoantibody activities in the serum. This corresponds, in physiology, to the well-documented positive selection of autoreactive B cells into the central IS.[5]

A variety of pathological deviations of these equilibria are known as autoimmune diseases (and we expect many more to be identified, as the molecular characterization of cell surface receptors progresses). Classical examples are hyperthyroidism due to Ig-mediated stimulation of thyroid cells upon binding to the TSH receptor (or Graves Disease), and Myasthenia Gravis, due to the inhibition of acetylcholine receptor function by antibody binding at the endplates. Antibodies to the hormone and/or the respective receptor, however, have been described in many other diseases (e.g., both anti-insulin and anti-insulin receptor antibodies coexist in diabetes). These dysfunctions constitute, for us, the major argument in support of the notion that such dynamic equilibria, this motion in shape space, are an integral part of vertebrate physiology, as postulated here. This has actually been demonstrated experimentally, by the repeated isolation of natural antibodies binding to various receptors (TSH, acetyl-choline, insulin) in normal healthy individuals. Most interestingly, and likely as a result of the particular dynamics of the four-way interactions (hormone, antihormone antibody, hormone receptor, and antireceptor antibody), very often to the surprise and dismay of the clinical immunologist, there is little or no correlation between the titers of a particular antireceptor antibody and the clinical state or the severity of symptoms of the patient. The classical example is again Myasthenia Gravis, where isolated antireceptor antibodies can actually transfer the typical clinical symptoms to naive animals, and yet, many normally active individuals (particularly in the families of patients) contain in their serum high titers of antireceptor antibodies, often higher than in the symptomatic patients.

The very existence of autoimmune conditions is, therefore, another very good illustration of motion in shape space, the result of the two-way exchange between the IN and the somatic environment. Such pathologies would not exist unless a disarray in the network could not go as far as inducing a drastic cellular change in a selected cell class. Autoimmunity is not a simple elevation of titers of specific idiotypes, but it is better understood as a true network dysfunction[10,16,22]: it is the entire idiotypic morphology that is put into motion in this case, albeit in a pathological direction. It follows that the converse must be the case: a global co-definition is the normal condition, and it has a limited robustness and flexibility to maintain the entire bodily identity.

PROPOSED EXPERIMENTS

We come now to the crunch: Can we translate our arguments further into concrete, feasible experiments? Recall that on our hypothesis, there are two distinct types of event: (1) on a short time scale, there is influence at a distance in shape space between otherwise unrelated molecular events, i.e., the manifestation of the IN's dynamics; (2) on a longer time scale, there is the construction of the routes of influence determining which molecular events will be related to which others, i.e., the manifestations of the IN's metadynamics.

Correspondingly, there are two types of experiments, short-term and long-term. The priority is to establish short-term experiments, both because of expediency (they are intrinsically more convenient to perform), and because what long-term experiments can measure is essentially *changes* in short-term performance so that identification of the latter is a prerequisite. What follows is not, of course, an exhaustive list, but some examples to render the ideas more tangible.

One interesting kind of experiment would consist in changing the somatic environment at one or two points (by inhibiting the expression of one specific self component for example) and to follow the dynamics of readjustment of the entire IN in the form of its global profile. Another interesting experiment is to exploit the recently available RAG-1 and RAG-2 "knock out" mice that lack Ig's. If such mice are raised in an antigen-free environment, we would expect that their rate of mortality should not be very different from those raised in normal conditions. This kind of experimental model can provide a real measure of the tightness of the putative global regulation given by the IN to the entire organism.

On the whole, the IN-mediated self-assertion opens new perspectives into clinical research, because in principle every molecular profile in the body should be accessible to regulation via the network. To learn how to intervene in this global regulation is of paramount importance, and a good model could be to study the dynamics of a protein, such as renin or insulin, in normal conditions or in the absence of the IN, or even after selective manipulation of the respective V-region reactivities in the central IS or simply after blocking its binding to the respective receptors.

REFERENCES

1. Avrameas, S. "Natural Antibodies." *Immunol. Today* **12** (1991): 153–159.
2. Bersini, H., and Varela, F. "The Immune Recruitment Mechanism: A Selective Evolutionary Strategy." In *Genetic Algorithms, Proc. XIVth International Conference*, edited by R. K. Belew and L. S. Booker, 520–526. San Mateo, CA: Morgan Kauffman, 1991.
3. Cohen and Young. "Immunol. Today." **12** (1991): 105–110.
4. Coutinho, A. "Beyond Clonal Selection and Network." *Immunol. Rev.* **110** (1989): 63–87.
5. Coutinho, A., A. Bandeira, P. Pereira, D. Portnoi, D. Holmberg, A. C. Martinez, and A. Freitas. "Selection of Lymphocyte Repertoires: The Limits of Clonal Versus Network Organization." *Cold Spring Harbor Symposia in Quantitative Biology* **51** (1989): 159–170.
6. De Boer, R. J., I. G. Kevrekidis, and A. S. Perelson "A Simple Idiotypic Network Model with Complex Dynamics." *Chem. Eng. Sci.* **45** (1990): 2375-2382.
7. De Boer, R. J., and A. S. Perelson "Size and Connectivity as Emergent Properties of a Developing Immune Network." *J. Theor. Biol.* **149** (1990): 381–424.
8. De Boer, R. J., L. A. Segel, and A. S. Perelson. "Pattern Formation in One- and Two-Dimensional Shape Space Models of the Immune System." *J. Theor. Biol.* **155** (1992): 195–233.
9. Freitas, A. A., B. Rocha, and A. Coutinho. "Lymphocyte Population Kinetics in the Mouse." *Immunol. Rev.* **91** (1986): 5–37
10. Kaveri, S.-V., H. Dietrich, and V. Kazatchkin. "Intravenous Immunoglobulins (IVIg) in the Treatment of Autoimmune Diseases." *Clin. Exp. Immunol.* **86** (1992): 92–198.
11. Lundqvist, I., A. Coutinho, F. Varela, and D. Holmberg. "Evidence for the Functional Dynamics in an Antibody Network." *Proc. Natl. Acad. Sci USA* **86** (1989): 5074–5078.
12. Maturana, H., and F. Varela. *The Tree of Knowledge*. Boston: Shambhala, 1987.
13. Nobrega, A., M. Haury, A. Grandien, A. Sundblad, and A. Coutinho. "Analysis of Natural Antibody Reactivities: The 'Immunculus' and the Natural History of the Individuals Molecular Composition." In *Natural Antibodies*, edited by Y. Schoenfeld and D. A.Isenberg. (1992): in press.
14. Perelson, A. S., and G. Oster. "Theoretical Studies of Clonal Selection: Minimal Antibody Repertoire Size and Reliability of Self–Non-Self Discrimination." *J. Theor. Biol.* **81** (1979): 645–670.
15. Segel, L. A., and A. S. Perelson. "Shapespace: An Approach to the Evaluation of Cross-Reactivity Effects, Stability, and Controllability in the Immune System." *Immun. Lett.* **22** (1989): 91–100.

16. Stewart, J., F. Varela, and A. Coutinho. "The Relationship Between Connectivity and Tolerance as Revealed by Computer Simulation of the Immune Network: Some Lessons for an Understanding of Autoimmunity." *J. Autoimmunity Suppl.* **2** (1989): 15–23.

17. Stewart, J., and F. Varela. "Dynamics of a Class of Immune Networks. II. Oscillatory Activity of Cellular and Humoral Components." *J. Theor. Biol.* **144** (1990): 103–115.

18. Stewart, J., and F. Varela. "Morphogenesis in Shape Space: Elementary Meta-Dynamics of Immune Networks." *J. Theor. Biol.* **153** (1991): 477–498.

19. Stewart, J. "The Immune System in an Evolutionary Perspective." In *Theoretical and Experimental Insights into Immunology, NATO ASI Series,* edited by A. Perelson and G. Weisbuch, 27–49. Berlin: Springer-Verlag, 1992.

20. Varela, F., A. Coutinho, B. Dupire, and N. M. Vaz. "Cognitive Networks: Immune, Neural and Otherwise." In *Theoretical Immunology,* edited by A. S. Perelson, 359–374. Santa Fe Institute Studies in the Sciences of Complexity, Proc. Vol. II. Reading, MA: Addison-Wesley, 1988.

21. Varela, F., and A. Coutinho. "Second Generation Immune Networks." *Immunol. Today* **12** (1991): 159–167.

22. Varela, F., A. Anderssen, G. Dietrich, A. Sundblad, D. Holmberg, M. Kazatchkine, and A. Coutinho. "The Population Dynamics of Natural Antibodies in Normal and Autoimmune Individuals." *Proc. Natl. Acad. Sci. USA* **88** (1991): 5917–5921.

23. Vaz, N. M., and F. Varela. "Self and Non-Sense: An Organism-Centered Approach to Immunology." *Med. Hypoth.* **4** (1978): 231–267.

Rob J. de Boer, Jan D. van der Laan, and Pauline Hogeweg
Bioinformatica RUU, Padualaan 8, 3584 CH Utrecht, The Netherlands

Randomness and Pattern Scale in the Immune Network: A Cellular Automata Approach

INTRODUCTION

The immune system is a beautiful example of a complex information processing system. The complexity of the immune system is comparable to that of the nervous system. Both systems are comprised of a large number of different cell types communicating via the production of stimulatory or inhibitory molecules. Several of these cell types form connected networks of billions of different nodes. In the neural network the nodes are fixed and communicate via electrical signals; in the immune system the nodes recirculate and communicate via molecules (e.g., antibody) or cell-to-cell contacts. An important property of both systems is "learning." During its early life the immune system learns to discriminate between self and nonself. Additionally, the immune system has a form memory which is known as "immunity": secondary immune responses are usually different from primary responses.

Immune network theory[21] postulates that the functioning of the immune system is based upon network interactions. The assumption is that the lymphocyte clones comprising the immune system are capable of activating and/or inhibiting each other. Interactions between lymphocytes are based upon the binding of their receptors. Lymphocytes are recirculating blood cells, which are formed in the bone marrow and which die in the periphery. The specificity of each lymphocyte clone is determined by its receptor molecule. Each clone is unique because each receptor has a unique variable region. Typical estimates for the number of different lymphocyte clones in mice and men are 10^6 and 10^9 different specificities. The collection of different specificities is usually called the immune repertoire.

A major difference between the immune system and the nervous system is the turnover of the nodes in the network. In a neural network the neurons typically live for years while they respond on a time scale of seconds. In the immune network, the lifetime and the response time of clones both have a time scale of days. The high turnover of network nodes in the immune system is a consequence of a huge production of novel clones in the bone marrow. The bone marrow production suffices to replace the entire immune system in just a few days.[11]

Recent views of the immune network posit that only part of the immune system is involved in network interactions.[1,4,19,34,35] These authors argue that the immune network is little involved with immune reactions to foreign antigens. Rather the network is autonomously active and/or responds to the environment of self-antigens. Further, these authors argue that the network is most pronounced during early life. This early immune network seems to play a role in the selection of the immune repertoire.[22] One possible mechanism of selection is that the activation of lymphocytes by network interactions precludes the death of the lymphocytes thus allowing the clone to be maintained in the immune repertoire. In this paper we study this hypothesis and show that such an immune repertoire self-organizes into clusters of activated and suppressed specificities. The size of these clusters turns out to depend on the bone marrow production. Thus, the main interest of the present paper is the *scale* of the patterns in the immune repertoire.

CELLULAR AUTOMATA

The immune network model that we develop below is a cellular automaton (CA). CAs were introduced in the late forties[32] as a general formalism for the study of complex systems (see Bak[2]). A CA consists of a collection of cells or automata that are ordered in a regular pattern, which is usually a rectangular grid (see Wolfram[38] or Toffoli and Margolus[31] for reviews). The neighborhood in a CA is usually formed by the four or eight cells surrounding each cell on the grid. Each automaton has a finite set of states and a next-state function that provides the next state as a function of the current state and the neighborhood. All cells behave according to the same rule and usually change to their next state synchronously. Because CAs are comprised of local automata in a spatial embedding, they have been viewed as "computing matter." They have been used to define a variety of enjoyable "artificial worlds."

A classic example of an artificial world CA is the game of "Life" by J. H. Conway.[2,12,13] Life has become famous because of its simple rules and complex behavior. In the game of Life, automata are on (i.e., "alive") or off (i.e., "dead"). At each time-step every automaton responds to the state of its local environment. Automata become alive whenever they have exactly three live neighbors. Automata remain alive when they have two or three live neighbors. Automata die when there are fewer or more neighbors alive. Thus, in this artificial world, being "overcrowded" or being too "lonely" leads to death. The most important result of the game of Life is that extremely simple rules may allow for an enormous complexity. In general, CAs have been classified into different categories on the basis of predictability[38] and complexity.[24] CAs like Life belong to the unpredictable classes and are capable of universal computation.[38]

In biology, CA models have been used to study spatial processes in general. Examples of this are reaction-diffusion-type systems,[3,14] ecological systems,[17] and immune systems.[18] The immune network CA that we develop below is based upon a rule similar to the game of Life: clones are maintained (i.e., are alive) when their stimulation is sufficient but not too high.

BITPLANE APPROACH

In our modeling approach we make use of the parallelism of the CA. The states of a binary CA, i.e., an automaton with black/white, on/off, or one/zero states, can be stored in bitplanes. This speeds up the simulations enormously because most computers have fast algorithms for manipulating bitplanes (bitplanes are used for writing the screen and for windows). A more detailed description of the bitplane approach can be found in Appendix I.

VOTING RULES

As an example illustrating the main principles of CAs, we discuss here majority voting. In a voting rule CA, every cell in the grid is a binary state automaton. For a majority vote, the next state of a cell is the state in which the majority of its neighbors and itself are in. Thus, the next state of each automaton is determined by a vector I of nine bits. If the sum of this vector is larger than four, i.e., if $\sum I \geq 5$, the majority is black, and the next state will be black. Conversely, if $\sum I < 5$, the majority is white, and the next state will be white. Starting with a random initial configuration, in which 50% of the cells is black and 50% is white (see Figure 1(a)), a voting rule CA yields a stable pattern of small white and black regions (see Figure 1(b)). This takes only a few generations (here 29). The most stable attractors of a voting rule CA are the two global patterns in which all cells have the same state (i.e., all white or all black). These attractors are never attained because the system gets stuck in smaller scale patterns of black and white regions. Locally, the boundaries between these areas are stable.

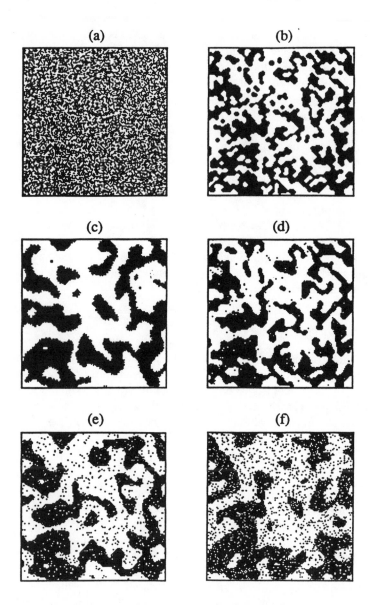

(a)   (b)

(c)   (d)

(e)   (f)

FIGURE 1   The effect of annealing or randomness on the scale of patterns in a CA with majority voting rules. This CA consists of 140×140 bits and is simulated using periodic boundary conditions. Thus it takes the shape of a torus where the upper and lower, and the left and right, boundaries are connected. (a) initial random distribution, (b) stable configuration attained after 29 majority voting steps, (c) snapshot after 50 voting steps with annealing, and (d–f) snapshots after 50 voting steps with 1, 5, and 10% random "errors."

The scale of these patterns can be increased by simulated annealing and/or by randomness. Simulated annealing is a general technique for avoiding local optima.[23] In the context of the CA, annealing introduces "errors" at the critical point of the next-state function.[36] Thus, if there is a profound majority, say, $\sum \mathbf{I} \geq 6$ or $\sum \mathbf{I} \leq 3$, the majority is chosen. However, if there is a majority of just one vote (i.e., $\sum \mathbf{I} = 4$ or $\sum \mathbf{I} = 5$), then the next state will be the minority. The increase in the scale of the pattern as a result of this type of annealing is shown in Figure 1(c): the black and white areas become much larger. In the long end, the system usually attains one of the two global attractors, i.e., only white or only black cells. Similar results can be obtained by introducing random "errors" in the transition rule. Figure 1(d–f) show an increase of the scale of the pattern if the minority is chosen with a low probability. Additionally, an increase in randomness increases the scale of the pattern: in Figure 1(d–f), the probability to follow the minority is 1, 2, and 4%, respectively.

The process of pattern formation by simulated annealing in voting rules is further illustrated in Figure 2. Small-scale structures can been seen to emerge from an initial, random distribution of black and white. As time proceeds the scale of these patterns increases. The pattern in Figure 2(i) largely consists of one big black island in a white sea (be reminded of the periodic boundary conditions). Because of its concave boundaries, this island will eventually disappear.

Annealing affects the boundaries between the black and white regions. Due to the annealing, curved boundaries become unstable and the system straightens its boundaries. This results in the erosion of "capes" and the filling of "bays." The increase of scale by introducing randomness in the voting rule CA can also be explained this way. Random errors will also make curved boundaries unstable.

Summarizing, the scale of the pattern increases both with the degree of randomness and with time.

IMMUNE NETWORKS

The enormous recruitment of novel clones from the bone marrow is one of the most salient characteristics of the immune system.[8,33] We have argued above that in the immune network, the decay and the growth of established clones, and the recruitment of novel clones, occur on the same time scale. In terms of modeling, this means that one cannot make an equilibrium assumption for either of the two processes, i.e., recruitment or growth, and that both have to be studied in combination.[8]

Thus, we have previously developed an asynchronous CA in which both processes were accounted for and could be scheduled at any time scale.[8] In this asynchronous CA, the scale of the pattern increases with the rate of random recruitment. Because of the similarity between this result and the effect of randomness

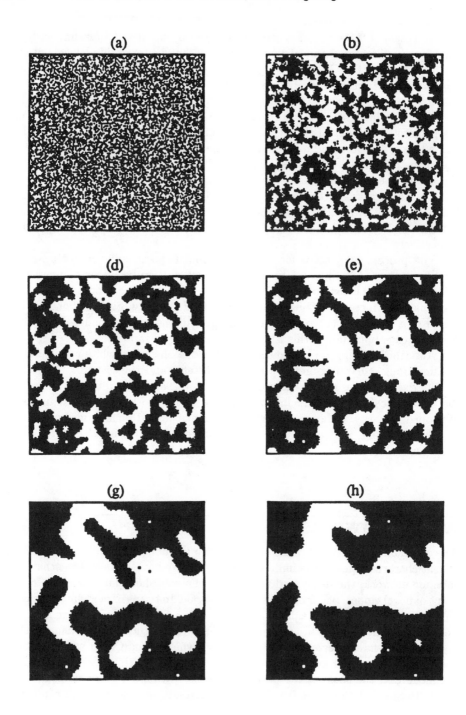

(c)

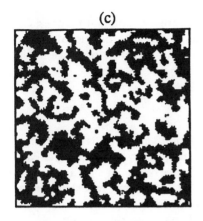

(f)

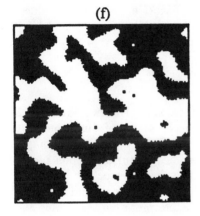

(i)

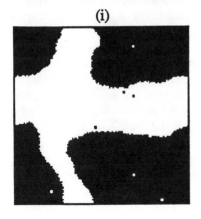

FIGURE 2 The evolution of voting rule patterns by annealing. Periodic boundary conditions were used. Starting with an initial random distribution (a), we perform majority voting with annealing, and make snapshots at the following time steps: $t = 5$ (b), $t = 10$ (c), $t = 20$ (d), $t = 40$ (e), $t = 80$ (f), $t = 160$ (g), $t = 320$ (h), and $t = 640$ (i).

in voting rule systems (see Figure 1), we here compare our immune network CA with voting rule systems. We simplify our previous CA by ignoring clonal growth and by making it parallel. Thus, for each clone, we only record its presence in the repertoire.

SHAPE SPACE

An elegant concept for the modeling of the immune network is the "shape space."[26,28,29] Because each lymphocyte carries a unique receptor molecule, each clone can be characterized by a unique shape. The degree of complementarity of the receptors determines the affinity of the binding interaction. Thus, in a shape space model, a vector $\mathbf{x}$ describes the shape of each receptor. In principle, $\mathbf{x}$ may account for several molecular characteristics such as the shape of protuberances, the hydrophobicity, and/or the charge. For reasons of simplification, one usually reduces $\mathbf{x}$ to an abstract low-dimensional shape variable. The first shape space model[28,29] was one-dimensional; later models[9] were two-dimensional.

THE IMMUNE NETWORK CA

Here we study a two-dimensional shape space by means of a two-dimensional CA. We implement complementarity by considering two planes of complementary shapes, i.e., the "plus" plane with shapes $\mathbf{x}_+$ and the "minus" plane with shapes $\mathbf{x}_-$. A possible interpretation of the two planes would be to think of the plus planes as protuberances and the minus planes as clefts. The shape vector $\mathbf{x}_+$ could represent the height and the diameter of a protuberance whereas $\mathbf{x}_-$ could represent the depth and the diameter of a cleft. An alternative interpretation of the plus and the minus planes is as molecules with a positive or negative charge, respectively.

In the CA formalism this means that the neighborhood of a shape $\mathbf{x}_+$ corresponds to the local neighborhood of $\mathbf{x}_-$ in the minus plane. In our CA we use a weighted 7×7 neighborhood. The Figure 3 shows the neighborhood that a single cell evokes at the complementary plane. The perfect match, $\mathbf{x}_+ = \mathbf{x}_-$, is counted 17 times (see Appendix II). This neighborhood weighting is roughly Gaussian and is generated by just 15 parallel shifts in our bitplane approach (see Appendix II). Note that nine shifts are required for the standard 3×3 neighborhood (see Appendix I). The weighted sum of the neighborhood, which we call the "field," or h, is the total stimulation of each shape.

Figure 3 shows that $0 \leq h \leq 225$. The field h determines whether or not a clone will be maintained according to the window automaton proposed by Neumann and Weisbuch[25]

$$w(h) = \begin{cases} 1, & \text{if } \theta_1 \leq h \leq \theta_2; \\ 0, & \text{otherwise.} \end{cases} \tag{1}$$

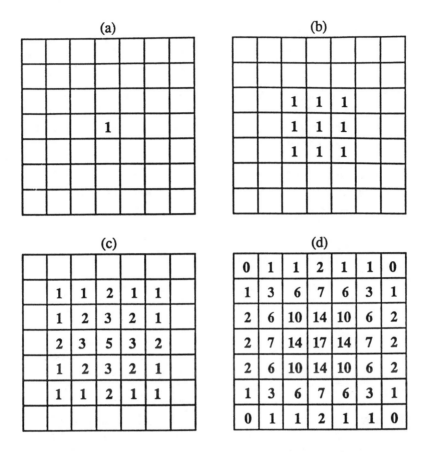

FIGURE 3 The weighted neighborhood of the immune network CA. We show how the neighborhood (d) is obtained from a single bit (a). First, the bit evokes a standard 3 × 3 Moore neighborhood (this requires eight parallel shifts). Second, the Moore neighborhood is shifted in the four diagonal directions. This gives a weighted 5 × 5 neighborhood (c). Third, the weighted neighborhood is shifted in the two horizontal and two vertical directions. This results in the final weighted 7 × 7 neighborhood (d).

The reason for choosing a window automaton is that experimental data and mathematical models of B lymphocyte activation suggest that the degree of activation is a log bell-shaped function of the stimulus.[27] Most recent immune network models are based upon a continuous log bell-shaped function. This class of models is the topic of several recent reviews.[6,10,34]

Finally we implement bone marrow recruitment. Recruitment is a random process which allows a fixed percentage P_R of all shapes to become present. The rate of recruitment, P_R, is varied throughout the analysis. The limits of the window $w(h)$ were fixed to $\theta_1 = 1$ and $\theta_2 = 80$. We have studied the CA for other values of the thresholds and have always found similar patterns. The reason for choosing a low value for θ_1 is to allow for easy network activation. The large value of θ_2 reflects our intuition that clones can only be suppressed by the combined effect of several other clones.

The bitplane implementation of our immune network CA is explained in more detail in Appendix II. The CA consists of two shape planes of 140×140 bits. We here use fixed boundary conditions. This implies that there is no contribution to the field from outside the shape space.

Here, we summarize the steps as follows:

(1) Initialization: make 10% of the shape planes present.
(2) Neighborhood: calculate h, the sum of the weighted neighborhood.
(3) Recruitment: allow a percentage P_R of all shapes to become present.
(4) Selection: eliminate all shapes for which $w(h) = 0$.
(5) Time step: go to step two.

Since the recruitment makes shapes present only after the fields have been determined, this algorithm assumes that only those clones that have been maintained for at least one time step can contribute to the field. This is reasonable because it is unlikely that the small clones emerging from the bone marrow are immediately able to activate the other clones in the network.

RESULTS

Visualization of results is a notorious problem in network models. Conversely, in bitplane CAs this is straightforward because each plane of the CA can be presented by displaying the bitplane as a black and white pixel. Thus, two bitplanes of 140×140 pixels provide a complete description of the state of our CA. Further, by color coding, different bitplanes can be combined into one color display. We here use a color table of four colors. Red means present in the Shape$_+$ bitplane, green means present in the Shape$_-$ bitplane, black means absent from both planes, white means present in both planes. Now, one 140×140 picture completely describes the state of the CA.

An example of one particular network after 400 time steps is shown in Plate 1(a) (see color insert for plates). The pattern consists of irregular red and green areas. The borderlines of the areas are densely populated. The red and green areas fit almost exactly into each other. A tiny empty region separates borderlines of different color. The red and green areas are almost exclusive: in panel (a) we find hardly any white areas corresponding to presence in both planes. Within each area the pattern is noisy having scattered red and green pixels. Apart from the noise, some small green clusters occur in the large red areas. Each of the green clusters is surrounded by a heavy red borderline. (The same is, of course, true for the green areas.) Because of their circular appearance we call these borderlines around the small areas the "atolls." The pattern that we find, i.e., large areas dominated by one color and containing atolls of the same color, is of a much larger scale than the 7×7 pixel neighborhoods of our CA. Thus, the pattern that is formed cannot directly be derived from the local next-state function of the CA. Instead, the pattern emerges by self-organization.

The immunological interpretation of these patterns is that large areas in shape space are predisposed toward one class of antibodies (e.g., toward a cluster of positively charged molecules). Within such a cluster the other class of molecules is scattered at a low frequency.

The ability of a region to mount an immune response (to either self or nonself antigens) is defined by the field in that region. The absence of clones in a region may have two reasons: too little or too much stimulation. Thus, in Plate 2(a) we plot the field, h_-, of the Shape$_-$ plane. This corresponds to the field of the green pattern as it is generated by the red pattern. In Plate 2, red means that the field is suppressive, i.e., $h_- > \theta_2$, black means a lack of stimulation, i.e., $h_- < \theta_1$, and green stands for a stimulatory field, i.e., $w(h_-) = 1$. The suppressive (red) areas look like a blurred version of the red borderlines of Plate 1(a). Thus, the heavy borderlines evoke a suppressive field. The field is stimulatory (i.e., green in Plate 2(a)) in the majority of large green area of Plate 1(a). Thus, although the repertoire is noisy within each area, the field is stimulatory almost everywhere. Conversely, in the large areas that are mainly occupied by red, the field tends to be suppressive. This is only a tendency because several small spots of stimulatory fields exist in the red areas (compare Plate 1(a) with Plate 2(a)). The clones located in these small spots sustain the simulation of the noisy parts of the large areas. Areas with a lack of stimulation (i.e., black in Plate 2(a)) are rare. This implies that the network roughly covers the shape space.[7,9]

The distribution of the field strengths in Plate 2 allows us to investigate the hypothesis that only part of the immune system is involved in the network.[4,19,34,35] Thus, we consider the introduction of a foreign antigen at a random location in shape space. This area in shape space may or may not be under the influence of the network. Stimulation with any antigen corresponds to a local increase of the field. In this model an effective immune response to such an increase of the field requires that this area in shape space becomes more densely populated. Such a response is only possible if that area is not yet suppressed by the network. Thus, in Plate 2, an immune response can only come from a black or green area. A response from a black area, where $h_- < \theta_1$, would correspond to a classical immune response of that part of the system that is not functionally connected to the network. A response from a green area, where $\theta_1 \leq h_- \leq \theta_2$, would correspond to a reaction from the network. Because the black areas are rare, immune reactions from clones functionally connected to the network are expected to be most abundant in this model.

The stability of the immune repertoire, i.e., of the pattern shown in Plate 1(a), is studied by making time slices. These are shown in Plate 3(a). Each horizontal line in Plate 3(a) corresponds to a horizontal slice through the center of the shape space of Plate 1(a). At every time step we obtain one line of pixels of Plate 3(a). In Plate 3 time increases from top to bottom. The panel is square because we only show the last 140 time steps of the simulation (i.e., the panel is 140×140 pixels). In such a time plot, any moving pattern, e.g., a so-called traveling wave, shows up as a diagonal line. Instead, steady patterns show up as vertical lines, and noise shows up as noise.

Thus, inspection of Plate 3(a) reveals that the pattern is stable, i.e., most lines are vertical. Only part of the pattern is noisy, most of the noise being confined to small regions within the large areas. The borderlines between the red and green areas are all stable.

RECRUITMENT

The effect of the rate of recruitment, i.e., of P_R, on the immune repertoire is shown in Plate 1. In each panel we show the repertoires as they are attained after 400 time steps. From panel (a) to (i), the rate of recruitment varies from $P_R = 1$ to 90. Visual inspection of the panels reveals the impact of the random recruitment on the scale of the pattern. As P_R increases, the areas grow in size and merge together to form connected patterns that span the entire shape space. For instance, in Plate 1(g) the shape space is largely one red cluster. Further, the patterns become smoother because the borderlines loose their irregular appearance. Within the areas we see more structure arising because the areas become covered with atolls.

The rate of recruitment P_R not only influences the spatial scale of the pattern, it also influences its temporal scale. On Plate 3 the dynamic behavior of the shape space is shown as a function of P_R. When the rate of recruitment is small (e.g., panels (a)–(c), $P_R = 1, 2, 4$) vertical lines dominate. This indicates that the boundaries between the areas are stable. Within the areas the pattern is noisy. As P_R increases, i.e., $P_R \geq 8$, some boundaries become unstable and fewer vertical lines are observed. Note that the main reason for having fewer boundaries, either stable or unstable, is the increase in the scale of the pattern when P_R increases.

In the time series, the atolls appear as triangular or drop-shaped patterns with the top pointing downwards (e.g., panels (f) and (g)). This indicates that the atolls appear suddenly but are unstable, and decrease in size and disappear. The typical lifetime of an atoll is 15 to 30 time steps. High rates of recruitment, i.e., $P_R = 60 - 90$ (panels (g)–(i)), give rise to periodic behavior. In the time plots one observes alternating black and white lines. Thus, presence in both planes (i.e., white) oscillates with absence from both planes (i.e., black). The fact that clones from both planes are recruited in these spots means that the field is stimulatory for both planes. Such an overlap in the stimulatory fields is to be expected because we have observed above that, in the areas predisposed toward one of the planes, small stimulatory spots exist for the other plane.

Because of the high rate of recruitment, clones from both planes invade such a stimulatory area at a high rate. Because novel (i.e., unstimulated) clones do

Color Plates

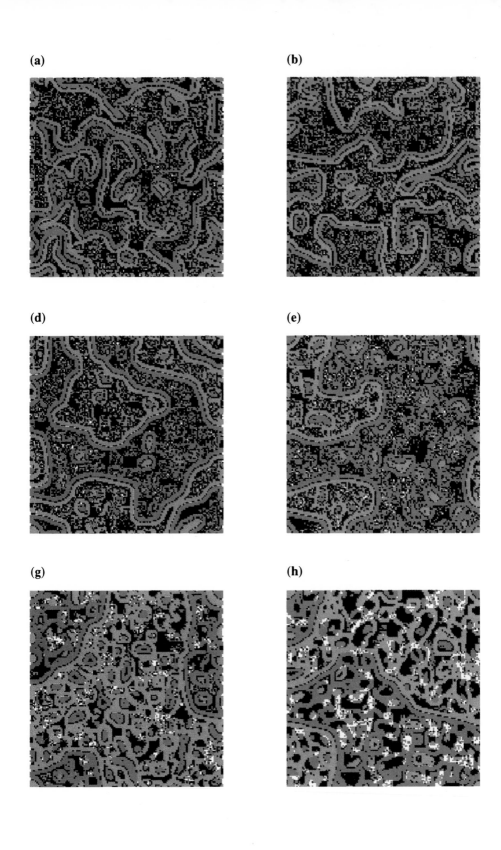

(c)

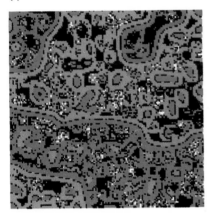

(f)

PLATE 1

The effect of random recruitment on the repertoire. The nine panels show the presence of clones in the plus and minus planes after 400 time steps. The clones present in the plus plane are colored red; those present in the minus plane are colored green. White pixels denote that a site is occupied in both planes, whereas black indicates absence from both planes. The nine panels show the repertoire for $\theta_1 = 1$, $\theta_2 = 80$, $P_R = 1, 2, 4, 8, 20, 40, 60, 80, 90$. Fixed boundary conditions are used throughout in the immune CA.

(i)

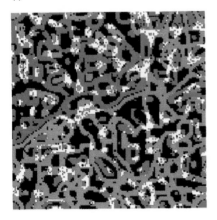

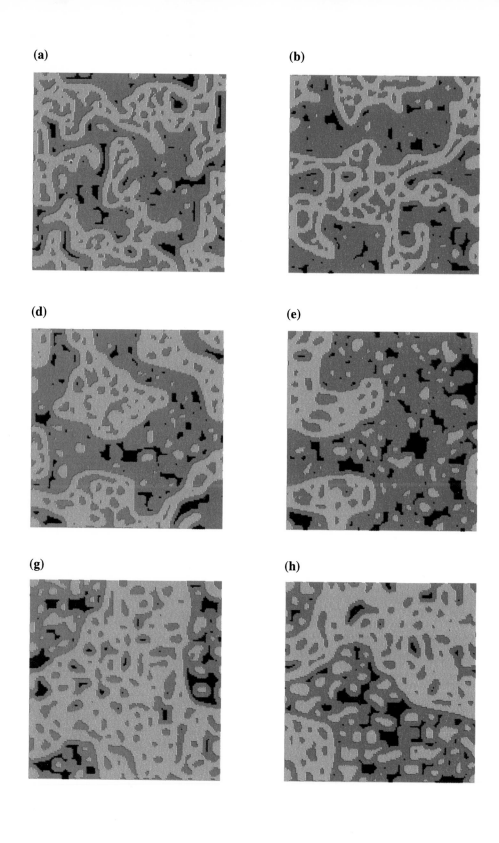

(c)

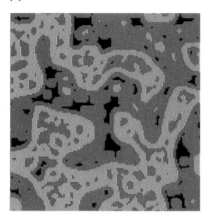

PLATE 2

Fields as a function of recruitment. The field, h_-, of the minus plane (i.e., of the green pattern in Plate 1) is color coded in three categories. Black means little or no stimulation, i.e., $h_- < \theta_1$. Green denotes a stimulatory field, i.e., $\theta_1 \leq h_- \leq \theta_2$. Red indicates a suppressive field, $h_- > \theta_2$. The nine panels show the fields for $\theta_1 = 1$, $\theta_2 = 80$, $P_R = 1, 2, 4, 8, 20, 40, 60, 80, 90$.

(f)

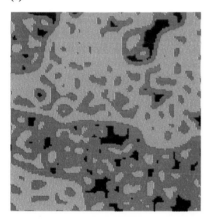

(i)

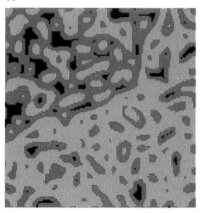

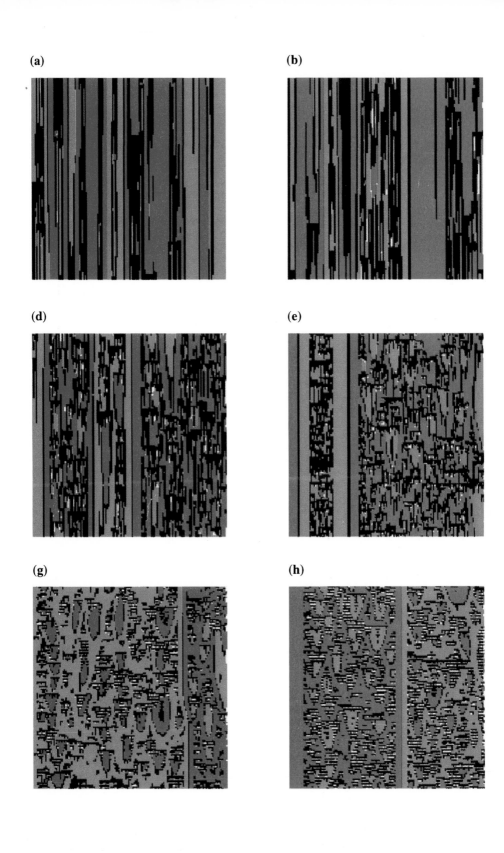

(c)

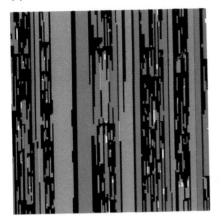

PLATE 3

Time series of a slice through the shape space. The panels are composed of 140 horizonal lines that correspond to horizontal slices through the center of the shape space at subsequent time steps. The slices correspond to the last 140 time steps of the simulation; time increases from top to bottom. The color coding is the same as in Plate 1. The nine panels show the time series for $\theta_1 = 1$, $\theta_2 = 80$, $P_R = 1, 2, 4, 8, 20, 40, 60, 80, 90$.

(f)

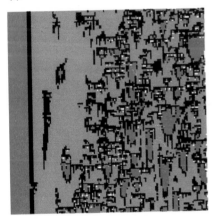

(i)

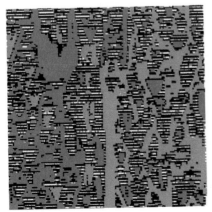

(a)

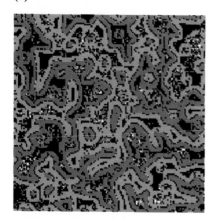

PLATE 4

The ontogeny of the pattern. For $\theta_1 = 1$, $\theta_2 = 80$, $P_R = 30$, Plate 4 shows five snapshots of the immune repertoire taken at day 15, 30,..., 75. Looking at the boundaries between the red and green areas, one can see that "capes" become eroded and that "bays" become filled. The color coding is the same as in Plate 1.

(b)

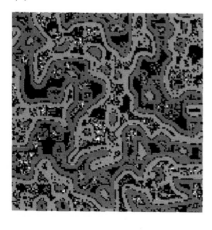

(c)

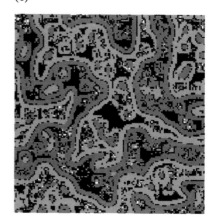

(d)

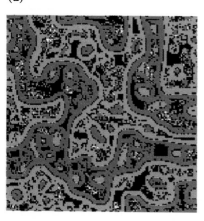

(e)

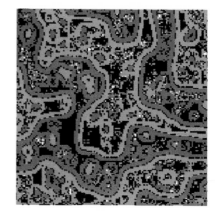

not contribute to the field, these novel clones do not interact and persist for one time step. This appears as a white line in the plot. Because the high recruitment has generated a high density, they suppress each other the next time step. As a consequence they all disappear. This appears as a black line in the plot. This process also explains why in Plate 1 we observe a higher density of black and white regions: these are sustained by the oscillation.

The effect of the rate of recruitment on the fields is shown in Plate 2. Since the distribution of the fields follows that of the clones closely, the effect of P_R in Plate 2 is comparable to that in Plate 1. Additional effects are that increasing P_R increases the suppression, i.e., the part of the shape space covered in red. This reduces the responsivity of the system to external stimuli. A similar effect has been observed before in models based upon bitstrings.[7]

Summarizing, visual inspection of the results suggests that the system self-organizes on two spatial scales. Increasing the recruitment increases the scale of the large areas. Conversely, within the areas, a high rate of recruitment generates a smaller scale pattern of atolls. On the temporal scale the atolls become short lived when the rate of recruitment is high.

ONTOGENY OF THE PATTERN

A possible explanation for the increase in scale as a function of the rate of random recruitment is the similar effect that randomness has in voting rule systems (see Figure 1). In systems with simulated annealing and/or randomness, boundaries become fluid and small areas merge into larger ones. The temporal effects of annealing in voting rule systems are shown in Figure 2. This revealed that the patterns start at small scale and slowly increase in scale during subsequent generations.

We can test whether similar effects play a role in our immune network CA by studying the early evolution, i.e., the ontogeny, of the pattern. Thus, for $P_R = 30$, Plate 4 shows five snapshots of an immune repertoire taken after $15, 30, \ldots, 75$ time steps. This shows that in our immune network the scale of the patterns is increasing by a very similar process. Looking at the boundaries between the red and green areas, one can see that "capes" become eroded and that "bays" become filled. Thus, the effect of recruitment in the immune network is comparable to that of annealing or randomness in voting rule systems.

FILTERING BY MAJORITY VOTES

By defining a quantitative measure of scale, we here make our intuitive observations on the scale of the pattern a little more substantial. A simple scale variable is the ratio between border length and surface area (which is minimized by large discs). A problem with our patterns is that they are noisy. Red areas are not really red, and green areas are not really green. Thus, we resort to filtering techniques.

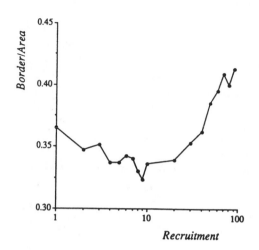

FIGURE 4   Scale, i.e., the border/area ratio, as a function of recruitment P_R. Each point corresponds to an average of the two fields of five simulations with $\theta_1 = 1, \theta_2 = 80$. For each simulation we take the minus plane after 400 time steps, normalize it, and perform majority voting until stabilization. For the filtered pattern the total number of bits in the borderlines and the total number of bits in areas are counted. The ratio of the two is plotted as a measure of scale.

An interesting technique for stressing the difference between regions is majority voting. Areas in which one of the colors dominates can be filled with that color by repeated majority votes. However, this procedure will only work if that dominant color occupies more than 50% of the sites. The average density in each of the shape planes is between 30% to 40% coverage. Further, the coverage increases somewhat with recruitment. Thus we first normalize the patterns to a coverage of 50%. This has two effects. First, areas in which one color dominates become covered with that color for more than 50%. Second, the coverage becomes independent of the rate of recruitment.

Thus, for each of the two shape planes, we count the coverage, and add random bits until the coverage is 50%. Actually, the normalization technique is (1) to calculate how many bits are required for attaining a coverage of 50%, (2) to make a random plane filled with that number of bits, and (3) to perform the logical OR operation on the original plane and the random plane. We then take this normalized version of the two planes of the immune CA as the initial configurations of a majority voting CA. This is expected to eliminate the noise and to connect some of the areas. We observed that this procedure preserves most of the characteristics of the original patterns (not shown). Subsequently, we calculate the border/area ratio of the pattern. A parallel technique for finding the borderlines is to filter out a plane of all bits that have less than eight neighbors. By counting the number of bits in the plane of borderlines and the number of bits in the plane subjected to normalization and voting, we know the total border length and the total surface area. We express scale as the ratio of the two, i.e., border/area.

We study the influence of the rate of recruitment on this measure of pattern scale. Thus, for $P_R = 1, 2, \ldots, 10, 20, \ldots, 90$, five simulations were carried out. Each dot in Figure 4 is the average border/area ratio of the two fields of the five simulations, i.e., each dot is based upon ten samples. The curve connecting these points has a minimum around $P_R \approx 9$. Visually, the patterns indeed seem to be at the largest scale in this range of recruitment rates. Because of the emergence of many atolls, there is a steep increase of the border/area ratio at high rates of recruitment.

Biological data suggest that the rate of recruitment is very high. In the adult mouse, it is estimated that the production in the bone marrow amounts to 2 to 5×10^7 B cells per day.[11] This is sufficient for replacing the entire B-cell population in just a few days. Thus, if we assume that each day 20% of the repertoire can be replaced, an estimate for the probability of recruitment is $P_R = 20$. In this parameter range the patterns have a large scale (see Figure 4 and Plate 1(e)).

DISCUSSION

The model that we have developed is similar to a model developed by Stewart and Varela.[30] They also consider a two-dimensional shape space model in which clones are present or absent, and also implement complementarity by considering two planes of shapes. In their model, clones are randomly introduced in one of the planes and are maintained whenever they receive enough but not too much stimulation. The two-dimensional patterns that they report resemble those reported here. With our CA model we generalize the results of Stewart and Varela[30] and show that the scale of pattern depends on the rate of recruitment.

The network hypothesis that we have investigated here postulates that only part of the immune system is functionally connected to the immune network.[4,19,34,35] These ideas are (partly) based upon the percentage of activated lymphocytes in animals raised in "antigen-free" conditions.[20] Thus, it is estimated that 10–20% of the lymphocytes are functionally connected to the network, and that 80–90% remain disconnected.

In our model network we do find regions of the shape space that are functionally disconnected from the network. These regions correspond to the black areas in Plate 2 in which the field is low (i.e., $h_- < \theta_1$). The major problem with this result is, however, that the percentage of functionally disconnected clones is very small. Similar results have been obtained before with other network models.[7,8,30] Generically, the repertoires of the model networks are determined by a roughly complete coverage of shape space.[7] The networks are expected to encroach on areas that are poorly stimulated.

These results raise two questions. First, how can one explain the experimental observations that 80–90% of the clones remain functionally disconnected from the network? Second, what part of the immune system is responsible for the immune reactions, and for immunological memory?

With respect to the first question, we could use our results on the large parts of shape space that tend to be suppressed. Cells in these regions are not stimulated to grow and could be measured experimentally as nondividing lymphocytes.

Experimentally, such cells are interpreted as functionally disconnected cells. From our work, however, we know they are connected to the network by suppressive interactions, and cannot respond in an immune reaction. Another explanation that would account for a large fraction of apparently functionally disconnected clones is that the network might be composed of a special class of B cells, i.e., the Ly1 or CD5 B cells[16] which predominates during early life. The conventional B cells, which appear later, have several different properties. It is an open question whether or not the conventional B cells connect to the network. If they do not, they may account for the 80–90% nondividing cells.

With respect to the second question on the origin of immune reactions and memory, we interpret our results on the large stimulated areas that are only sparsely inhabited by clones. In these regions the presence of an antigen will allow more clones to be recruited and to respond in a primary immune reaction. This corresponds to a shift in the repertoire. If such a shift were to persist after the removal of the antigen, the network would account for memory. A secondary exposure with the same antigen would lead to a more vigorous immune reaction because that part of shape space would be densely populated. For our immune network CA, this speculation can be tested by extending our CA with an additional plane for the antigens. This distributed form of memory, i.e., one based upon repertoire shifts, has already been observed in previous models.[7] A more localized form of memory has been accounted for by network models with multiple stable points.[5,6,37]

The immune repertoire in this CA model, and that of previous network models,[7,8,9] is highly clustered. All clones in a cluster tend to have the same field and tend to be in the same state. This suggests that one could develop novel models in terms of "super clones" representing an entire cluster. In fact, such a super clone would correspond to the immunological concept of an "idiotype" which also comprises several clones. For such a super clone, the rate of recruitment corresponds to the average of many populations. Thus, it becomes realistic to model recruitment as a continuous, instead of a stochastic, process. This suggests that many earlier models that are based upon a continuous recruitment are realistic.

Our simple immune network CA has demonstrated that such clusters are likely to emerge because random recruitment increases the fluidity of the repertoire.

ACKNOWLEDGEMENTS

We thank Maarten C. Boerlijst and Alan S. Perelson for helpful discussions. This work is supported in part by grants to the Santa Fe Institute, including core funding from the John D. and Catherine T. MacArthur Foundation, the National Science Foundation (PHY-8714918), and the U.S. Department of Energy (ER-FG05-88ER25054).

APPENDIX I: CELLULAR AUTOMATA AS BITPLANES

One of the main properties of most CAs is their synchronicity or parallelism. This allows one to develop specialized hardware for fast CA simulations. Toffoli and Margolus[31] developed a Cellular Automata Machine (i.e., the CAM) that is based upon hardware bitplanes. In our bitplane approach we borrow a lot of their ideas. Thus, in our CAs, all operations are defined on the level of (software) bitplanes, and run for all individual bits simultaneously.

Because most computers use bitplanes for updating the screen, the bitplane approach has two advantages. First, the simulations are fast due to efficient algorithms in the system software. Second, it is fast and straightforward to visualize the state of the CA by copying a bitplane to the screen. Note that piles of bitplanes can be converted to color pixels. Visualization is a major advantage because one can actually watch the behavior of the CA and see its patterns emerge.

Operations on the level of bitplanes correspond to logical operations (e.g., OR, AND, and XOR). For example, one can perform the AND operation on two bitplanes to obtain a new bitplane that is black in those positions where both bitplanes are black. Note that whenever rules become complex, the translation into a sequence of logical operations may become quite complicated. Another operation is the SHIFT, which yields a copy of a bitplane with all positions shifted in some direction. Shift operations are helpful for getting neighborhood information. For instance, shifting a bitplane one bit to the right yields a novel bitplane with in each cell the states of its left neighbor. Shift operations require a specification of the boundaries of the CA. The conventional procedure is to use periodic boundaries that wrap the CA in the form of a torus. In our immune CA we assume fixed boundary conditions which means that all cells outside the boundaries are absent.

CAs with more than two states simply require more bitplanes. This is very common, e.g., the sum of neighbors in voting rule CAs ranges from zero to nine. We represent integer numbers in a pile of bitplanes by encoding it in the standard binary representation. Arithmetic operations on piles of bitplanes are also defined in terms of logical operations on the individual bitplanes (see Goldberg and Robson[15] for a more detailed explanation of counting by logical operations on bitplanes).

Random values are also obtained in parallel by running a random generator CA, i.e., the NOISE-BOX, described by Toffoli and Margolus.[31] At every generation the NOISE-BOX generates two planes with a random distribution of 50% ones and 50% zeros. Other percentages are obtained by running the NOISE-BOX for a few time steps and performing logical operations on several random planes.

By showing the code for the majority voting rule, we here illustrate the basic principles of bitplane CAs. To this end we define an *ad hoc* formal language. In this language, variables in normal typeface denote planes, variables in boldface denote piles of planes, and variables in italics denote integers. We have a library of several functions:

1. **Shift(Pile,***i***,***j***)** returns a copy of the **Pile** shifted i positions horizontally and j positions vertically.
2. **Sum(Pile,Pile)** returns a pile which is the arithmetic sum of its two arguments.
3. Random(i) returns a random Plane in which a percentage of i of the bits are black.
4. LargerEQ(**Pile**,i) returns a plane with ones for those positions where the integer encoded in the Pile is $\geq i$.

Having defined the library functions, the following function retrieves the sum of the 3×3 neighborhood of each cell in a plane in parallel:

```
GetNeighborhood(World) {
    N ← World                            initialize the neighborhood by the center cell
    N ← Sum(N, Shift(World, −1, −1))     add the NorthEast neighbor
    ⋮                                    shift eight times in total
    N ← Sum(N, Shift(World, 1, 1))       add the SouthWest neighbor
    GetNeighborhood ← N                  return the sum
}
```

Having defined the neighborhood function, the program is simple. It consists of an iteration loop that calls the neighborhood function n times:

```
World ← Random(50)                       initialize with 50% random bits
Repeat n Times {
    World ← LargerEQ(GetNeighborhood(World), 5)
    Display(World)                       show the results on the screen
}
```

APPENDIX II: THE IMMUNE SHAPE SPACE CA

In the immune network CA, we require a large number of planes and piles for modeling the different parts of the system. The complementary shape space planes are represented in the two bitplanes Shape$_+$ and Shape$_-$. These planes are both initialized with a random distribution of 10% ones, i.e., 10% of the shapes is present. The field, i.e., the network stimulation, is retrieved by a GetWeightedNeighborhood function (see the code below). The result of this function (i.e., the weighted neighborhood) is stored in the piles **Field$_+$** and **Field$_-$**. These piles are then filtered twice by the routine LargerEQ(). This returns a plane (i.e., Field$_+$ and Field$_-$, respectively) with ones at those positions where the weighted sum in the pile is between θ_1 and θ_2. Thus, Field$_+$ and Field$_-$ correspond to the stimulatory fields.

The recruitment is realized by first creating two random planes (i.e., Marrow$_+$ and Marrow$_-$) with a certain percentage, P_R, of ones. The Marrow planes are added to the Shape planes by an OR operation on both planes. An AND operation on the new Shape plane and the corresponding Field plane determines which clones are to be maintained. Each time step the results are displayed on the computer screen.

For illustrative reasons we translate this formalism in our illustrative formal language. Below, the code for the Shape$_+$ plane is listed. The code for the Shape$_-$ plane is very similar.

The program is based upon a GetWeightedNeighborhood function, which is in turn based upon the GetNeighborhood function defined in Appendix I. The main idea of the GetWeightedNeighborhood function is that shifting the standard neighborhood pile of a plane increases the size of the neighborhood and provides a weighting with respect to the distance. By repeating such a shift, one increases the neighborhood. Here we repeat the shift twice.

```
GetWeightedNeighborhood(Shape) {
    N ← GetNeighborhood(Shape)
    P ← N                              initialize with the standard neighborhood
    P ← Sum(P, Shift(N, −1, −1))                        now shift four times
    P ← Sum(P, Shift(N, 1, −1))
    P ← Sum(P, Shift(N, −1, 1))
    P ← Sum(P, Shift(N, 1, 1))
    Q ← P                              initialize with the shifted neighborhood
    Q ← Sum(Q, Shift(P, −1, 0))                        shift another four times
    Q ← Sum(Q, Shift(P, 0, −1))
    Q ← Sum(Q, Shift(P, 1, 0))
    Q ← Sum(Q, Shift(P, 0, 1))
    GetWeightedNeighborhood ← Q                        return the weighted sum
}
```

In terms of these functions, the program for the plus planes in the immune network CA becomes

```
Shape₊ ← Random(10)
Repeat 400 Times {
    Field₊ ← GetWeightedNeighborhood(Shape₋)
    Field₊ ← LargerEQ(Field₊, θ₁) AND NOT LargerEQ(Field₊, θ₂+1)
    Marrow₊ ← Random(P_R)
    Shape₊ ← (Shape₊ OR Marrow₊) AND Field₊        determine next state
    Display(Shape₊)
}
```

REFERENCES

1. Avrameas, S. "Natural Antibodies: From 'Horror autotoxicus' to 'Gnothi seauton.'" *Immunol. Today* **12** (1991): 154–159.
2. Bak, P. "Self-Organized Criticality and Gaia." This volume.
3. Boerlijst, M. C., and P. Hogeweg. "Spiral Wave Structure in Prebiotic Evolution: Hypercycles Stable Against Parasites." *Physica D* **48** (1991): 17–28.
4. Coutinho, A. "Beyond Clonal Selection and Network." *Immunol. Rev.* **110** (1989): 63–87.
5. De Boer, R. J., and P. Hogeweg. "Memory but No Suppression in Low-Dimensional Symmetric Idiotypic Networks." *Bull. Math. Biol.* **51** (1989): 223–246.
6. De Boer, R. J. "Recent Developments in Idiotypic Network Theory." *Neth. J. Med.* **39** (1991): 254–262.
7. De Boer, R. J., and A. S. Perelson. "Size and Connectivity as Emergent Properties of a Developing Immune Network." *J. Theor. Biol.* **149** (1991): 381–424.
8. De Boer, R. J., P. Hogeweg, and A. S. Perelson. "Growth and Recruitment in the Immune Network." In *Theoretical and Experimental Insights into Immunology*, edited by A. S. Perelson, G. Weisbuch, and A. Coutinho. New York: Springer, 1992 (in press).
9. De Boer, R. J., L. A. Segel, and A. S. Perelson. "Pattern Formation in One- and Two-Dimensional Shape Space Models of the Immune System." *J. Theor. Biol.* **155** (1992): 295–333.
10. De Boer, R. J., A. U. Neumann, A. S. Perelson, L. A. Segel, and G. W. Weisbuch. "Recent Approaches to Immune Networks." In *Proceedings First European Biomathematics Conference*, edited by V. Capassso and P. Demongeot. Berlin: Springer, 1992 (in press).
11. Freitas, A. A., B. Rocha, and A. A. Coutinho. "Lymphocyte Population Kinetics in the Mouse." *J. Immunol.* **91** (1986): 5–37.
12. Gardner, M. "Mathematical Games: The Fantastic Combinations of John Conway's New Solitaire Game 'Life.'" *Sci. Am.* **223(4)** (1970): 120–123.
13. Gardner, M. "Mathematical Games: On Cellular Automata, Self-Reproduction, the Garden of Eden and the Game 'Life.'" *Sci. Am.* **224(2)** (1971): 112–117.
14. Gerhardt, M., H. Schuster, and J. J. Tyson. "A Cellular Automaton Model of Excitable Media Including Curvature and Dispersion." *Science* **247** (1990): 1563–1566.
15. Goldberg, A., and D. Robson. *Smalltalk-80, the Language and Its Implementation*, 412–413. Redwood City, CA: Addison-Wesley, 1983.
16. Herzenberg, L. A., A. M. Stall, P. A. Lalor, C. Sidman, W. A. Moore, D. R. Sparks, and L. A. Herzenberg. "The Ly-1 B Cell Lineage." *Immunol. Rev.* **93** (1986): 81–102.

17. Hogeweg, P. "Cellular Automata as a Paradigm for Ecological Modeling." *Appl. Math. Comp.* **27** (1988): 81–100.
18. Hogeweg, P. "Local T–T and T–B Interactions: A Cellular Automaton Approach." *Immunol. Lett.* **22** (1989): 113–122.
19. Holmberg, D., Å. Andersson, L. Carlson, and S. Forsgen. "Establishment and Functional Implications of B-Cell Connectivity." *Immunol. Rev.* **110** (1989): 89–103.
20. Hooykaas, H., R. Benner, J. R. Pleasants, and B. S. Wostmann. "Isotypes and Specificities of Immunoglobulins Produced by Germ-Free Mice Fed Chemically Defined 'Antigen-Free' Diet." *Eur. J. Immunol.* **14** (1984): 1127–1130.
21. Jerne, N. K. "Towards a Network Theory of the Immune System." *Ann. Immunol. (Inst. Pasteur)* **125C** (1974): 373–389.
22. Kearney, J. F., and M. Vakil. "Idiotype-Directed Interactions During Ontogeny Play a Major Role in the Establishment of the Adult B Cell Repertoire." *Immunol. Rev.* **94** (1986): 39–50.
23. Kirkpatrick, S., C. D. Gelatt, and M. P. Vecchi. "Optimization by Simulated Annealing." *Science* **220** (1983): 671–680.
24. Langton, C. G. "Life at the Edge of Chaos." In *Artificial Life II*, edited by C. G. Langton, C. Taylor, J. D. Farmer, and S. Rasmussen. SFI Studies in the Sciences of Complexity, Vol. X, Part Two, 41–91. Redwood City, CA: Addison-Wesley, 1991.
25. Neumann, A. U., and G. Weisbuch. "Window Automata Analysis of Population Dynamics in the Immune System." *Bull. Math. Biol.* **54** (1991): 21–44.
26. Perelson, A. S., and G. F. Oster. "Theoretical Studies on Clonal Selection: Minimal Antibody Repertoire Size and Reliability of Self–Non-Self Discrimination." *J. Theor. Biol.* **81** (1979): 645–670.
27. Perelson, A. S. "Some Mathematical Models of Receptor Clustering by Multivalent Ligands." In *Cell Surface Dynamics: Concepts and Models*, edited by A. S. Perelson, C. DeLisi, and F. M. Wiegel, 223–276. New York: Marcel Dekker, 1984.
28. Segel, L. A., and A. S. Perelson. "Computations in Shape Space: A New Approach to Immune Network Theory." In *Theoretical Immunology*, Part Two, edited by A. S. Perelson. SFI Studies in the Science of Complexity, Vol. III, 321–343. Redwood City, CA: Addison-Wesley, 1988.
29. Segel, L. A., and A. S. Perelson. "Shape-Space Analysis of Immune Networks." In *Cell to Cell Signalling: From Experiments to Theoretical Models*, edited by A. Goldbeter, 273–283. New York: Academic Press, 1989.
30. Stewart, J., and F. J. Varela. "Morphogenesis in Shape-Space. Elementary Meta-Dynamics in a Model of the Immune Network." *J. Theor. Biol.* **153** (1991): 477–498.
31. Toffoli, T., and N. Margolus. *Cellular Automata Machines. A New Environment for Modeling.* Cambridge, MA: MIT Press, 1987.

32. Ulam, S. "Random Processes and Transformations." *Proc. Intl. Congr. Math.* **2** (1952): 264–275.

33. Varela, F. J., A. Coutinho, B. Dupire, and N. N. Vaz. "Cognitive Networks: Immune, Neural, and Otherwise." In *Theoretical Immunology*, Part Two, edited by A. S. Perelson. SFI Studies in the Science of Complexity, Vol. III, 359–375. Redwood City, CA: Addison-Wesley, 1988.

34. Varela, F. J., and A. Coutinho. "Second Generation Immune Networks." *Immunol. Today* **12** (1991): 159–166.

35. Varela F. J., A. Coutinho, and J. Stewart. "What is the Immune Network for?" This volume.

36. Vichniac, G. Y. "Cellular Automata Models of Disorder and Organization." In *Disordered Systems and Biological Organization*, edited by E. Bienenstock, F. Fogelman Soulié, and G. Weisbuch, 1–20. Berlin/Heidelberg: Springer-Verlag, 1986.

37. Weisbuch, G. "A Shape Space Approach to the Dynamics of the Immune System." *J. Theor. Biol.* **143** (1990): 507–522.

38. Wolfram, S. "Universality and Complexity in Cellular Automata." *Physica D* **10** (1984): 1–35.

III. Species and Societies: Communication Among Multiple Agents

Per Bak
Department of Physics, Brookhaven National Laboratory, Upton, NY 11973

Self-Organized Criticality and Gaia

INTRODUCTION

The philosophy behind Lovelock's Gaia hypothesis[16] is that life on earth forms an integrated entity. The individual species constitute each other's environment. One cannot think of the species competing in a preexisting background. The climate, the ozone layer, etc. are all generated and regulated by biology; the individual species in isolation did not just passively adapt to all of this, but must have coevolved with the environment. When evolution started, the environment must have been very different, maybe like the surface of Mars, and Earth as we see it now merely reflects a snapshot in an ever-continuing evolutionary process.

Raup[19] has made the interesting observation that biology is intermittent. Most of the time there is very little activity in the evolutionary process. However, this quiescence is interrupted by episodic events at all scales. There are a few large catastrophic events, such as the extinction of the dinosaurs, but there is also a spectrum of smaller events. Gould[10] has used the term "punctuated equilibrium" to describe this intermittency. This is a simple relation between the number of small and large events which indicates a common origin. If the large events were caused by exogenous cataclysmic forces, such as meteorites, one would expect the number of large events to be uncorrelated with the number of small events, which presumably would be caused by different mechanisms.

A few years ago, Kurt Wiesenfeld, Chao Tang, and I[1,4] demonstrated that large interactive dynamical systems typically self-organize into a globally correlated "critical" state, far out of equilibrium. The critical state is highly sensitive, in the sense that a small local event can lead to large avalanches of activity spreading throughout the system. The activity propagates like a critical chain reaction and there may be avalanches of all sizes. For a simple analogy, think of a nuclear chain reaction where the amount of fissionable material is precisely at the point where an explosion becomes possible. Self-organized critical systems tune themselves to this critical point, in contrast to the nuclear chain reaction where one has to carefully prepare the system in order to make it critical. The theory provided a natural explanation of a number of physical and geophysical intermittent phenomena, including earthquakes,[2,18] volcanic eruptions,[9] solar flares,[17] noise in electronic circuits, and even economics.[5]

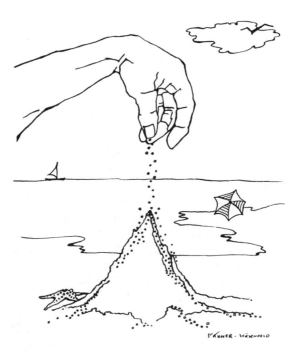

FIGURE 1 Evolving sandpile.

Indeed, biological evolution can be seen as a dynamical process in which many species interact. I shall argue that biology can indeed be thought of as one huge interactive, evolving ecosystem which exhibits self-organized criticality (SOC). The extinction events are coevolutionary avalanches, which are endogenous and unavoidable features of the critical state. The extinctions cannot be thought of as noise in an otherwise smooth biological process; evolution is an intrinsically intermittent process. The individual complexity of the various species is a consequence of the global connectivity. The various species in the SOC state would be globally connected, and evolving together, precisely as envisioned in the Gaia theory.

The case for self-organized criticality in biology has not quite been made yet; we are at the starting point of our study. Since a complete, realistic model of a complex system like biology is completely out of the question, we have to study grossly oversimplified models in the hope of catching the essential features. The models range from caricatures of biology such as Conway's "Game of Life,"[8,11,12,13] to models involving interacting spin glasses and interacting Kauffman models of rough fitness landscapes. All these models can be prepared in a critical state, but the extend to which the criticality must be tuned is not clear. In any case, Raup's observations can be taken as empirical evidence. But, first, a brief prelude on the ideas of self-organized criticality.

ON SELF-ORGANIZED CRITICALITY

The phenomenon of self-organized criticality appears to be quite universal and provides a unifying concept for large-scale behavior in systems with many degrees of freedom. It complements the concept of chaos wherein simple systems with a small number of degrees of freedom can display quite complex behavior.

The prototypic example of self-organized criticality is a sandpile (Figure 1). Adding sand to, or tilting, an existing heap will result in the slopes increasing precisely to a critical value where an additional grain will give rise to unpredictable behavior. If the slope is too steep, there would be a single large avalanche collapsing the pile to a flatter and more stable configuration. On the other hand, if it were less steep the new sand will just accumulate locally to make the pile steeper. The critical state is an attractor of the dynamics. One might start either from a completely flat pile, where there would be no avalanches in the beginning, or from a very steep pile which would instantly collapse like a sandcastle, and eventually reach the critical state. At the critical slope, the distribution of avalanches has no overall scale (there is no typical size of an avalanche). The large fluctuations provide a feedback fast mechanism returning the slope to the critical one, whenever the system is modified, for instance by using different types of sand.

The most spectacular accomplishment of the SOC theory is probably its explanation of the famous Gutenberg-Richter law for the distribution of earthquakes,[2,18] with the earthquakes playing the role of sand slides. The Gutenberg-Richter law says roughly that every time there is one earthquake of magnitude 7, there are 10 earthquakes of magnitude 6, 100 earthquakes of magnitude 5, and so on. Raup's observation of an intermittent pattern of extinction events might be thought of as a Gutenberg-Richter law for the dynamics of evolution: every time there is one event where 10% of the species of earth become extinct (or are created), there are 10 events where 1% is affected and 100 events where 0.1% are affected, and so on.

Self-organized criticality seems to go against the grain of classical mechanics. According to that picture of the world, things evolve towards the most stable state, i.e., the state with lowest energy. So all that classical mechanics says about the sandpile is that it would be more stable if spread out in a flat layer. That is true, but utterly unhelpful for understanding real sandpiles. The most important property of self-organized criticality is its robustness with respect to noise and any modifications of the system: this is necessary for the concept to have any chance of being relevant to realistic complicated systems.

An observer who studies a specific area of the pile can easily identify the mechanisms that cause sand to fall, and he or she can even predict where there will be activity in the near future. To a local observer, large avalanches would remain unpredictable, however, because they are consequences of the total history of the entire pile. No matter what the local dynamics are, the avalanches would mercilessly persist at a relative frequency that cannot be altered. The criticality is a global property of the sandpile. For an interesting discussion of the sandpile metaphor of the global ecology, I recommend Al Gore's book *"Earth in the Balance."*[14]

TOY MODELS OF BIOLOGY
SANDPILES AS A METAPHOR FOR LIFE

I don't want to insult the intelligence of the reader too much by suggesting that the sandpile is a realistic model of evolution; it is merely supposed to describe a dynamical principle. Note, however, the following analogy: in the beginning, when the pile is flat, the system is that of many non-interacting grains of sand. The individual grains do not "see" each other in the sense that the motion of a grain at one point does not affect grains far away. Once the system reaches the critical state, we don't talk about *many* individual grains, but about one coherent sandpile, where motion of grains at any part might soon affect grains at any other part. There are global avalanches despite the fact that the individual toppling rules are local. Similarly, in biology, evolution has taken the earth from a system of *many* non-interacting molecules to *one* biology of globally interacting species. Self-organized criticality provides a simple general mechanism for persistent nonequilibrium behavior. Lovelock[16] has pointed out that in our environment, for instance, the oxygen concentration exists in a persistent state, far out of equilibrium.

FIGURE 2   Stationary configuration of the "Game of Life." Live individuals are indicated by black squares. The pluses ("+") are blinkers oscillating between vertical and horizontal bars of three live cells.

In the critical state the individual grains only barely support each other. The structure of the critical pile is quite complex (it is a fractal) compared to a flat beach. The local complex structure is not robust to changes, but the critical pile is. The individual species are not robust, but biology is. The species try continuously to improve their fitness by modifying their genetic code, but avalanches originating all over the system tend to prevent this from happening.

CONWAY'S "GAME OF LIFE"

If self-organized criticality is indeed so ubiquitous, then it might appear in the simplest cellular automata models, like the ones discussed by De Boer et al. in this book. There is nothing fundamentally deep in using cellular automata: they are simply used for computational convenience since they are much simpler than, for instance, coupled differential equations. Thus motivated, we[3] studied the famous "Game of Life" automaton, invented by John Conway.[8,11,12,13]

Life may be thought of as a model of a simple society of living organisms. The organisms live on a two-dimensional square lattice, of size $L \times L$. They can be either "live" or "dead." Their fate depends on the states of the eight neighbors, at the up, down, left, right, and four corner positions. The states of all organisms are updated in parallel. A live organism remains live as long as the number of live neighbors is not too small or too large. If the number of live neighbors is larger than three, it dies of overcrowding at the next time step; if the number is less than two, it dies of loneliness. A new individual is born at a dead site which has precisely three live neighbors.

If the system is initiated at a random configuration of live and dead individuals, it will come to rest after a while in a configuration of static clusters, and simple periodic states, "blinkers" (Figure 2). Most of the analysis of cellular automata is strictly restricted to the behavior of this transient behavior, not of particular interest to us. The system is now perturbed by adding or subtracting randomly a single live individual. This leads to an avalanche of extinction and creation of individuals by simply updating according to the rules above; eventually the system will come to rest in a new static configuration. This procedure is repeated again and again. Eventually the system appears to organize itself into a statistically stationary state with avalanches of all sizes, i.e., the society has become globally connected.

The size of an avalanche is measured as the number of births and deaths following a single perturbation. To see that the distribution follows a Gutenberg-Richter law, we have counted how many avalanches there are of each size. Figure 3 shows

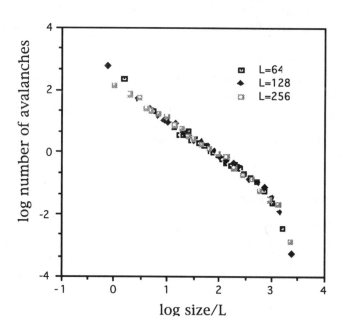

FIGURE 3 Histogram of sizes of avalanches in the "Game of Life" for various linear dimensions L of the lattice. The linear behavior over three decades indicates criticality. Finite-size scaling of the variables condenses all curves to a single one.

histograms for these distributions, plotted on a logarithmic scale, for several sizes L of lattices with closed boundary conditions. The linear behavior over three decades indicates a power-law distribution. The cutoff at large avalanches depends on the size of the system. This is seen by plotting the histograms for various L in terms of a rescaled coordinate size/L, rather than the size itself. The rescaling corresponds to L-dependent rigid shifts of the curves, and makes the curves for various L fall on a single curve: we say that the system obeys finite-size scaling. Only critical systems obey finite-size scaling. In contrast, if the system were subcritical, the avalanches would have a cutoff which is independent of the size of the system, i.e., there would not be any large events no matter how large the system is, and one would *not* have to shift the curves in order to have them coalesce. The biggest catastrophic avalanches for the 256×256 system involved as many as 10 million events following a single perturbation. Another indication of criticality is the fact that the scaling depends on the type of boundary conditions, such as whether the boundaries are open or periodic. For different boundary conditions, one would have to rescale, the axis with different powers of L.

Our paper[3] has become somewhat controversial, with other work[7] indicating a possible size-independent cutoff for lattices larger than $L = 100$, but the finite scaling result shown above seems to refute this. The computational demands for establishing equilibrium for larger lattices are prohibitively large, so we cannot guarantee that there is not an intrinsic cutoff for $L > 500$. If there is an intrinsic cut-off, the criticality would not be self-organized, but accidental.

Thus, the question is whether "life" organized itself to the critical state, or Conway accidentally did it? In any case, it is quite interesting that John Conway, in his search for models exhibiting local complex structures arrived at a model which at the same time is at (or near) criticality. This suggests an intimate relation between *criticality* and *local complexity* in self-organizing systems. Maybe complexity and criticality are synonymous concepts. Maybe one can only identify objects as complex after defining the self-consistent functional integration of these objects into an extended outer world. A car cannot be characterized as a complex object before one has defined its environment with roads, gas stations, and the people that fit into it.

KAUFFMAN'S INTERACTING DANCING FITNESS LANDSCAPES

We consider an ecology of interacting species. The properties of each species is defined by its genetic code $T = (011000111101\ldots)$. The fitness is some function of this code. Note that we are not talking about the fitness of individuals, but the collective fitness of an entire species represented by a single code. This is a tremendous oversimplification of biology. A species can only be present or absent, there is no room for variations in the number of individuals belonging to each species. In physics, we have successfully utilized similar discrete up-down Ising models to describe phenomena where the variables, in principle, should be continuous. The fitness function can be thought of as a more or less rough landscape. The species try to improve their fitness by walking around step by step in this landscape by flipping genes, one at a time. After a few steps the fitness may reach a local maximum, from where no single step can improve the fitness. There may be some other better maximum further away, but the species have no means of getting there. The state is frozen.

However, following Kauffman and Johnsen,[15] the fitness of each species depends not only on its own genetic code, but also on the properties, i.e., the genetic codes of other species: we have an interactive ecology. In other words, the fitness landscape of each species depends on the other species, and may be deformed by their actions. As each species tries to increase its fitness by changing its genetic code T, the other species do the same, and at the end of the day you see a new landscape in which to continue your adaptation. According to Kauffman, the key to understanding biology is the understanding of such "interacting dancing landscapes." One can think of the "Game of Life" as a special very simple interacting landscape where there is only one gene with two alleles characterizing each species.

Formally, the situation is depicted in Figure 4. A number of species, represented by the square boxes, live together in the world ecology. Each species has a number

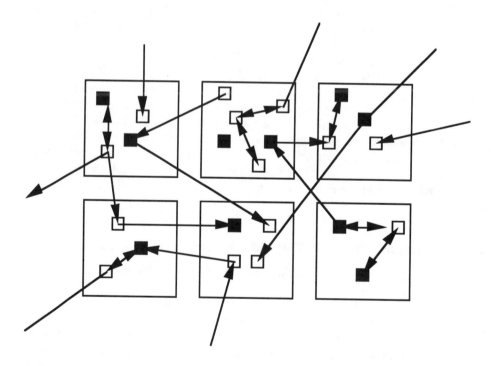

FIGURE 4 Interacting species. The arrows represent symmetric couplings amongst the genes of each species, and asymmetric coupling between genes of different species.

of genes, represented by small boxes, which can be in either of two states, black or white, or 1 and 0. Each gene looks at a number of other genes within itself, and outside itself, in order to decide its fitness, and changes its value if it can benefit by doing so, disregarding the possible harm that this might inflict on other species looking at this particular gene.

Interacting genes are depicted as arrows pointing from one gene to another. The difference between "self" and "nonself" is that the appropriate fitness is the total fitness of all genes within each box. Thus, the gain in fitness is the gain of the flipping gene, minus the loss of the interacting genes within the same box. The interactions between genes are thus symmetric within the box, as represented by arrows pointing in both directions. In the language of physics, the internal interactions may be represented by a Hamiltonian, or energy, which is minimized in stable states. Maximization of fitness corresponds to minimization of energy. In contrast, a flip is made irrespectively of the possible loss or gain of other species, so the arrows between species are unidirectional. This asymmetric distinction between "self" and "nonself" is necessary for the evolution to take place, and prevent freezing into a local energy minimum.

ASYMMETRIC SPIN GLASS MODELS

How does one implement this picture in a specific model? Our investigation of this vastly complex problem is yet at its infancy. The models that we have studied are highly unsatisfactory in several different ways, mostly because of our inability to capture the essential features in biology, while keeping the models at a numerically tractable level. Analytical work seems out of the question at present, and all conclusions are extremely tentative.

The simplest way of implementing a random fitness landscape is a spin glass model: The individual bits can be thought of as magnetic spins, interacting through random positive and negative interactions, which can assume values of plus and minus one. The contribution to the fitness from each spin is a sum of its interaction with all the interacting spins. Such models have many equilibrium states, that is, valleys where the energy cannot be further reduced by single-flip operations. The various co-existing species correspond to various local energy minima (fitness maxima). This model simulates the behavior of a single box in Figure 1. The roughness of the landscape in spin space can be varied by changing the number of spins interacting with a given spin.

In our model, each spin in the ith species, $T^i = (S_1^i, S_2^i, \ldots, S_N^i)$, $S = \pm 1$, sees C external spins in species T^j with $j \neq i$ in addition to K internal spins. The fitness of the individual T^j becomes

$$F_i = \sum_{\langle i_1, i_2 \rangle} J(i, i_1, i_2) S_{i_1}^i S_{i_2}^i + \sum_j \sum_{i_1} \sum_{i_2} J(i, j, i_1, i_2) S_{i_1}^i S_{i_2}^j .$$

The first term is the internal fitness. The interactions are nonzero for pairs $\langle i_1, i_2 \rangle$ chosen such that each spin interacts with K random other spins within the same species. By varying K, one can vary the roughness of the landscape. The second term is the contribution to the fitness from other species Tj. Each spin $S_{i_1}^i$ interacts with C randomly chosen spins in other species. The interactions are zero for noninteracting spins, and random functions in the interval $[-1, 1]$ for the interactive spins. The asymmetry manifests itself by the fact that $J(i, j, i_1, i_2)$ is different from $J(j, i, i_2, i_1)$: if the spin $S_{i_1}^i$ sees the spin $S_{i_2}^j$, the inverse is not necessarily true, and even if it is, the interactions are different since they are independently chosen random functions. The interactions J represent the chemistry or physics of the world; these functions are predetermined. The space of possible configurations is enormous: for M species each with N genes, the number of possible states is 2^{NM}. Thus, for $M = 64$ and $N = 16$, a situation which we have studied, there are 10^{300} states.

The model is updated by successively checking if flipping a spin will increase the fitness of the species to which it belongs. If it does, the spin is flipped whether or not it lowers the fitness of other species looking at the particular species. Starting from random initial configurations, one of two different patterns may arise. For a large number C of external interactions compared with the number K of internal interactions, that is, $C \gg K$, the dynamics never comes to rest. The landscape changes faster than the species can equilibrate. This is usually, somewhat misleadingly, called the chaotic state. If $C \ll K$, the situation is similar to that of the noninteracting spin glass: the system comes to rest in a state where there is at most some local activity where a small number of spins may evolve in a periodic pattern (like the "blinkers" in the "Game of Life.") This is called the frozen state. It has been suggested that evolution might take place at the phase transition between the two states, at the "edge of chaos." But the all important question is how the system gets there without any tuning of parameters. The system must self-organize to the critical state. In order to get criticality, it is not enough to specify the local rule: the driving mechanism and the boundaries are essential in defining the outcome of the evolutionary process.

Armed with the insight achieved from the study of sandpiles, the "Game of Life," etc. I suggest the following scenario:

We focus on values of the parameters where the initial transient leads to a *frozen* state. After the system comes to rest, it is perturbed by a single random mutation, i.e., a spin flip which does not increase the fitness. This induces a coevolutionary avalanche, rather small in the beginning. When the system comes to a stop again, another mutation is made. The hope is that after a while, the system will be pumped up to a "poised" state, where yet another mutation may induce an avalanche of any size. Figure 5 shows the distribution of avalanche size measured as the total number of gene-flips induced by a single random mutation, once the system has built up to the statistically stationary state. The plot is for a total of 1 million avalanches in a system of 64 species, each with 12 genes. Each gene interacts with

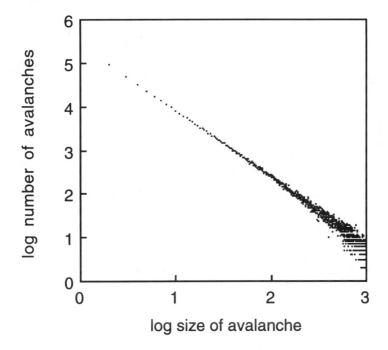

FIGURE 5   Distribution of one million coevolutionary avalanches in the asymmetric coupled spin glass model. Number of species $= 64$, $N = 12$, $K = 8$, $C = 4$. The linear behavior over more than three decades indicates criticality, but is it self-organized?

$K = 10$ internal genes and $C = 4$ external genes, so each species interacts roughly with $C \times N = 48$ other species. Before sampling the distribution, 100,000 avalanches were discarded to assure that the system had evolved into the stationary state. The linear behavior over three orders of magnitude in the log-log plot indicates that the distribution is a power law, $N(S) = S^{-\tau}$, with $\tau = 1.50 \pm 0.01$. The continuous drive by rare random mutations appears to have driven the subcritical biology to the critical point!

The scatter of points for large avalanches is statistical fluctuations due to the small number of large avalanches. In the model, as in Nature, the large catastrophic events are few and far between. The tendency of the points to fall below the line for large avalanches is probably due to a finite-size cutoff: the size of avalanches in SOC systems is generally cutoff by the finite number of constituents. The cutoff increases, hopefully to infinity, as the system size increases. It might be, however, that one may have to increase the number of interactions as the system size increases in order to maintain criticality.

KAUFFMAN'S NKC MODELS

Kauffman and Johnsen[15] have introduced another class of interactive models: the NKC models. The contribution to the fitness function from each gene is a *random* function of the state of that particular gene, the state of K other internal genes, and C external genes, in contrast to the bilinear interaction in the spin glass model. The general behavior appears to be the same for the two types of models.[6]

We have also studied models where the number of genes, N, within each species itself is a dynamical variable which can be adjusted to increase fitness. Thus, the model could be set up with only a single gene or molecule at each site, these being able to interact only locally. Eventually more and more genes are added to allow the species to benefit from more and more other species, until they reach the globally connected state.

In general, it is possible to have avalanches with very large cutoffs in rather broad ranges of parameters. We can typically vary parameters 50% and still have avalanches with cutoff of the order of 1,000. Studies on larger systems leave open the possibility that there is an intrinsic cutoff not related to the system size: this would indicate that the criticality must be tuned and is not self-organized. There is definitely much more work to do before the ideas of self-organized criticality may condense into a realistic theory of biology.

A CONCLUDING REMARK ON FITNESS AND THE SPEED OF EVOLUTION

From the perspective of a single species, a miraculous process has taken place. It finds itself situated in an incredibly ingenious fitness maximum, in a hostile and complicated world, which barely allows it to survive. Suppose we, for a while, freeze the outer world (i.e., the configuration of all other species), and try by engineering rather than by evolution to construct from scratch an equally fit species. This cannot be done by starting at a random configuration and simply relaxing it, since this will certainly carry it to a wrong and much less fit maximum. We would essentially have to systematically go through all configurations, i.e., the time for the evolutionary process would be exponentially large in the number of genes. This is, in fact, a major problem with the traditional Darwinian evolution where the most fit of random mutants are selected: it is much too slow to explain real evolution.

In contrast, the coevolution described here, where the individual species adapt slowly to a changing environment, without ever climbing high fitness barriers, constitutes a fast evolutionary process. What the individual sees as his superior fitness may better be characterized as a self-consistent integration into a complex system. Biology constructed the solution to the fitness problem together with the problem itself. It is much simpler to construct a complicated crossword puzzle by a coevolutionary process, than to solve it by trial and error.

ACKNOWLEDGMENTS

I am deeply indebted to Stuart Kauffman for numerous discussions on self-organized criticality, evolution, and many other related and not-so-related phenomena. This work was supported by the U.S. Department of Energy under contract DE-AC02-76-CH00016.

REFERENCES

1. Bak, P., C. Tang, and K. Wiesenfeld. "Self-Organized Criticality." *Phys. Rev. A* **38** (1987): 364–373.
2. Bak, P., and C. Tang. "Earthquakes as a Self-Organized Critical Phenomenon." *J. Geophys. Res. B* **94** (1989): 15635–15637.
3. Bak, P., K. Chen, and M. Creutz. "Self-Organized Criticality in the Game of Life." *Nature* **342** (1989): 780.
4. Bak, P., and K. Chen. "Self-Organized Criticality." *Sci. Am.* **264(1)** (1991): 46–53.
5. Bak, P., K. Chen, J. Scheinkman, and M. Woodford. "Self-Organized Criticality and Fluctuations in Economics." Working Paper 92-04-018, Santa Fe Institute, Santa Fe, NM, 1992.
6. Bak, P., and S. A. Kauffman. "Calculations." Unpublished.
7. Bennett, C., and M. Bourzutschky. "Life Not Critical." *Nature* **350** (1991): 468.
8. Berlekamp, E., J. Conway, and R. Guy. *Winning Ways for Your Mathematical Plays*, Vol 2. New York: Academic Press, 1982.
9. Diodati, P., F. Marchesoni, and S. Piazza. "Acoustic Emission from Volcanic Rocks: An Example of Self-Organized Criticality." *Phys. Rev. Lett.* **67** (1991): 2239–2242.
10. Eldridge, M., and S. J. Gould. *Models in Paleobiology*, edited by T. J. M. Schopf, 82–115. San Fransisco: Freeman, 1972.
11. Gardner, M. "Mathematical Games." *Sci. Am.* **223(4)** (1970): 120–124.
12. Gardner, M. "Mathematical Games." *Sci. Am.* **(5)** (1970): 116.
13. Gardner, M. "Mathematical Games." *Sci. Am.* **224(2)** (1971): 112.
14. Gore, A. *Earth in the Balance and the Human Mind.* Boston: Houghton-Mifflin, 1992.
15. Kauffman, S. A., and S. Johnsen. "Coevolution to the Edge of Chaos—Coupled Fitness Landscapes, Poised States, and Coevolutionary Avalanches." *J. Theor. Biol.* **149** (1991): 467–506.
16. Lovelock, J. E. *Gaia.* Oxford: Oxford University Press, 1979.

17. Lu, E. T., and R. J. Hamilton. "Avalanches and the Distribution of Solar Flares." *Astrophys. J.* **380** (1991): L89–L92.
18. Olami, Z., H. J. S. Feder, and K. Christensen. "Self-Organized Criticality in a Continuous Nonconserved Cellular Automaton Modelling Earthquakes." *Phys. Rev. Lett.* **68** (1992): 1244–1247.
19. Raup, D. M. "Biological Extinction in Earth History." *Science* **231** (1986): 528–1533.

Stuart A. Kauffman
Department of Biochemistry and Biophysics, School of Medicine, University of Pennsylvania, Philadelphia, PA 19104-6059 and Santa Fe Institute, 1660 Old Pecos Trail, Suite A, Santa Fe, NM 87501

Requirements for Evolvability in Complex Systems: Orderly Dynamics and Frozen Components

Abridged and amended from *Complexity, Entropy, and the Physics of Information*, edited by W. H. Zurek, 151–192. Santa Fe Institute Studies in the Sciences of Complexity, Proceedings Volume VIII. Redwood City, CA: Addison-Wesley, 1990.

This article discusses the requirements for evolvability in complex systems, using random Boolean networks as a canonical example. The conditions for crystalization of orderly behavior in such networks are specified. Most critical is the emergence of a "frozen component" of the binary variables, in which some variables are frozen in the active or inactive state. Such frozen components across a Boolean network leave behind *functionally isolated islands* which are not frozen.

Adaptive evolution or learning in such networks via near mutant variants depends upon the structure of the corresponding "fitness landscape." Such landscapes may be smooth and single peaked, or highly rugged. Networks with frozen components tend to adapt on smoother landscapes than those with no frozen component. In coevolving systems, fitness landscapes themselves deform due to coupling between coevolving partners. Conditions for optimal coevolution may include tuning of landscape structure for the emergence of frozen components among the coadapting entities in the system.

INTRODUCTION

The dynamical behavior of complex information-processing systems, and how this behavior may be improved by natural selection, or by other learning or optimizing processes, are issues of fundamental importance in biology, psychology, economics, and, not implausibly, in international relations and cultural history. Biological evolution is perhaps the foremost example. No serious scientist doubts that life arose from nonlife as some process of increasingly complex organization of matter and energy. Four billion years later we applaud organisms that evolved from simple precursors, that unfold in their own intricate ontogenies, that sense their worlds, categorize the states of those worlds with respect to appropriate responses, and in their interactions form complex ecologies whose members coadapt more or less successfully over ecological and evolutionary time scales. We suppose, probably rightly, that Mr. Darwin's mechanism, natural selection, has been fundamental to this astonishing story. We are aware that, for evolution to "work," there must be entities which in some general sense reproduce, but do so with some chance of variation. That is, there must be heritable variation. Thereafter, Darwin argued, differences will lead to differential success, culling out the fitter, leaving behind the less fit.

But, for at least two reasons, Darwin's insight is only part of the story. First, in emphasizing the role of natural selection as the Blind Watchmaker, Darwin and his intellectual heritors have almost come to imply that without selection there would be no order whatsoever. It is this view which sees evolution as profoundly historically contingent; a story of the accidental occurrence of useful variations accumulated by selection's sifting: evolution as the Tinkerer. But, second, in telling us that natural selection would cull the fitter variants, Darwin has implicitly assumed that successive cullings by natural selection would be able to *successively accumulate* useful variations. This assumption amounts to presuming what I shall call *evolvability*. Its assumption is essential to a view of evolution as a tinkerer which cobbles together ad hoc, but remarkable, solutions to design problems. Yet "evolvability" is not itself a self-evident property in complex systems. Therefore, we must wonder what may be the construction requirements which permit evolvability, and whether selection itself can achieve such a system.

Consider the familiar example of a standard computer program on a sequential von Neumann universal Turing machine. If one were to randomly exchange the order of the instructions in a program, the typical consequence would be a catastrophic change in the computation performed.

Try to formulate the problem of evolving a *minimal program* to carry out some specified computation on a universal Turing machine. A minimal program is one in which the program is encoded in the shortest possible set of instructions and, per-haps, initial conditions in order to carry out the desired computation. The length of such a minimal program defines the *algorithmic complexity* of the computation. Ascertainment that a given putative minimal program is actually minimal cannot, however, in general be carried out. Ignore for the moment the problem of ascer-tainment, and consider the following: Is the minimal program itself likely to be evolvable? That is, does one imagine that a sequence of minimal alterations in highly compact computer codes could lead from a code which did not carry out the desired computation to one which did?

I do not know the answer; nevertheless, it is instructive to characterize the obstacles. Doing so helps define what one might mean by "evolvability." In order to evolve across the space of programs and achieve a given compact code to carry out a specified computation, we must first be able to ascertain that any given program actually carries out the desired computation. Think of the computation as the "phenotype," and the program as the "genotype." For many programs, it is well known that there is no short cut to "seeing the computation" carried out other than running the program and observing what it "does." That is, in general, given a program, we do not know what computation it will perform by any shorter process than observing its "phenotype." Thus, to evolve our desired program, we must have a process which allows candidate programs to exhibit their phenotypes, and then have a process which chooses variant programs and "evolves" towards the target minimal compact program across some defined program space. Since programs and, if need be, their input data can be represented as binary strings, we can represent the space of programs in some high-dimensional Boolean hyperspace. Each vertex is then a binary string, and evolution occurs across this space toward the desired minimal target program.

Immediately we find two problems. First, can we define a "figure of merit" which characterizes the computation carried out by an arbitrary program—defines its phenotype—and which can be used to compare how "close" the phenotype of the current program is to that of the desired target program? This requirement is important since, if we wish to evolve from an arbitrary program to one which computes our desired function, we need to know if alterations in the initial program bring the program closer to, or further from, the desired target program. The distri-bution of this figure of merit or, to a biologist, "fitness" across the space of programs defines the "fitness landscape" governing the evolutionary search process. Such a fitness landscape may be smooth and single peaked, with the peak corresponding to the desired minimal target program, or may be very rugged and multipeaked. In the latter case, typical of complex combinatorial optimization problems, any local evolutionary search process is likely to become trapped on local peaks. In general, in such tasks, attainment of the global optimum is an intractable prob-lem, and an evolutionary search will not attain the global optimum in reasonable time. Thus, the second problem with respect to evolvability of programs relates to how rugged and multipeaked the fitness landscape is. The answers are not known, but the intuition is clear. The more compact the code becomes, the more *violently* the computation carried out by the code changes at each minimal alteration of the code. That is, long codes may have a variety of internal sources of redundancy which allows small changes in the code to lead to small changes in the computation. By definition, a minimal program is devoid of such redundancy. Thus, inefficient redundant codes may occupy a landscape which is relatively smooth and highly correlated in the sense that nearby programs have nearly the same fitness in carrying out similar computations. But, as the programs become shorter, small changes in the programs induce ever more pronounced changes in the phenotypes. That is, the landscapes become ever more *rugged* and *uncorrelated*. In the limit where fitness landscapes are entirely uncorrelated, such that the fitness of "one-mutant" neighbors in the space are random with respect to one another, it is obvious that the fitness of a neighbor carries no information to help choose good directions to move across the space in an evolutionary search for global, or at least good, optima. Evolution across fully uncorrelated landscapes amounts to an entirely random search process where the landscape itself provides no information about where to search.[40] In short, since minimal programs almost surely "inhabit" fully uncorrelated landscapes in program space, one comes strongly to suspect that *minimal programs are not themselves evolvable.*

Analysis of the conditions of evolvability, therefore, requires understanding: (1) what kinds of systems inhabit what kinds of "fitness landscapes"; (2) what kinds of fitness landscapes are "optimal" for coadaptive evolution; and (3) whether there may be selective or other adaptive processes in complex systems that might "tune" (1) and (2) to achieve systems which are able to evolve well.

Organisms are the paradigmatic examples of complex systems which patently have evolved, and hence now do fulfill the requirements of evolvability. Despite our fascination with sequential algorithms, organisms are more adequately characterized as complex *parallel-processing* dynamical systems. A single example suffices to make this point. Each cell of a higher metazoan such as a human harbors an identical, or nearly identical, copy of the same genome. The DNA in each cell specifies about 100,000 distinct "structural" genes, that is, those which code for a protein product. Products of some genes *regulate* the activity of other genes in a complex regulatory web which I shall call the genomic regulatory network. Different cell types in an organism (nerve cell, muscle cell, liver hepatocyte, and so forth) differ from one another because different subsets of genes are active in the different cell types. Muscle cells synthesize myoglobin, red blood cells hemoglobin. During ontogeny from the zygote, genes act in parallel, synthesizing their products and mutually regulating one another's synthetic activities. Cell differentiation, the production of diverse cell types from the initial zygote, is an expression of the parallel processing on the order of 10,000 to 100,000 genes in each cell lineage. Thus the metaphor of a "developmental program" encoded by the DNA and controlling ontogeny is more adequately understood as pointing to a parallel-processing genomic dynamical system whose dynamical behavior unfolds in ontogeny. Understanding development from the zygote, and the evolution of development, hence the *evolvability* of ontogeny, requires understanding how such parallel-processing dynamical systems might give rise to an organism, and be molded by mutation and selection.

Other adaptive features of organisms, ranging from neural networks to the anti-idiotype network in the immune system, are quite clearly examples of parallel-processing networks whose dynamical behavior, and changes with learning or with antigen exposure, constitute the "system" and exhibit its evolvability.

The hint that organisms should be pictured as parallel-processing systems leads me to focus the remaining discussion on the behavior of such networks and on the conditions for their evolvability. Central to this is the question of whether even random, disordered, parallel-processing networks can exhibit sufficiently ordered behavior to provide the raw material upon which natural selection might successfully act. This discussion therefore serves as an introduction to the following topics:

1. What kinds of random, disordered, parallel-processing networks exhibit strongly self-organized behavior which might play a role in biology and elsewhere?
2. What kinds of "fitness landscapes" do such systems inhabit?
3. What features of landscapes abet adaptive evolution?
4. Might there be selective forces which "tune" the structures of fitness landscapes by tuning the structure of organisms, and tune the couplings among fitness landscapes, such that coevolutionary systems of coupled adapting organisms coevolve "well?"

In the next section I discuss random Boolean networks as models of disordered dynamical systems. We will find that such networks can exhibit powerfully ordered dynamics. In the third section I discuss why such systems exhibit order. It is due to the percolation of a "frozen" component across the network. In such a component, the binary elements fall to fixed active or inactive states. The frozen component breaks the system into a percolating frozen region, and isolated islands, which continue to change but cannot communicate with one another. In the fourth section I discuss the evolvability of such Boolean networks, and show that networks with frozen components evolve on more correlated landscapes than those without frozen components. In the fifth section I discuss a new class of coupled spin-glass models for *coevolution* where the adaptive moves by one partner *deform* the fitness landscapes of its coevolutionary partners. We find that the analogue of frozen components re-merges in this coevolutionary context. We find that selective forces acting on individual partners can lead them to tune the structure of their own fitness landscapes, and their coupling to other landscapes, so to increase their own sustained fitness. In addition, we find that these same adaptive moves "tune" the entire coevolutionary system towards an optimal structure where all partners coevolve "well." Thus, we have a hint that selection itself may, in principle, achieve systems that have optimized evolvability.

DISCRETE DYNAMICAL SYSTEMS: INTRODUCING BOOLEAN DYNAMICAL NETWORKS

I have now asked what kinds of complex, disordered, dynamical systems might exhibit sufficient order for selection to have at least a plausible starting place, and whether such systems adapt on well-correlated landscapes. This is an extremely large problem which goes to the core of the ways that complex systems must be constructed so that improvements by accumulation of improved variants by mutation and selection, or any analogue of mutation and selection, can take place. We will not soon solve so large a problem. Yet we can make substantial progress. The immediate task is to conceive of a coherent way to approach such a vast question. In this section I shall try to confine this question into one such coherent approach by asking what kinds of "discrete" dynamical systems, whose variables are limited to two alternative states, "on" and "off," adapt on well-correlated landscapes.

Switching networks are of central importance in such an effort. I collect the reasons for this:

1. For many systems, the on/off "Boolean" idealization is either accurate, or the best idealization of the nonlinear behavior of the components in the system.
2. We are concerned with dynamical systems with hundreds or thousands of coupled variables. These might represent active or inactive genes coupled in a genetic regulatory cybernetic network,[28,29,30,31,32,34,35] the linked cellular and molecular components of the immune system and idiotype network,[27,41] the interacting polymers in an autocatalytic polymer system,[13,14,37] or the interacting neurons in a neural network.[24,51,54] The idealization to on/off switching elements allows us to actually study such enormously complex systems. The corresponding problems are often intractable by computer simulations using continuous equations.
3. We can *pose and answer* the following question: What are the construction requirements in very complex switching networks such that they spontaneously exhibit orderly dynamics by having small attractors? (We define attractors in the following section.)
4. The same properties which assure orderly dynamics, hence spontaneous order, simultaneously yield systems which *adapt on well-correlated fitness landscapes.*
5. Having identified these properties in idealized on/off networks, we will find it easier to begin to identify homologous properties in a wider class of continuous nonlinear dynamical systems.

POSITIVE COOPERATIVITY, SIGMOIDAL RESPONSE FUNCTIONS, AND THE ON/OFF IDEALIZATION

A short example demonstrates why a Boolean or on/off idealization captures the major features of many continuous dynamical systems. Many cellular and biochemical processes exhibit a response which changes in an S-shaped, or sigmoidal, curve as a function of altered levels of a molecular input.[52] For example, hemoglobin is a tetrameric protein. That means that four monomers are united into the functional hemoglobin molecule. Each monomer of hemoglobin binds oxygen. But the binding behavior of the four monomers exhibits *positive cooperativity*. Binding of oxygen by a first monomer increases the affinity of the remaining monomers for oxygen. This implies that the amount of oxygen bound by hemoglobin as a function of oxygen concentration increases, at first, faster than linearly as oxygen levels increase from a base level. But, at sufficiently high oxygen concentration, all four monomers have almost always bound an oxygen; thus, further increases in oxygen concentration do not increase the amount bound per hemoglobin molecule. The response saturates. This means that a graph of bound oxygen concentration as a function of oxygen tension is S-shaped, or sigmoidal. It starts by increasing slowly, becomes increasingly steep, then passes through an inflection and bends over, increasing more slowly again to a maximal asymptote.

Positive cooperativity and ultimate saturation in enzyme systems, in cell receptor systems, and the binding of regulatory molecules to DNA regulatory sites[1,45,52] and other places are extremely common in biological systems. Consequently, sigmoidal response functions are common as well.

The vital issue is to realize that even with a "soft" sigmoidal function, whose maximum slope is less than vertical, *coupled systems* governed by such systems, are properly idealized by on/off systems. It is easy to see intuitively why this might be. Consider a sigmoidal function graphed on a plane and, on the same plane, a constant or proportional response where the output response is equal to the input, i.e., the slope is 1.0. The sigmoidal function is initially *below* the proportional response. Thus a given input leads to even less output. Were that reduced output *fed back as the next input*, then the subsequent response would be even less. Over iterations the response would dwindle to 0. Conversely, the sigmoidal response becomes steep in its mid-range and crosses *above the proportional response*. An input above this critical crossing point leads to a greater than proportional output. In turn, were that output fed back as a next input, the response would be still greater than that input. Over iterations, the response would climb to a maximal response. That is, feedback of signals through a sigmoidal function tend to *sharpen to an all-or-none response*.[25,60] This is the basic reason that the "on/off" idealization of a flip-flop in a computer captures the essence of this behavior.

In summary, *logical switching systems* capture major features of a homologous class of nonlinear dynamical systems governed by sigmoidal functions, because such systems tend to *sharpen* their responses to *extremal values of the variables*. The logical or switching networks can then capture the logical skeleton of such continuous systems. However, the logical networks miss detailed features and, in particular, typically *cannot represent the internal unstable steady states* of the continuous system. Thus, Boolean networks are a caricature, but a good one, a very powerful idealization with which to think about a very broad class of continuous nonlinear systems, as well as switching systems in their own right. I stress that it is now well established that switching systems are good idealizations of many nonlinear systems.[25] But just how broad may be the class of nonlinear systems that are "homologous," in a useful sense, to switching networks remains a large mathematical problem.

THE STATE SPACE DYNAMICS OF AUTONOMOUS BOOLEAN NETWORKS

Boolean networks are comprised of binary, "on/off" elements. A network has N such elements. Each element is *regulated* by some of the variables in the network which serve as its "inputs." The *dynamical behavior* of each element, whether it will be active (1) or inactive (0) at the next moment, is governed by a *logical switching rule*, or *Boolean function*. The Boolean function specifies the activity of the regulated element at the next moment for each of the possible combinations of current activities of the input variables. For example, an element with two inputs might be active at the next moment if either one of the other or both inputs were active at the current moment. This is the Boolean "OR" function. Alternatively, the element might be active at the next moment only if both inputs were active at the present moment. This is the Boolean "AND" function.

Let K stand for the number of input variables regulating a given binary element. Since each element can be active or inactive, the number of combinations of states of the K inputs is just 2^K. For each of these combinations, a specific Boolean function must specify whether the regulated element is active or inactive. Since there are two choices for each combination of states of the K inputs, the total number of Boolean functions, F, of K inputs is

$$F = 2^{(2^K)}. \tag{1}$$

The number of possible Boolean functions increases rapidly as the number of inputs, K, increases. For $K = 2$ there are $(2^2)^2 = 16$ possible Boolean functions. For $K = 3$ there are 256 such functions. But by $K = 4$ the number is $2^{16} = 24336$, while for $K = 5$ the number is $2^{32} = 5.9 \times 10^8$. As we shall see, special subclasses of the possible Boolean functions are important for the emergence of orderly collective dynamics in large Boolean networks.

An autonomous Boolean network is specified by choosing which K elements will serve as the regulatory inputs for each binary element and assigning to each binary element one of the possible Boolean functions of K inputs. If the network has no inputs from "outside" the system, it is considered to be "autonomous." Its behavior then depends upon itself alone.

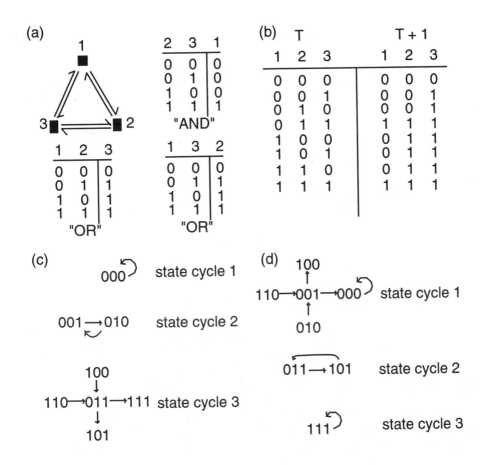

FIGURE 1 (a) The wiring diagram in a Boolean network with three binary elements, 1, 2, 3, each an input to the other two. One element is governed by the Boolean AND function, the other two by the OR function. (b) The Boolean rules of (a) rewritten showing for all $2^3 = 8$ states of the Boolean network at time T, the activity assumed by each element at the next time moment, $T + 1$. Read from left to right this figure shows, for each state, its successor state. (c) The state transition graph, or behavior field, of the autonomous Boolean network in (a) and (b), obtained by showing state transitions to successor states, (b), as connected by arrows, (c). This system has three state cycles. Two are steady states (000) and (111), the third is a cycle with two states. Note that (111) is stable to all single perturbations, e.g., to (110), (101), or (011), while (000) is unstable to all such perturbations. (d) Effects of mutating the rule of element 2 from OR and AND. From *Origins of Order: Self Organization in Evolution* by S. A. Kauffman. Copyright ©1993 by Oxford University Press, Inc. Reprinted by permission.

Figure 1(a) shows a Boolean network with three elements, 1, 2, and 3. Each element receives inputs from the other two. Element 1 is governed by the AND function, while 2 and 3 are governed by the OR function. The simplest class of Boolean networks is *synchronous*. All elements update their activities at the same moment. To do so, each element examines the activities of its K inputs, consults its Boolean function, and assumes the prescribed next state of activity. This is summarized in Figure 1(b). Here, I have rewritten the Boolean rules. Each of the 2^3 possible combinations of activities of the three elements corresponds to one *state* of the entire network. Each state, at one moment, T, causes all the elements to assess the values of their regulatory inputs, and, at a clocked moment, $T+1$, causes them to assume the proper next activity. Thus, at each moment, the system passes from a state to a unique successor state.

Over a succession of moments, the system passes through a succession of states, called a *trajectory*. Figure 1(c) shows these successions of transitions.

The first critical feature of autonomous Boolean networks is: since there is a finite number of states, the system must eventually reenter a state previously encountered; thereafter, since the system is deterministic and must always pass from a state to the same successor state, the system will *cycle repeatedly* around this *state cycle*. These state cycles are the *dynamical attractors* of the Boolean network. The set of states flowing into one state cycle or lying on it constitute the *basin of attraction* of that state cycle. The *length* of a state cycle is the number of states on the cycle, and can range from 1 (a steady state) to 2^N.

Any such network must have at least one state cycle attractor, but many have more than one, each draining its own basin of attraction. Further, since each state drains into only one state cycle, the set of state cycles are the dynamical attractors of the system, and their basins *partition the* 2^N state space of the system.

The simple Boolean network in Figure 1(a) has three state cycle attractors. Each is a discrete, *alternative, recurrent, asymptotic pattern* of activities of the N elements in the network. Left to its own, the system eventually settles down to one of its state cycle attractors and remains there.

The *stability* of attractors to minimal perturbation may differ. A minimal perturbation in a Boolean network consists of transiently "flipping" the activity of an element to the opposite state. Consider Figure 1(c). The first state cycle is a steady state, or state cycle of length one, (000) which remains the same over time. Transient flipping of any element to the active state, e.g., (100), (010), or (001), causes the system to move to one of the remaining two basins of attraction. Thus the (000) state cycle attractor is unstable to any perturbations. In contrast, the third state cycle is also a steady state (111). But it remains in the same basin of attraction for any single perturbation (011), (101), or (110). Thus, this attractor is stable to all possible minimal perturbations.

A structural perturbation is a permanent "mutation" in the connections or the Boolean rules in the Boolean network. Figure 1(d) shows the result of mutating the rule governing element 2 from the OR function to the AND function. As you can see, this alternation has not changed state cycle (000) or state cycle (111), but has altered the second state cycle. In addition, state cycle (000) which was an isolated state now drains a basin of attraction and is stable to all minimal perturbation, while (111) has become an isolated state and now is unstable to all minimal perturbations.

The following features of autonomous Boolean networks are of immediate interest:

1. The number of states around a state cycle, i.e., its length. The length can range from 1 state for a steady state to 2^N states.
2. The number of alternative state cycles. At least one must exist, but a maximum of 2^N might occur. These are the permanent asymptotic alternative behaviors of the entire system.
3. The sizes of the basins of attraction drained by the state cycle attractors.
4. The stability of attractors to a minimal perturbation, i.e., flipping any single element to the opposite activity value.
5. The changes in dynamical attractors and basins of attraction due to mutations in the connections or the Boolean rules. These changes will underlie the character of the *adaptive landscape* upon which such Boolean networks evolve by mutation of the structure and rules of the system.

Boolean networks are discrete dynamical systems. The elements are either active or inactive. A major difference between a continuous and a discrete deterministic dynamical system is that two trajectories in a discrete system can merge. To be concrete, Figure 1(c) shows several instances where more than one state converges upon the same successor state.

THE NK BOOLEAN NETWORK ENSEMBLE: CONDITIONS FOR ORDERLY DYNAMICS

In the present section, and in the next section, I summarize the behaviors of Boolean networks as a function of N, the number of elements in the net, and of K, the average number of inputs to each element in the net.

In order to assess the expected influence of these parameters, I have analyzed the *typical behavior* of members of the entire *ensemble of Boolean networks* specified by any values of the parameters N and K. I simplify and require that each binary element has exactly K inputs.

In order to analyze the typical behavior of Boolean networks with N elements, each receiving K inputs, it is necessary to sample at random from the ensemble of all such systems, to examine their behaviors, and to accumulate statistics. Numerical simulations to accomplish this, therefore, construct exemplars of the ensemble entirely at random. Thus, the K inputs to each element are chosen at random, then fixed, and the Boolean function assigned to each element are chosen at random, then fixed. The resulting network is a specific member of the ensemble of NK networks.

I stress, therefore, that NK Boolean networks are examples of *strongly disordered systems*.[5,7,8,16,17,18,19,28,29,34,35,36,57,58] Both the connections and Boolean functions are assigned at random. Were any such network examined, its structure would be a complex tangle of interactions, or "input wires," between the N components. The rule characterizing the behavior of one element will typically differ from its neighbors in the network. Such Boolean networks are spiritually similar to spin glasses, and the NK family of landscapes, described elsewhere[43,44] and below. Here, however, we generate networks with random wiring diagrams, and random "logic," and ask whether orderly behavior nevertheless emerges. Note that such behavior is occurring in a *parallel-processing network*. All elements compute their next activities at the same moment. If we find order in random networks, then "random" parallel networks with random logic has order despite an apparent cacophony of structure and logic.

MAJOR FEATURES OF THE BEHAVIOR OF RANDOM BOOLEAN NETWORKS

I report here briefly the behavior of random Boolean networks.

Table 1 summarizes the salient features for the following cases: $K = N$, $K > 5$, $K = 2$, and $K = 1$.

1. $K = N$. In these networks, each element receives inputs from all elements. Hence there is only one "wiring diagram" among the elements. Each element is assigned at random one of the 2^N Boolean functions. In these *maximally disordered systems*, the successor to each state is a completely *random choice* among the 2^N possible states.

   Table 1 shows that the lengths of state cycles average $0.5 \times 2^{(N/2)}$, that the number of state cycle attractors averages N/e, that state cycles are unstable to almost all minimal perturbations, and that state cycles are all totally disrupted by random replacement of the Boolean function of a single variable by another Boolean function.

   State cycle lengths of $0.5 \times 2^{(N/2)}$ are vast as N increases. For $N = 200$, the state cycles average $2^{100} = 10^{30}$. At a microsecond per state transition, it would require billions of times the history of the Universe to traverse the attractor. Here is surely a "big" attractor wandering through state space before finally returning. I will call such attractors, whose length increases *exponentially* as N increases, *"chaotic."* This does *not* mean that flow "on" the attractor is divergent, as in the low-dimensional chaos in continuous dynamical systems. A state cycle is the analogue of a one-dimensional limit cycle.

   Because the successor to each state is randomly chosen, each element is equally likely to assume either activity 1 or 0 at the next moment; hence, virtually all elements "twinkle" on and off around the long attractor.

   The number of cycles, hence basins of attractions, however, is small, N/e. Thus a system with 200 elements would have only about 74 alternative asymptotic patterns of behavior. This is already an interesting intimation of order even in extremely complex disordered systems. A number of workers has investigated this class of systems.[7,8,28,29,30,31,34,35,36,37,38,64] The stability of such attractors to minimal perturbations remains low.

2. $K > 5$. Networks in this class have an enormous number of alternative connection patterns among the N elements. As shown in Table 1, the essential feature of these systems is that their attractors remain "chaotic"; they increase in length exponentially as N increases. The exponential rate at which attractors grow is low for small values of K, and increases to $N/2$ as K approaches N. This implies that, even for $K = 5$, state cycle lengths eventually become huge as N increases. Similarly, along any such attractor, any element "twinkles" on and off around such an attractor.[36,37,38]

3. $K = 2$. Random Boolean networks with $K = 2$ inputs exhibit unexpected and powerful collective spontaneous order. As shown in Table 1, the expected length of state cycles is only $N^{1/2}$. Similarly, the number of alternative state cycle attractors is also $N^{1/2}$, while each state cycle is stable to almost all minimal perturbations, and mutations deleting elements, or altering the logic of single elements, alter dynamical behavior only slightly.[5,7,8,28,29,30,31,32,34,35,36,37,57,58] Each property warrants wonder. State cycles are only $\sqrt{N}$ in length. Therefore, a system of 10,000 binary elements, with $2^{10,000} = 10^{3000}$ alternative states, settles down and cycles among a mere 100 states. The attractor "boxes" behavior into a tiny volume 10^{-2998} of the entire state space. Here, if I may be forgiven delight, is indeed spontaneous order. At a microsecond per state transition, the system traverses its attractor in 100 microseconds, rather less than billions of times the history of the universe.

The number of alternative attractors is only $\sqrt{N}$. A system with 10,000 elements and 10^{3000} combinations of activities of its elements has only 100 alternative asymptotic attractor patterns of integrated behavior. Ultimately, the system settles into one of these.

Along these state cycle attractors, many elements are "*frozen*" into either the active or inactive value. I return to this fundamental property below. It governs the correlated features of the adaptive landscapes in these systems. More critically, this property points to a new principle of collective order.

Another critical feature of random $K = 2$ networks is that each attractor is stable to most minimal perturbations. Small state cycles are therefore correlated with *homeostatic return to an attractor after perturbation*.

In addition, we will find shortly that most "mutations" only alter attractors slightly. $K = 2$ networks adapt on highly correlated landscapes.

The previous properties mean that this class of systems simultaneously exhibit *small attractors, homeostasis, and correlated landscapes* abetting adaptation. Further, these results demonstrate that random parallel-processing networks exhibit order without yet requiring any selection.

TABLE 1 Properties of Random Boolean Nets for Different Values of K^1

	State Cycle Length	Number of State Cycle Attractors	Homeostatic Stability	Reachability Among Cycles After Perturbation
$K = N$	$0.5 \times 2^{N/2}$	N/e	Low	High
$K > 5$	0.5×2^{BN}	$\sim N \left[\frac{\log(\frac{1}{1/2 \pm \alpha})}{2} \right];$	Low	High
	$(B > 1)$	$\alpha = P_{(K)} - 1/2$		
$K = 1$	Short	Short	Low	High
$K = 2$	$\sqrt{N}$	$\sqrt{N}$	High	Low

^{1} Column 2: state cycle length is the median number of states on a state cycle. Column 3: number of state cycle attractors in behavior of one net. ($\alpha = P_K - 1/2$, where P_K is mean internal homogeneity of all Boolean functions on K inputs.) Column 4: homeostatic stability refers to the tendency to return to the same state cycle after transient reversal of the activity of any one element. Column 5: reachability is the number of other state cycles to which the net flows from each state cycle after all possible minimal perturbations, due to reversing the activity of one element.

4. $K = 1$. In these networks, each element has only a single input. The structure of the network falls apart into separate loops with descendent tails. If the network connections are assigned at random, then most elements lie on the "tails" and do not control the dynamical behavior, since their influence "propagates" off the ends of the tails. Each separate loop has its own dynamical behavior and cannot influence the other structurally isolated loops. Thus such a system is *structurally modular*. It is comprised by separate isolated subsystems. The overall behavior of such systems is the product of the behaviors of the isolated systems. As Table 1 shows, the median lengths of state cycles increase rather slowly as N increases, the number of attractors increases exponentially as N increases, and their stability is moderate. There are four Boolean functions of $K = 1$ input, "yes," "not," "true," and "false." The last two functions are constantly active, or inactive. The values in Table 1 assume that only the Boolean function "yes" and "not" are utilized in $K = 1$ networks. When all four functions are allowed, most isolated loops fall to fixed states, and the dynamical behavior is dominated by those loops with no "true" or "false" functions assigned to elements of the loop. Flyvbjerg and Kjaer[16] and Jaffee[26] have derived detailed results for this analytically tractable case.

The results summarized here are discussed elsewhere,[28,29,34,35,43] where I interpret the binary elements as *genes switching one another on and off*, and the Boolean network as modeling the cybernetic genetic regulatory network underlying ontogeny and cell differentiation. I interpret a *state cycle attractor of recurrent patterns of gene activity as a cell type* in the behavioral repertoire of the genomic regulatory system. Then:

1. The sizes of attractors map into a measure of how confined a pattern of gene expression corresponds to one cell type. The theory accounts for cell types as confined patterns of gene expression, and correctly predicts that the cell cycle time varies as about the square root of the DNA content per cell.

2. The number of attractors maps into the number of cell types in an organism. The theory predicts that the number of cell types in an organism should vary as about a square root function of the number of genes in an organism. This, too, is approximately true. Bacteria have two cell types, yeast two to four, and humans about 255.[1] If one assumes humans have on the order of 100,000 genes, then the square root of the expected number of cell types is about 317. The observed number of cell types as a function of DNA content, or estimated number of genes, is between a square root and linear function.[28,29,32,43]

3. The stability of attractors maps into homeostatic stability of cell types.

4. The number of attractors accessible by perturbing the states of activities of single genes maps into the number of cell types into which any cell type can "differentiate." Since the number is small compared to the total number of cell types in the organism, ontogeny must be, and is, organized around branching pathways of differentiation.

5. The overlap in the gene activity pattern between attractors then maps to the similarity of cell types in one organism. The predicted, and actual, differences in gene activities between two cell types is on the order of 5% to 10%. Thus higher plants have perhaps 20,000 genes, and two cell types typically differ in the activities of 1000 to 2000 genes.

6. A core of genes in the model systems falls to fixed active or inactive states, predicting a core of genes which shares the same activity patterns among all cell types of organism. Such a core, typically comprising 70% or more of the genes which are transcribed into heterogeneous nuclear RNA, is observed.

7. The alterations of attractors by mutations corresponds to evolution of novel cell types. Typical mutation in organisms affect the activities of a small fraction of the other genes. The same limit to the spread of "damage" occurs in Boolean networks in the canalizing ensemble.

The spontaneous order we have just uncovered in $K = 2$ networks and their generalizations underlies a serious hope to account for much of the order seen in the orderly coordinate behavior of genetic regulatory systems underlying ontogeny in the absence of selection. "Random" genetic programs can behave with order.

THE BALANCE BETWEEN SELF-ORGANIZATION AND SELECTION: SELECTIVE ADAPTATION OF INTEGRATED BEHAVIOR IN BOOLEAN NETWORKS

We have just considered which kinds of random, disordered Boolean networks exhibit highly ordered dynamics in the absence of selection or other organizing forces. Such spontaneous order suggests that many features of organisms might reflect such self-organization rather than the handiwork of selection. These features include the number of cell types in an organism, the stability of cell types, the restricted number of cell types into which each cell type can "differentiate," and, hence, the existence of branching pathways of differentiation in all multicellular organisms since the Paleozoic. Nevertheless, natural selection is always at work in actual biology. Other adaptive processes analogous to selection are at work in economic and cultural systems. Yet we have no body of theory in physics, biology, or elsewhere, that seeks to understand the ways selection and self-organization may interact. We have almost no idea of the extent to which selection can, or cannot, modify the self-organization exhibited in ensembles of systems such as oolean networks. For example, if properties of Boolean networks resemble those of real organisms, are we to explain those features of real organisms as a consequence of membership in the ensemble of Boolean regulatory systems per se, or does selection account for the features that we see? Or, more plausibly, both? We need to develop a theory exploring how selection acts on, and modifies, systems with self-ordered properties, and to understand the limits upon selection. If selection cannot avoid those properties of complex systems which are generic to vast ensembles of systems, then much of what we see in organisms is present, despite selection, not due to it. If so, a kind of physics of biology is possible.

Let me be clear about the question that I want to ask. Boolean networks exhibit a wide range of properties. We want to investigate whether adaptive evolution can attain Boolean networks with some desired property. More generally, we wonder how the structure of Boolean networks governs the structure of their fitness landscapes for any such property, how the structure of such landscapes governs the capacity of evolutionary search to evolve across the space of networks toward those with desired properties, and whether selection can markedly change the properties of networks, from those generic to the ensemble in which evolution is occurring. Among the obvious properties of parallel-processing Boolean networks, the attractors of such systems commend themselves to our attention. A central question therefore is whether an adaptive process which is constrained to pass via fitter one-mutant or "few mutant" variants of a network, by altering the input connections between elements in a net, and the logic governing individual elements in the net, can "hill climb" to networks with desired attractors.

Notice that, as in the space of sequential computer programs, we confront a *space of systems.* Here the space is the space of NK Boolean networks. Each network is a one-mutant neighbor of all those networks which differ from it by altering a single connection, or a single Boolean function. More precisely, each network is a one-mutant neighbor of all those which alter the beginning or end of a single "input" connection, or a single "bit" in a single Boolean function.

In considering program space I defined a *fitness landscape* as the distribution over the space of the figure of merit, consisting in a measurable property of those programs. This leads us to examine the statistical features of such fitness landscapes, including its correlation structure, the numbers of local optima, the lengths of walks to optima via fitter one-mutant variants, the number of optima accessible from any point, and so forth. Similarly, in considering adaptation in Boolean network space, any specific measurable property of such networks yields a fitness landscape over the space of systems. Again we can ask what the structure of such landscapes looks like.

I shall choose to *define* the fitness of a Boolean network in terms of a *steady target pattern* of activity and inactivity among the N elements of the network. This target is the (arbitrary) goal of adaptation. Any network has a finite number of state cycle attractors. I shall define the *fitness of any specific network* by the *match* of the target pattern to the closest state on any of the net's state cycles. A perfect match yields a normalized fitness of 1.0. More generally, the fitness is the fraction of the N that matches the target pattern.

In previous work, Kauffman and Levin[40] studied adaptive evolution on fully uncorrelated landscapes. More recently,[43,44] my colleagues and I introduced and discussed a spin-glass-like family of rugged landscapes called the NK model. In this family, each site, or spin, in a system of N sites, makes a fitness contribution which depends upon the site, and upon K other randomly chosen sites. Each site has two alternative states, 1 or 0. The fitness contribution of each site is assigned at random from the uniform distribution between 0.0 and 1.0 for each combination of the 2^{K+1} states of the $K + 1$ sites which bear on that site. The fitness of a given configuration of N site values (e.g., 110100010) is defined as the mean of the fitness contributions of each of the sites. Thus, this model is a kind of K-spin spin glass, in which an analogue of the energy of each spin configuration depends, at each site, on interactions with K other sites. In this model, when $K = 0$, the landscape has a single peak, the global optimum, and the landscape is smoothly correlated. When K is $N - 1$, each site interacts with all sites, and the fitness landscape is fully random. The limit corresponds to Derrida's random-energy spin glass model.[4,5,6] Two major regimes exist, K proportional to N, and K of order 1. In the former, landscapes are extremely rugged, and local optima fall toward the mean of the space as N increases. In the latter, there are many optima, but they do not fall toward the mean of the space as N increases. For $K = 2$, the highest optima cluster near one another.

Such rugged landscapes exhibit a number of general properties. Among them, there is a "universal law" for long-jump adaptation. In "long jump" adaptation, members of an adapting population can mutate a large number of genes at once, hence jump a long way across the landscape at once. Frame-shift mutations are examples. In long-jump adaptation the waiting time to find fitter variants doubles after each fitter variant is found; hence, the mean number of improvement steps, S, grows as the logarithm base 2 of the number of generations. Further, there is a complexity catastrophe during adaptation via fitter one-mutant variants on sufficiently rugged landscapes which leads, on those landscapes, to an inexorable decrease in the fitness of attainable optima as the complexity of the entities under selection increases. A similar complexity catastrophe applies to an even wider class of rugged landscapes in the long-jump limit. There, the fitness attained after a fixed number of generations dwindles as complexity increases. Finally, some landscapes, namely those in which the number of epistatic interactions, K, remains small, retain high optima as N increases. These landscapes have "good" correlation structures.

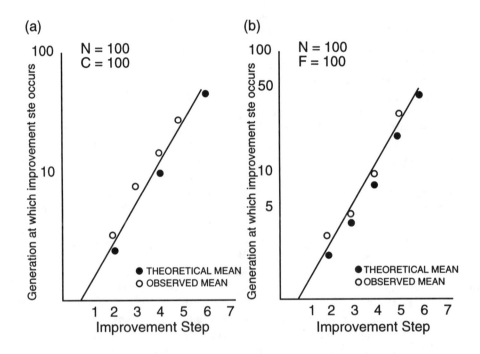

FIGURE 2 Tests of the "Universal Law" for long-jump adaptation. Figures show cumulative number of improvement steps following (a) mutations of half the connections in $K = 2$ and $N = 100$ element Boolean nets in each member of the population—except for a "current best" place holder—plotted against the logarithm of the generation at which the improvement occurred. Each walk yields a sequence of generations at which an improvement step arose. Mean of observed values are plotted, as well as theoretical expectations. In (b) $1/4$ of all "bits" in the Boolean functions within each member of the population of $N = 100$ networks were reversed at each generation as a "long jump" mutation in the logic of the network. From *Origins of Order: Self Organization in Evolution* by S. A. Kauffman. Copyright ©1993 by Oxford University Press, Inc. Reprinted by permission.

Together, these properties identify twin limits to selection as complexity increases. In smooth landscapes, as N increases, the fitness differentials between one-mutant neighbors dwindle below a critical level, *a mutation error catastrophe sets in*. Selection cannot hold an adapting population at fitness peaks, and the population falls inexorably to lower fitness values and typically "melts" from the peak into vast reaches of the space of systems. Conversely, in very rugged landscapes, the complexity catastrophe sets in. As systems become more complex, the conflicting constraints implied by high K, leads to ever poorer local optima and to trapping into small regions of the highly rugged landscape.

Here we are dealing with adaptation in the *coordinated dynamical behavior of Boolean networks*. It is not obvious that the same generic features and limitations will be found. But they are.

LONG-JUMP ADAPTATION IN $K=2$ NETWORKS CONFIRMS THE "UNIVERSAL LAW"

Figure 2(a) and (b) show numerical results of "long jump" adaptation in Boolean nets with $N = 100$ binary elements, each receiving $K = 2$ inputs. The population consists of 20 networks, located "at" the current fittest network found in the adaptive search process. At each "generation," each of 20 networks mutated 25% of the "bits" in its N Boolean functions (b), or 50% of the connections were mutated (a). The attractors of the grossly mutated nets were tested for their match against a predefined target pattern. If a fitter net was found, the entire population of 20 "hopped" to that fitter net and searched via long-jump mutations from that new site in network space on the next generation. If no fitter net was found on that generation, the search repeated with 20 new long jumps from the current best fit network.

Figure 2(a) and (b) compare, with the numerical simulation data, the expectation that the cumulative number of improved variants should increase as log 2 of the number of generations. The agreement is extremely close.

The range of applicability of the "Universal law" for long-jump adaptation, closely related to the theory of records,[15] is not yet clear, but seems to be broad.

THE COMPLEXITY CATASTROPHE OCCURS IN LONG-JUMP ADAPTATION

The complexity catastrophe occurs not only in NK landscapes, but in long-jump adaptation in Boolean networks. That is, as N increases, *long-jump adaptation achieves substantially less fit* networks at any fixed generation. To test this, adaptation was carried out in the long-jump limit in which half the connections among binary variables in $K = 2$ input nets were mutated in all members of a population except one which was left at the current best network. As N increases from 20 to 100, the fitness achieved after 100 generations declines from 0.88 to 0.67. The difference is statistically significant. Thus, as N increases, the fitness clearly falls after a fixed number of generations. In principle it falls toward 0.5, the mean fitness of networks in the space.

This result is of considerable interest. As in the NK landscape family and the Traveling Salesman problem, so in Boolean nets the rate of finding improved variants depends upon the mutant search range and on how well it matches the correlation structure of the landscape. Often, search via fitter two-mutant variants is better than via one-mutant variants. But, in the limit of long jumps on rugged landscapes, the rate of improvement slows to the log 2 law, and the *complexity catastrophe sets in*. Thus, long-jump adaptation is a progressively worse adaptive strategy as the complexity of Boolean networks, N, increases.

BOOLEAN NETWORK SPACE IS FULL OF LOCAL OPTIMA WHICH TRAP ADAPTIVE WALKS

Rugged landscapes have many local optima. These trap adaptive walks. In the current case we are asking whether Boolean networks can adapt via mutation and selection to have a specific pattern of activities among the N binary elements as a steady-state attractor. Note first that no mathematical constraint foredooms such an effort. Any network will fit the bill if each element that is to be "active" in the target pattern is "active" for all patterns of inputs, while all elements that are inactive in the target pattern respond by being inactive to all input patterns. The constantly active rule is the Tautology Boolean function. The constantly inactive rule is the Contradiction Boolean function. It follows that adaptation by altering single bits in Boolean functions can in principle achieve such a network.

Figure 3(a) shows the results of adaptive walks via fitter one-mutant, two-mutant, and five-mutant variants of Boolean networks. The number of mutants refers to the number of bits altered in the net's Boolean functions. As before, a population of 20 nets is adapting from the best net found in the current generation. Figure 3(b) shows similar phenomena when 1, 2, or 5 connections in the networks are mutated.

Note the following general features.

First, improvement is rapid at first, then slows, and typically appears to stop. Walks become arrested on *local optima*. The fact that improvement slows shows that the fraction of fitter mutant neighbors dwindles as optima are approached.

Second, walks always stop at local optima which are *well below the global optimum*. Trapping is rife in network space. This has a critical consequence.

ADAPTIVE WALKS TYPICALLY CANNOT ACHIEVE ARBITRARY PATTERNS OF ACTIVITIES ON ATTRACTORS

Adaptation via fitter variants in network space becomes grossly hindered by the rugged structure of the landscape. Walks become frozen into small regions of the

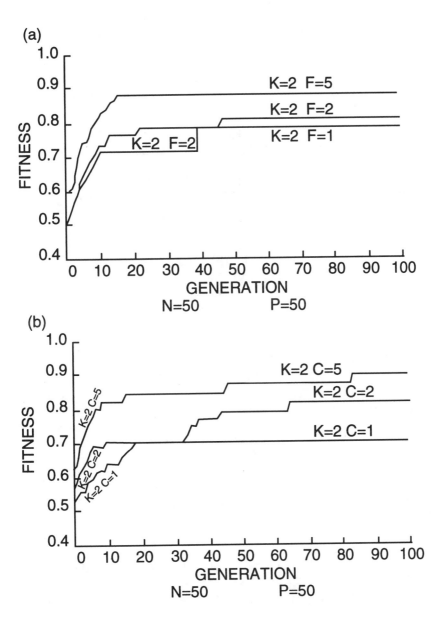

FIGURE 3   (a) Adaptation via fitter 1, 2, and 5 mutant variants in $K = 2$ networks. Mutations altered the "bits" within Boolean functions in the networks of the adapting population. (b) Same as (a) except that 1, 2, and 5 of the connections were mutated in networks. From *Origins of Order: Self Organization in Evolution* by S. A. Kauffman. Copyright ©1993 by Oxford University Press, Inc. Reprinted by permission.

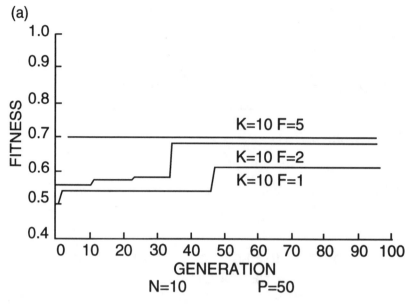

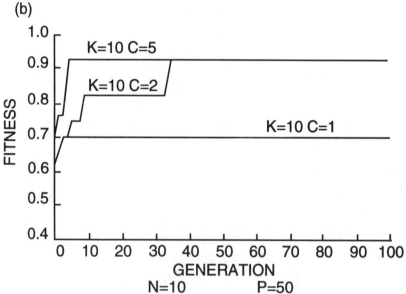

FIGURE 4   (a) As in Figure 3, except that $K = 10$. (b) Same as (a), except that the connections were mutated in networks. From *Origins of Order: Self Organization in Evolution* by S. A. Kauffman. Copyright ©1993 by Oxford University Press, Inc. Reprinted by permission.

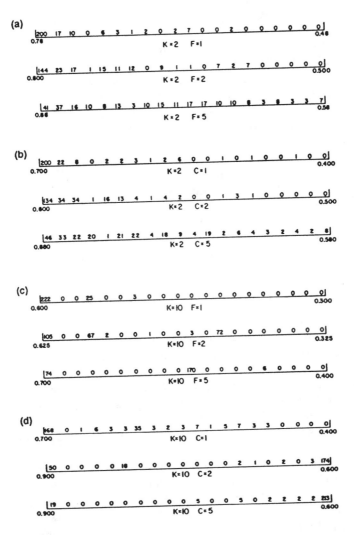

FIGURE 5 The fitness of 1, 2, and 5 mutant variants of the fittest network found after adaptive hill climbing in $K = 2$ networks (a)–(b), or $K = 10$ networks (c)–(d). "F" refers to mutants to "bits" in Boolean functions, "C" to mutations of connections. The fractional numbers at the ends of each line refer to the fitness values, while the numbers above the line refer to the number of times that the value of fitness was found. From *Origins of Order: Self Organization in Evolution* by S. A. Kauffman. Copyright ©1993 by Oxford University Press, Inc. Reprinted by permission.

space. Any intuition which we may have harbored that mutation and selection alone could "tune" attractors to arbitrary patterns of behavior appears to be wrong.

Such problems are very complex combinatorial optimization tasks, and selection confronts enormous problems moving successfully in such spaces. Generally, one cannot get to "there" from "here." These limitations appear to be very important. In the introduction to this article, I described a number of examples of networks which exhibit adaptation. This adaptation often pertains to the attractors of the parallel-processing networks that are interpreted in biological contexts ranging from memories in neural networks to cell types in genetic networks. In general, learning or adaptation is imagined to occur by altering couplings among the network components to achieve "desired" attractors. The present results suggest that this may typically be extremely difficult, or impossible. If so, then either alternative means to search rugged adaptive landscapes in network spaces must exist, or adaptation and learning does not achieve arbitrary attractors. One suspects that the latter possibility is more plausible.

Note that adaptation via two-mutant and five-mutant variants is more rapid and reaches higher optima than adaptation via fitter one-mutant variants in the same time. Thus the correlation structure favors search at slightly longer distances.

$K=2$ NETS ADAPT ON A MORE CORRELATED LANDSCAPE THAN DO $K=10$ NETS

Our results above show that $K = 2$ networks have highly orderly global dynamics, with small stable attractors. Networks in which each element has more inputs per element, $K = 10$, have chaotic attractors which increase in length exponentially as N increases. Mutations cause more drastic alterations in attractors in $K = 10$ nets; hence $K = 10$ nets should adapt on more rugged landscapes than should $K = 2$ nets. Figures 4(a) and (b) show adaptive walks for $K = 10$ networks. Because cycle lengths are long, we studied small networks. The same basic features were found as for $K = 2$ nets.

Figure 5(a)–(d) compares the ruggedness of fitness landscapes in $K = 2$ and $K = 10$ nets. It shows the fitness of the one-mutant, two-mutant, and five-mutant variants of the best network found after 100 generations of adaptation in $K = 2$ and $K = 10$ networks. The salient feature is that, in $K = 2$ nets, the one-mutant neighbors of the best net have nearly the same fitness. The landscape is highly correlated. This impression is confirmed by looking at the two- and five-mutant variants. The spread in fitness only increases slightly. In contrast, for $K = 10$ nets, the spread in fitness is wider and increases rapidly as one-mutant to five-mutant variants are examined. Thus $K = 10$ networks adapt on a very much more rugged landscape than do $K = 2$ networks.

TABLE 2 Mean fitness in $K = 2$ nets attained after 100 and 200 generations for $N = 20$ and $N = 100$ nets[1]

N	Generations	
	100	200
20	0.90	0.91
100	0.78	0.79

[1] Means are averages of 20 nets.

$K=2$ NETS EXHIBIT THE COMPLEXITY CATASTROPHE, BUT SLOWLY

Table 2 depicts the fitness of optima attained after 100 and 200 generations of adaptation in $K = 2$ networks of $N = 20$ and $N = 100$. The important result is that as N increases, the fitness appears to have reached a local optimum after 200 generations. Nevertheless, the fitness *decreases* as N increases. For $N = 20$ networks at 200 generations mean fitness is .90, while for $N = 100$ networks at 200 generations mean fitness has fallen to .78.

This means that even though $K = 2$ networks adapt on well-correlated, "good" landscapes, they cannot avoid the complexity catastrophe. Presumably, the fitness attained will ultimately be hardly better than chance, 0.5. On the other hand, comparison with long-jump adaptation for the same class of $K = 2$ networks suggest that in the long-jump limit the rate of decrease of fitness as N increases is faster. Thus adaptation via near neighbors on the correlated $K = 2$ fitness landscape does not fall prey to the complexity catastrophe as rapidly as would occur were the landscape fully uncorrelated.

These results strongly suggest that $K = 2$ nets adapt on *better landscapes* than do $K = 10$ nets with respect to selection for attractors which match a desired steady-state pattern.

A general summary of our results is that the features of adaptive landscapes found for evolution in sequence spaces,[40,44] and in the NK family of landscapes,[43] extends to adaptation in the integrated dynamical behavior of Boolean networks. It was not obvious that the same features would be found, since sequence space and landscapes over proteins might be very different than fitness landscapes over spaces of dynamical systems, with respect to their attractors. Nevertheless, similar features are found. Landscapes are rugged and multipeaked. Adaptive processes typically become trapped on such optima. The "long jump" law obtains. Most importantly, as the complexity of the entities under selection increases, here the number of binary switching variables in a disordered Boolean network, the attainable optima again fall toward the mean of the space. We do not know at this stage just how general is the complexity catastrophe that limits the power of selection when operating on complex systems, but it appears likely to be a powerful factor in evolution. Finally, Boolean networks of different connectivities, $K = 2$ and $K = 10$, clearly adapt on radically different landscapes. The capacity to attain and maintain high fitness depends upon landscape structure, mutation rate, and coevolutionary couplings of landscapes. It follows that dynamical systems in different classes, constructed in different broad ways, can have very different capacities to adapt. Tentatively, it appears that Boolean nets of low connectivity are likely to adapt more readily than those of high connectivity.

Among the themes to be investigated in understanding the relation between self-organization and selection is the extent to which selection can achieve systems whose behavior is very *untypical* of those in the ensemble in which adaptive evolution is occurring. In the current context, can selection achieve networks with *short stable attractors* when operating on Boolean networks with $K = 20$ inputs and $N = 10,000$? The answer is unknown. But, since the generic properties of this class of random Boolean networks includes attractors which scale exponentially in N, and are grossly unstable to minimal perturbations, one doubts strongly that selection could achieve such systems within the $N = 10,000$, $K = 20$ ensemble. But, if the structure of such networks governs the sizes and stability of their attractors, it also governs the ruggedness of the fitness landscapes upon which they evolve. If selection can "tune" K in such networks, or bias the choice of Boolean functions in such networks, then selection can change the ensemble being explored by evolution. Such changes would tune the landscape structure of the systems, hence their evolvability. The fact that the $K = 2$ ensemble fits so many features of organisms, and that organisms are themselves now clearly evolvable, suggests that this ensemble may itself have been achieved by selection, in part to achieve evolvability. In the next section we turn to ask what features of fitness landscapes and the couplings between landscapes, such that landscapes deform as partners adapt, abet coevolution.

COEVOLUTIONARY TUNING OF LANDSCAPE STRUCTURE AND COUPLING TO ACHIEVE SYSTEMS ABLE TO COEVOLVE SUCCESSFULLY: FROZEN COMPONENTS AND SELF-ORGANIZED CRITICALITY

The results above show that different classes of disordered dynamical systems, different ensembles of Boolean networks, adapt on fitness landscapes of different degrees of ruggedness. We now turn to sketch a broad further topic and some initial insight into it. In real biological evolution, the adaptive evolution of members of one species occurs in the context of other species. Development of a stickier tongue by the frog lowers the fitness of the fly, and also alters the fly's fitness landscape.

Coevolution is a story of fitness landscapes which are coupled together such that moves by one coevolutionary partner cause the fitness landscapes of its partners to deform more or less drastically. It is a story of coupled, dancing landscapes. On a fixed fitness landscape, there is the analogue of a potential function: the fitness at each point. In coevolution, no such potential function is present. Thus we can frame the following questions: (1) How are fitness landscapes coupled? (2) What kinds of couplings between landscapes allows the partners to dance happily and typically achieve "high fitness?" (3) Might there be evolutionary processes which alter the couplings among landscapes and the landscape structure of each partner, such that the entire system coevolves "well" or optimally in some sense?

Answers are not known, of course. I described briefly some preliminary work carried out with my colleague, Sonke Johnsen, using the spin-glass-like NK model of fitness landscapes. As noted briefly above, the NK model consists of N spins, each in two states, 1 or 0. Each spin makes a "fitness contribution" to the "organism" which depends upon the value at that spin site, and at K other randomly chosen sites. In our coevolutionary model, we consider a system with S organisms, one for each of S species. Each species interacts with R neighboring species. Each site in each species makes a fitness contribution which depends upon K sites within that species member, and "C" sites in each of the R species with which it interacts. The fitness contribution of each site therefore depends upon $K + 1 + R$ sites, each of which can be in the 1 or 0 state. The model assigns to each site a fitness contribution at random from the uniform interval between 0.0 and 1.0, for each of the 2^{K+1+R} combinations of these site values. In an extension to the model, each species also interacts with an external world of N sites, of which W affect each of the species' own sites. Thus, the oevolutionary model is a kind of coupled spin system. Each species is represented by a collection of N spins. Spins are K coupled within each species, and C coupled between species. The fitness of any species, whose current state is given by the values of its N spins, depends upon the states of those spins and those in its R neighbors which impinge upon it, and is the mean of the fitness contribution of the species' own N sites.

Consider a "square" 10×10 ecosystem with 100 species, each of which interacts with its four neighbors. Corner species interact with only two neighbors, edge species interact with three neighbors. Each species "plays" in turn by flipping each of its N spins, one at a time, and ascertaining if any one-mutant variant is fitter than the current spin configuration of that species. If so, the species randomly chooses one of the fitter variants and "moves" there. Each of the 100 players play in turn, in order. One hundred plays constitutes an ecosystem generation. After each ecosystem generation, a species may have changed spin configuration, or may not have changed. If the species changed, color it blue. If it remained fixed, color it red. Over time the system will continue to change unless all members stop changing, and the whole system becomes frozen in a "red" state. Such a state corresponds to a local, so-called Nash, equilibrium in game theory. Each player is at a local (one-mutant) optimum consistent with the local opima of its R neighbors.

Recall that increasing K increases the ruggedness of these NK landscapes. We find the following remarkable result: When K is large relative to $R \times C$, then over ecosystem generations, frozen red regions form, grow, and percolate across the ecosystem. At first these red frozen components leave behind blue islands of species which continue to undergo coevolutionary change. Eventually, the entire system becomes frozen in a red Nash equilibrium. In short, *frozen components* recur on this larger scale of coupled spin systems, in direct analogy with those found in Boolean networks. The number of ecosystem generations required for the frozen component to spread across the ecosystem increases dramatically when K is less than $R \times C$. Indeed, the systems behaved chaotically for very long periods.

Tuning the parameters of the coupled spin model, N, K, C, R, S, and the number of sites which can "flip" or mutate at once in each species, not only tunes the mean time to reach a Nash equilibrium, but also tunes the *mean fitness* of the coevolving partners. While full results are not yet available, it appears than in any model ecosystem, there is an *optimal value* of K. When K is too small relative to $R \times C$, the landscape of each partner is too "smooth," and the effects of altering a site internal to a species upon its fitness is too small with respect to the impact of site alterations in other species to withstand those exogenous perturbations to landscape structure. The waiting time to reach the frozen Nash equilibrium is long, the system is chaotic prior to reaching a Nash equilibrium and thus sustained fitness is low. Conversely, if K is too high, Nash equilibria are rapidly attained, but the high K value implies many conflicting constraints; thus the fitness of the local optima which comprise the Nash are low. Again, sustained fitness is low. An optimal value of K optimizes the waiting time to find Nash equilibria such that the sustained fitness is itself optimized. The optimal value of K appears to correspond to the phase transition between the chaotic and the ordered regimes.

It is also important that an evolutionary process guided by natural selection acting on members of individual species may lead partners to "tune" K to the optimum. For each partner in a system, where each has a suboptimal or overoptimal K value, any single partner improves its own sustained fitness by increasing or lowering its K value toward the optimal value. Thus natural selection, acting on members of individual species to tune the ruggedness of their own fitness landscapes, may optimize coevolution for an entire coupled system of interacting adapting species. Real coevolution confronts not only adaptive moves by coevolving partners, but exogenous changes in the external "world" impinging upon each partner. The coupled NK landscape model suggests that if each partner is occasionally shocked by a change in its external world, then sustained fitness may be optimized by increasing K slightly. In this case, the coevolving system as a whole tends to restore the red frozen Nash equilibria more rapidly in the face of external perturbations which destabilize the system.

Finally, it has been of interest to study the distribution of coevolutionary avalanches unleashed by changing the external "world" of species when the entire system is at a frozen Nash equilibrium. Small and large avalanches of coevolutionary change propagate across the system. To a first approximation, when the K value is optimized to maximize sustained fitness at the phase transition between order and chaos, the distribution of avalanche sizes appears to be linear in a log-log plot, suggesting a power law distribution. If so, the self-optimized ecosystem may harbor a self-organized critical state of the kind recently investigated by Bak[2] in other contexts. Interestingly, the distribution of such avalanches in these model ecosystems mirrors the distribution of extinction events in the evolutionary record.[53]

These results are first hints that coevolving systems may tune the structure of their internal landscapes and the coupling between landscapes, under the aegis of natural selection, such that the coupled system coadapts well as a whole. No mean result this, if true.

SUMMARY

What kinds of dynamical systems harbor the capacity to accumulate useful variations, hence evolve? How do such systems interact with their "worlds" in the sense of categorizing their worlds, act upon those categorizations, and evolve as their worlds with other players themselves evolve? No one knows. The following is clear. Adaptive evolution, whether by mutation and selection, or learning, or otherwise, occurs on some kind of "fitness landscape." This follows because adaptation or learning is some kind of local search in a large space of possibilities. Further, in any coevolutionary context, fitness landscapes deform because they are coupled. The structure and couplings among landscapes reflect the kinds of entities which are evolving and their couplings. Natural selection or learning may tune both such structures and couplings to achieve systems which are evolvable.

A further point is clear. Complex, parallel-processing Boolean networks that are disordered can exhibit ordered behavior. Such networks are reasonable models of a large class of nonlinear dynamical systems. The attractors of such networks are natural objects of interest. In the present article I have interpreted attractors as "cell types." But equally, consider a Boolean network receiving inputs from an external world. The attractors of a network are the natural classifications that the network makes of the external world. Thus, if the world can be in a single state, yet the network can fall to different attractors, then the network can categorize that state of the world in alternative ways and respond in alternative ways to a single fixed state of the external world. Alternatively, if the world can be in alternative states, yet the network fall to the *same* attractor, then the network categorizes the alternative states of the world as identical, and can respond in the same way. In brief, and inevitably, nonlinear dynamical systems which interact with external worlds classify and "know" their worlds.

Linking what we have discussed, and guessing ahead, I suspect that if we could find natural ways to model coevolution among Boolean networks, which received inputs from one another and external worlds, we would find that such systems tune their internal structures and couplings to one another so as to optimize something like their evolvability. An intuitive bet is that such systems would achieve internal structures in which the frozen components were nearly melted. Such structures live on the edge of chaos, in the "liquid" interface suggested by Langton,[49] where complex computation can be achieved. In addition, I would bet that couplings among entities would be tuned such that the red frozen Nash equilibria are tenuously held to optimize fitness of all coevolving partners in the face of exogenous perturbations to the coevolving system. But a tenuous frozen component in a coevolutionary context would be a repeat of "the edge of chaos" on this higher level. Perhaps such a state corresponds to something like Bak's self-organized critical state. It would be exciting indeed if coadaptation in mutually categorizing dynamical systems tended to such a definable state, for the same principles might recur on a variety of levels in biology and beyond.

REFERENCES

1. Alberts, A., D. Bray, J. Lewis, M. Raff, K. Roberts, and J. D. Watson. *Molecular Biology of the Cell.* New York: Garland, 1983.
2. Bak, P., C. Tank, and K. Wiesenfeld. "Self-Organized Criticality." *Phys. Rev. A.* **38(1)** (1988): 364–374.
3. De Arcangelis, L. "Fractal Dimensions in Three-Dimensional Kauffman Cellular Automata." *J. Phys. A. Lett.* **20** (1987): L369–L373.
4. Derrida, B., and H. Flyvberg. "Multivalley Structure in Kauffman's Model: Analogy with Spin Glasses." *J. Phys. A.: Math. Gen.* **19** (1986): L1003–L1008.
5. Derrida, B., and Y. Pomeau. "Random Networks of Automata: A Simple Annealed Approximation." *Biophys. Lett.* **1(2)** (1986): 45–49.
6. Derrida, B., and D. Stauffer. "Phase-Transitions in Two-Dimensional Kauffman Cellular Automata." *Europhys. Lett.* **2(10)** (1986): 739–745.
7. Derrida, B., and H. Flyvberg. "The Random Map Model: A Disordered Model with Deterministic Dynamics." *J. Physique* **48** (1987): 971–978.
8. Derrida, B., and H. Flyvberg. "Distribution of Local Magnetizations in Random Networks of Automata." *J. Phys. A. Lett.* **20** (1987): L1107–L1112.
9. Eigen, M. "New Concepts for Dealing with the Evolution of Nucleic Acids." In *Cold Spring Harbor Symposia on Quantitative Biology*, Vol. LII, 307–320. New York: Cold Spring Harbor Laboratory, 1987.
10. Eigen, M., and P. Schuster. *The Hypercycle, A Principle of Natural Self-Organization.* New York: Springer-Verlag, 1979.
11. Erdos, P., and A. Renyi. *On the Random Graphs 1*, Vol. 6. Debrecar, Hungary: Inst. Math. University Debreceniens, 1959.

12. Erdos, P., and A. Renyi. "On the Evolution of Random Graphs." *Math. Inst. Hung. Acad. Sci.* **5** (1960).
13. Farmer, J. D., K. S. Kauffman, and N. H. Packard. "Autocatalytic Replication of Polymers." *Physica* **22D** (1986): 50–67.
14. Farmer, J. D., N. H. Packard, and A. Perelson. "The Immune System, Adaptation, and Machine Learning." *Physica* **22D** (1986): 187–204.
15. Feller, W. *Introduction to Probability Theory and Its Application*, Vol. II, 2nd edition. New York: Wiley, 1971.
16. Flyvberg, H., and N. J. Kjaer. "Exact Solution of Kauffman's Model with Connectivity One." *J. Phys. A.* **21(7)** (1988): 1695–1718.
17. Fogelman-Soulie, F. "Frustration and Stability in Random Boolean Networks." *Discrete Appl. Math.* **9** (1984): 139–156.
18. Fogelman-Soulie, F. Ph.D. Thesis, Université Scientifique et Medical de Grenoble, 1985.
19. Fogelman-Soulie, F. "Parallel and Sequential Computation in Boolean Networks." *Theor. Comp. Sci.* **40** (1985).
20. Gelfand, A. E., and C. C. Walker. *Ensemble Modeling.* New York: Dekker, 1984.
21. Glass, L., and S. A. Kauffman. "Co-Operative Components, Spatial Localization and Oscillatory Cellular Dynamics." *J. Theor. Biol.* **34** (1972): 219–237.
22. Harary, F. *Graph Theory.* Reading, MA: Addison-Wesley, 1969.
23. Hartman, H., and G. Y. Vichniac. In *Disordered Systems and Biological Organization*, edited by E. Bienenstock, F. Fogelman-Soulie, and G. Weisbuch. Heidelburg: Springer-Verlag, 1986.
24. Hopfield, J. J. "Neural Networks and Physical Systems with Emerging Collective Computational Ability." *Proc. Natl. Acad. Sci. USA* **83** (1982): 1847.
25. Hopfield, J. J., and D. W. Tank. "Collective Computation with Continuous Variables." In *NATP ASI Series, Disordered Systems and Biological Organizations*, Vol. F20, edited by E. Bienenstock et al. Berlin: Springer-Verlag, 1986.
26. Jaffe, S. "Kauffman Networks: Cycle Structure of Random Clocked Boolean Networks." Ph.D. thesis, New York University, 1988.
27. Jerne, N. K. "Idiotypic Networks and Other Preconceived Ideas." *Immunol. Rev.* **79** (1984): 5–24.
28. Kauffman, S. A. "Homeostasis and Differentiation in Random Genetic Control Networks." *Nature* **224** (1969): 177–178.
29. Kauffman, S. A. "Metabolic Stability and Epigenesis in Randomly Connected Nets." *J. Theor. Biol.* **22** (1969): 437–467.
30. Kauffman, S. A. "Cellular Homeostasis, Epigenesis and Replication in Randomly Aggregated Macromolecular Systems." *J. Cybernetics* **1(1)** (1971): 71–96.
31. Kauffman, S. A. "Gene Regulation Networks: A Theory for Their Global Structure and Behavior." In *Current Topics in Developmental Biology 6*,

edited by A. Moscana and A. Monroy, 145–182. New York: Academic Press, 1971.

32. Kauffman, S. A. "The Large-Scale Structure and Dynamics of Gene Control Circuits: An Ensemble Approach." *J. Theor. Biol.* **44** (1974): 167–190.

33. Kauffman, S. A. "Development Constraints: Internal Factors in Evolution." In *Developmental Evolution*, edited by B. C. Goodwin, N. Holder, and C. G. Wylie, 195–225. Cambridge: Cambridge University Press, 1983.

34. Kauffman, S. A. "Pattern Generation and Regeneration." In *Pattern Formation*, edited by G. M. Malacinski, and S. V. Bryant, 73–102. New York: Macmillan, 1984.

35. Kauffman, S. A. "Emergent Properties in Random Complex Automata." *Physica* **10D** (1984): 145–156.

36. Kauffman, S. A. "Autocatalytic Sets of Proteins." *J. Theor. Biol.* **119** (1986): 1–24.

37. Kauffman, S. A. "A Framework to Think About Regulatory Systems." In *Integrating Scientific Disciplines*, edited by W. Bechtel, 165–184. Dordrecht, The Netherlands: Martinus Nijhoff, 1986.

38. Kauffman, S. A. "Boolean Systems, Adaptive Automata, Evolution." In *Disordered Systems and Biological Organization*, edited by E. Bienenstock, F. Fogelman-Soulie, and G. Weisbuch, 338–360. Berlin: Springer-Verlag, 1986.

39. Kauffman, S. A., and R. G. Smith "Adaptive Automata Based on Darwinian Selection." *Physica* **22D** (1986): 68–82.

40. Kauffman, S. A., and S. Levin "Towards a General Theory of Adaptive Walks on Rugged Landscapes." *J. Theor. Biol.* **128** (1987): 11–45.

41. Kauffman, S. A., E. D. Weinberger, and A. S. Perelson. "Maturation of the Immune Response via Adaptive Walks on Affinity Landscapes." In *Theoretical Immunology, Part One*, edited by. A. S. Perelson, 349–382. Santa Fe Institute Studies in the Sciences of Complexity, Proc. Vol. II. Reading, MA: Addison-Wesley, 1988.

42. Kauffman, S. A., and E. D. Weinberger. "Application of NK Model to Maturation of Immune Response." *J. Theor. Biol.*, in press.

43. Kauffman, S. A. *Origins of Order: Self-Organization and Selection in Evolution*. Oxford: Oxford University Press, 1993.

44. Kauffman, S. A., and D. Stein "Application of the NK Model of Rugged Landscapes to Protein Evolution and Protein Folding." Abstract AAAS Meeting on Protein Folding, June, 1989.

45. Koshland, D. E., J. "Evolution of Catalytic Function." *Cold Spring Harbor Symposia on Quantitative Biology*, Vol. LII, 1–8. New York: Cold Spring Harbor Laboratory, 1987.

46. Kurten, K. E. "Correspondence Between Neural Threshold Networks and Kauffman Boolean Cellular Automata." *J. Phys. A: Math. Gen.* **21** (1988): 615–619.

47. Kurten, K. E. "Critical Phenomena in Model Neural Networks." *Phys. Lett. A* **129(3)** (1988): 157.

48. Lam, P. M. "A Percolation Approach to the Kauffman Model." *J. Stat. Phys.* **50(5/6)** (1988): 1263–1269.
49. Langton, C. G. "Artificial Life." In *Artificial Life*, edited by C. G. Langton, 1–47. Santa Fe Institute Studies in the Sciences of Complexity, Proc. Vol. VI. Reading, MA: Addison-Wesley, 1989.
50. Maynard-Smith, J. "Natural Selection and the Concept of a Protein Space." *Nature* **225** (1970): 563.
51. McCulloch, W. S., and W. Pitts. "A Logical Calculus of the Ideas Immanent in Nervous Activity." *Bull. Math. Biophys.* **5** (1943): 115–133.
52. Monad, J., J.-P. Changeux, and F. Jacob. "Allosteric Proteins and Cellular Control Mechanisms." *J. Mol. Biol.* **6** (1963): 306.
53. Raup, D. M. "On the Early Origins of Major Biologic Groups." *Paleobiology* **9(2)** (1983): 107–115.
54. Rummelhart, D. E., J. L. McClelland, and the PDP Research Group. *Parallel Distributed Processing: Explorations in the Microstructure of Cognition*, Vols. I and II. Cambridge, MA: Bradford, 1986.
55. Schuster, P. "Structure and Dynamics of Replication-Mutation Systems." *Physica Scripta* **26B** (1987): 27–41.
56. Stanley, H. E., D. Staufer, J. Kertesz, and H. J. Herrmann. *Phys. Rev. Lett.* **59** (1987).
57. Stauffer, D. "Random Boolean Networks: Analogy with Percolation." *Philosophical Mag. B* **56(6)** (1987): 901–916.
58. Stauffer, D. "On Forcing Functions in Kauffman's Random Boolean Networks." *J. Stat. Phys.* **40** (1987): 789.
59. Stauffer, D. "Percolation Thresholds in Square-Lattice Kauffman Model." *J. Theor. Biol.*, in press.
60. Walter, C., R. Parker, and M. Ycas. *J. Theor. Biol.* **15** (1967): 208.
61. Weisbuch, G. *J. Phys.* **48** (1987): 11.
62. Weisbuch, G., and D. Stauffer. "Phase Transition in Cellular Random Boolean Nets." *J. Physique* **48** (1987): 11–18.
63. Weisbuch, G. *Complex Systems Dynamics: An Introduction to Automata Networks*, translated by Sylvie Ryckebusch. Santa Fe Institute Studies in the Sciences of Complexity, Lecture Notes Volume II. Redwood City, CA: Addison-Wesley, 1991. French language version *Dynamiques de Systems Complexes*. Paris: InterEditions, 1989.
64. Wolfram, S. "Statistical Mechanics of Cellular Automata." *Rev. Mod. Phys.* **55** (1983): 601.

L. E. H. Trainor
University of Toronto, Toronto, Ontario, Canada M5S 1A7

Modeling the Behavior of Ant Colonies as an Emergent Property of a System of Ant-Ant Interactions

1. INTRODUCTION

Social insects organized in colonies present a particular fascination for the observer and a particular challenge for the experimenter in trying to account for the group behavior in contradistinction to the behavior of individuals in the group.

The present article reviews a model of Gordon[7] and Trainor et al.[13] of the harvester ant species *Pogonomyrmex barbatus* based on observations taken in the field over a period of years by Gordon.[2,3,4,5] In the second section, the experimental observations of Gordon are surveyed and important parameters identified relating to colony behavior; in the third section, aspects of the modeling problem are identified and a perspective developed; in the fourth section, the main features of the model are established; section 5 deals with results and an evaluation of the success of the model in accounting for the experimental observations; section 6 gives a critical review of the problem of relating model parameters to experimental parameters and shows how recent work of Torres and Trainor[11] helps to put the problem in

Thinking About Biology, Eds. W. D. Stein and F. J. Varela, SFI Studies in the
Sciences of Complexity, Lect. Note Vol. III, Addison-Wesley, 1993   **303**

perspective. Finally, section 7 gives a review of the situation and comments on future possible developments.

2. EXPERIMENTAL OBSERVATIONS

In what follows, colony behavior is taken to mean the distribution of workers in time among the various tasks undertaken by the colony. The more traditional approach to understanding colony behavior (e.g., Oster and Wilson[10]) has been to assume a caste structure for the colony,[10] i.e., to assume that the colony is composed of distinct groups of workers, with workers within a group specialized to doing a particular task. In the Oster-Wilson model,[10] the caste organization "should provide the key to the ecology of social insects, in so far as the insects differ from their solitary counterparts. We may regard many of the principal processes of colony life. . .as subordinate to the evolution of caste. We postulate them to be the enabling devices by which labor is allocated and by which the colony as a whole precisely adjusts its relationship to the nest environs."

Gordon takes issue with this view based on recent empirical results, and particularly on the basis of her own field observations.[5] She argues that colony behavior doesn't depend simply on caste distributions, both because individuals switch tasks under external perturbations and because the intensity of colony activities changes; moreover, there is considerable variation from colony to colony, and colony dynamics change with maturation. Gordon emphasizes that while it is necessary to understand the colony dynamics at a deeper level and to appreciate that colony activities are carried on in parallel rather than series, it is essential to recognize that task allocation is distributed. In *Pogonomyrmex barbatus*, ants carrying out different tasks at any one time are not morphologically different; rather, they are closely identical.

This suggested to Gordon et al. an analogy with neural networks,[7] in which individual neurons are identical, but groups of interacting neurons lead to organizational states of specialized kinds on the basis of the parallel/distributed processing attributes of such a system.

In the field results[2,3,4,5] ants are observed to switch activity categories on the basis of external perturbations in ways which are nonobvious and nonlinear. The experimental observations are limited to counting the number of ants in the various activity categories for a discrete sequence of times, including both before and after a given perturbation. One can, in principle, also infer something about the dynamics from observing the "daily round,"[2] the sequence of activities and activity levels which occur while the colony is active on any particular day.

3. MODELING FEATURES

According to Gordon,[4,5] the study of colony behavior requires one to consider two broad categories of ants—active and inactive—as well as the four active categories—patrollers P, foragers F, nest maintenance workers N, and midden workers M. We shall use these letters P, F, N, and M to signify the appropriate categories, and on occasion (when specified) to represent the numbers of ants in these categories at any one time. Similarly, we use small case letters p, f, n and m to refer to the numbers of inactive ants in these four categories at any one time. In effect, then, one has eight categories for the ants in the colony (not counting internal workers which are probably younger and remain in the nest at all times, engaged in other activities).

The basic problem then is to devise a model of interacting identical ants (units) which can be in any one of eight activity categories, with colony behavior as an emergent property of the system. Just as individual neurons in a net do not possess wisdom, memory, etc., but these are only attributes of the network as a whole, so we assume individual ants do not possess the knowledge for colony organization and behavior, but these are emergent properties of the whole network of interacting ants.

We assume, moreover, that an individual ant "behaves" according to what category it is in and interacts likewise. Thus an ant in the foraging category will forage, and in interaction with other ants will behave as a forager. When an ant switches to the patroller category, it behaves as a patroller and interacts with other ants as though it were a patroller. For simplicity, we assume that all ants in a given activity category behave identically and interact with other ants identically.

It is appropriate to remark at this point on a subtle aspect of the model presented here suggested by the experimental observations. It appears that ants can switch from one category to any other, though some switches are more probable than others. This implies that each individual ant is capable of carrying on any of the activities that take place in the colony. Presumably this capacity is an intrinsic one (genetic) and not an emergent one. What is emergent[1] is the behavior (and organization) of the colony as a whole, arising out of the interactions between its identical units. For example, the proportion of ants that will be foraging under given circumstances at any one time is not hierarchically determined, but is determined by the collective interactions of the ants in the colony as a whole. In this sense the behavior of the colony is an emergent property.

---

[1] An example of an emergent property in a physical system is the appearance of superfluid behavior in a collective of helium atoms under appropriate circumstances. Individual atoms do not, in any sense, possess a superfluid property, but their interactions lead to the superfluid behavior in the collective. Similarly, individual neurons do not possess consciousness, but under appropriate conditions a network of them may. For further discussion on emergence, see Gordon[6] and Trainor.[12]

We model the colony as a highly interactive ant network of the Hopfield type[8] with two important differences. In the Hopfield model, there are only two activity categories, off and on neurons (or up and down spins in the analogous spin model). In our case, we must allow for eight activity categories; moreover, whereas, in the Hopfield model, interactions between units are fixed at the outset and remain fixed thereafter, in our case the interactions that are fixed are category interactions only—as two ants (units) change categories, the interaction between them changes likewise.

How one proceeds to model the eight category aspects is not unique. The choice made here is to use a binary decision model, somewhat analogous in this respect to the Boolean genetic networks of Kauffman.[9] We take the first binary decision to be whether an ant is active or not according to some appropriate Hebbian rule (see below). The second binary decision is to distinguish the forager/patroller subclass from the nest maintenance/midden worker subclass. Finally, the third decision is to distinguish forager from patroller and nest maintenance from midden worker. As has been previously pointed out by Kauffman,[9] such a binary description implies certain equalities in the interaction scheme, which may not be borne out by experiment. Alternatively, experiment may eventually show that there is a more appropriate subclass division than (patroller/forager) and (nest maintenance/midden worker).

We keep to the binary representation at this point, however, for two reasons. It is basically simple and easy to apply, and it corresponds to binary network models which have been explored extensively in the literature. Moreover, its predictions can be easily tested experimentally (see section 6). It can be replaced with analogous but more cumbersome models if, in time, experiment warrants. At the present stage one is more interested in exploring whether, in principle, colony behavior can be explained by emergence from parallel/distributed systems than whether the detailed model presented here is the most appropriate choice.

4. MODEL DETAIL

We ascribe, to each ant, three separate on-off switches which are independently activated by three distinct "fields" due to all other ants acting on the particular ant in question, a positive field value giving rise to an "on state" and a negative field value giving rise to an "off state," in close analogy to the situation in a magnetic system, where individual magnets line up with the local field direction. We thus consider a triad of binary vectors (a_k, b_k, c_k) for the kth ant, each vector taking on the on/off values of ± 1. The triad determines which activity state an ant is in, in accord with the scheme:

$$
\begin{aligned}
(1,1,1) &= \text{active patroller}, P \\
(-1,1,1) &= \text{inactive patroller}, p \\
(1,1,-1) &= \text{active forager}, F \\
(-1,1,-1) &= \text{inactive forager}, f \\
(1,-1,1) &= \text{active nest maintenance worker}, N \\
(-1,-1,1) &= \text{inactive nest maintenance worker}, n \\
(1,-1,-1) &= \text{active midden worker}, M \\
(-1,-1,-1) &= \text{inactive midden worker}, m
\end{aligned}
\tag{1}
$$

Thus, the field associated with a determines whether an ant is active or inactive; the field associated with b determines whether an ant is in the patroller/forager subclass or the maintenance/midden workers subclass. Finally, the field associated with c distinguishes between patroller and forager, and between nest maintenance and midden worker in the respective subclasses.

In the model, we make the assumption that ants interact with one another pairwise and that the ant network is fully connected. The first assumption that the ants interact pairwise seems justifiable from experiment since it is observed that the communication link is the exchange of chemical pheromones between the antennae of two ants on passing each other in pursuit of their separate activities. The second assumption implies that each ant interacts with every other ant engaged in one of the eight activity categories in the colony so that all ants affect a decision on the part of any individual ant to change or not to change from one activity category to another. This is probably unrealistic, but is partially justified by the heavy traffic and intense level of exchange which takes place between ants at the entrance to the nest so that in any substantial time interval, every pair of ants has had an opportunity to share in pheromone exchange (i.e., to interact). In the model, no interaction between pairs of ants would correspond to a zero matrix element in the interaction matrices.

Let the three fields acting on the states (spins) a, b and c be denoted by E, L, and K, respectively. We take the fields acting on ant i due to all other ants at any given time to be

$$
E_i = \sum_{\substack{j=1 \\ j \neq i}}^{N} \alpha_{ij} a_j
\tag{2}
$$

$$
L_i = \sum_{\substack{j=1 \\ j \neq i}}^{N} \beta_{ij} b_j
\tag{3}
$$

$$
K_i = \sum_{\substack{j=1 \\ j \neq i}}^{N} \gamma_{ij} c_j
\tag{4}
$$

where N is the total number of ants distributed over the eight activity categories and the matrix elements α_{ij}, β_{ij}, and γ_{ij} are the interaction strengths of the three fields connecting ant i and ant j. Self-interactions are excluded so that the diagonal elements α_{ii}, β_{ii}, and γ_{ii} are taken to be zero. If each ant were individual and different, there would be as many as $N(N-1)/2$ independent matrix elements in each of the three interaction matrices. However, since it is assumed that all ants in the same category behave in the same way, many matrix elements are the same. In fact, there are only $8^2 = 64$ independent elements in each of the three matrices, representing the interactions between ants in the various possible pairs of activity categories.

The same remark holds true for β_{ij} and γ_{ij} and for all category pairs.

The dynamics then proceeds as follows: an ant, say, the ith, is selected at random from among the total of N ants. The fields acting on ant i are then calculated in accordance with Eqs. (2), (3), and (4) and switching occurs or does not occur for the vectors a_i, b_i, and c_i depending on whether the state is already lined up with the local field or not. For example, if $a_i = -1$ and E_i is calculated to be positive, this state switches to $a_i = +1$.

Similar considerations apply to vectors b_i and c_i with respect to their fields L_i and K_i, respectively.

Thus, beginning with any initial state of the system, that is, with ants $1, 2, 3, \ldots, N$ distributed among the eight categories, an ant is chosen randomly and assessed for switching, and the new category assigned according to Eq. (1). This process is repeated and the category populations noted as change occurs, until the system reaches a stable state (fixed populations), or sets into a limit cycle of repeated population states, or reaches whatever persistent asymptotic behavior.

5. RESULTS

The detailed behavior of the system and, in particular, what asymptotic states occur depends both on the initial starting state (initial activity populations) and on the matrix elements of the α, β, γ matrices. As shown in Gordon,[7] it is straightforward to choose matrix elements which ensure that a simple attractor state exists corresponding to equal populations in all eight categories when the matrix elements are taken to be symmetric, which seems justifiable. Nonetheless, the system is not easily analyzable as to the structure of its attractor space, for reasons to be discussed below.

In the Hopfield[8] model, the interactions between units (neurons or whatever) are fixed at the outset and remain fixed throughout the dynamics. This feature allows Hopfield to construct a quadratic "energy" function, which decreases monotonically as the dynamics proceeds. This feature guarantees the existence of attractors and attractor basins in the "energy landscape" of state space; indeed, one can even choose landscapes (attractors as memory states). The same is not necessarily true here. Because ants change categories, the effective interactions between ant pairs alter as these change categories. Nonetheless, in practice, the system does appear to have stable states with quadratic energy minima, but the approach to these is not necessarily monotonic. This general situation is discussed in more detail in section 6.

The parallel distributed model of Gordon et al. presented here does seem to exhibit several characteristics of observed colony behavior. Using simple decision rules based on accumulated interactions between ants, it produces global changes in the numbers of ants in the various activity categories.

Changes in the number of ants in one category changes the number active in the other categories. In the models an initial increase or decrease in the number active in some given category propagates to other categories, until eventually the system settles into a stable state. In observed ant colonies, perturbations have a similar effect: colonies change the number engaged in various tasks, including the perturbed ones, and eventually return to some normal baseline state. As with experimental observations, after a large perturbation, the system may well settle down to a stable state which is different from the stable state obtaining before the perturbation. The model also shares with experimental observations a nonlinear behavior, in the sense that the results of two successive perturbations do not combine linearly. For example, if successive perturbations increase the number of foragers and decrease the number of midden workers, one cannot deduce the pattern of category populations from carrying out these perturbations separately.

6. RELATING THEORY AND EXPERIMENT

A particular problem with the above model is that the dynamics is determined by the choice of matrix elements in the three matrices α, β, and γ. As indicated above, the number of matrix elements is greatly reduced since all ants in the same category interact identically. Assuming symmetric matrices and c ($c = 8$) categories, the number of independent matrix elements is given by

$$\frac{3c(c+1)}{2} = 108. \tag{4}$$

However, the relationship of these matrix elements to experimental parameters is not clear. In the scheme below, due to Torres and Trainor,[11] it is shown how a close, convenient correspondence can be established.

A second problem with the model described above is that the dynamics do not necessarily imply a monotonically decreasing, quadratic energy function of the convenient type devised by Hopfield.[8] Again this problem has been successfully reformulated by Torres and Trainor[11] and is presented below.

Torres and Trainor replace the Hebbian dynamics used by Gordon[6] et al. and Hopfield[8] by the following scheme. They define a hypothetical "fitness" function or alignment function for each spin/field pair. For a randomly selected ant, say ℓ, the fitness is defined as the product $a_\ell E_\ell$ of the spin and associated local field E_ℓ as defined in Eq. (2). If the transition $a_\ell \to -a_\ell$ increases the fitness, the spin changes, otherwise not. Similar considerations apply to b_ℓ and c_ℓ and their associated local fields L_ℓ and K_ℓ from Eqs. (3) and (4). Again $b_\ell \to -b_\ell$ only if this increases the fitness $b_\ell L_\ell$ and $c_\ell \to -c_\ell$ only if this increases the fitness $c_\ell K_\ell$. Otherwise, these states remain the same. With these decision rules and the quadratic energy function defined in Eq. (7) below, one can then show that the energy decreases monotonically, so that attractors exist and the system eventually reaches stable states.

Following Torres and Trainor[11] we define, for the jth ant, a column vector $\vec{a}$, whose components are the polarities a_j, b_j, and c_j. We further define a set of 3×3 symmetrical matrices, one for each ant pair (j, k), as follows:

$$
A^{(jk)} = \begin{bmatrix} A_{11}^{(jk)} & A_{12}^{(jk)} & A_{13}^{(jk)} \\ A_{21}^{(jk)} & A_{22}^{(jk)} & A_{23}^{(jk)} \\ A_{31}^{(jk)} & A_{32}^{(jk)} & A_{33}^{(jk)} \end{bmatrix} \tag{5}
$$

such that $A^{(jk)} = A^{(kj)}$ for all $j, k = 1, 2, \ldots, N$ and with the property

$$
A^{(jk)} = A^{(\ell m)} \tag{6}
$$

if $\vec{a}_j = \vec{a}_\ell$ and $\vec{a}_k = \vec{a}_m$ (i.e., if the corresponding ants are in the same category).

We now define an energy function

$$
H = -\frac{1}{2} \sum_{\substack{j,k=1 \\ (j \neq k)}}^{N} \vec{a}_j^T A^{(jk)} \vec{a}_k \tag{7a}
$$

where $\vec{a}^T$ is the transpose of $\vec{a}$. We further define the field parameters i_n, being α, β and γ, as follows:

$$
\begin{aligned}
\alpha_{jk} &= A_{11}^{(jk)} + A_{12}^{(jk)} \frac{b_k}{a_k} + A_{13}^{(jk)} \frac{c_k}{a_k}, \\
\beta_{jk} &= A_{21}^{(jk)} \frac{a_k}{b_k} + A_{22}^{(jk)} + A_{23}^{(jk)} \frac{c_k}{b_k}, \\
\gamma_{jk} &= A_{31}^{(jk)} \frac{a_k}{c_k} + A_{32}^{(jk)} \frac{b_k}{c_k} + A_{33}^{(jk)} .
\end{aligned} \tag{7b}
$$

With this choice one can then show that for each random choice of ant and field assessment, the decision rules lead to $\Delta H \leq 0$, so that H is monotonically decreasing and, hence, a suitable choice of energy function.

From Eq. (6) above we can replace the $A^{(jk)}$ matrices (5) by a set $A^{(\sigma\delta)}$ where σ and δ run over the c categories ($c = 8$). It follows that H can be rewritten

$$H = \sum_{\sigma,\delta=1}^{c} M_{\sigma\delta} P_\sigma P_\delta - \sum_{\sigma=1}^{c} M_{\sigma\sigma} P_\sigma \qquad (8)$$

where

$$M_{\sigma\delta} = -\frac{1}{2}\vec{a}_\sigma^T A^{(\sigma\delta)} \vec{a}_\delta \qquad (9)$$

and $\vec{a}_\sigma = \vec{a}_j$ for ant j in the σ category. Moreover, we designate the population in the σ category by P_σ.

A correspondence between matrix elements and experimental observations can then be made as follows. Let a stable set of populations be denoted by $\{P_\sigma^*\}$. Such a set must correspond to a minimum of H and, hence, satisfy the set of equations

$$\frac{\partial H}{\partial P_\sigma} = 0 \text{ at } P_\sigma = P_\sigma^*, \sigma = 1, 2, \ldots, c \qquad (10a)$$

which yields

$$2\sum_{\delta=1}^{c} M_{\sigma\delta} P_\delta^* - M_{\sigma\sigma} - \theta = 0, \sigma = 1, 2, \ldots, c . \qquad (10b)$$

Here θ is a Lagrange multiplier corresponding to the condition

$$\sum_\delta P_\delta^* = N . \qquad (11)$$

If the quantities P_σ^* (stable populations) can be obtained from experimental observations, the set of Eq. (10b) represent $c = 8$ equations for the unknown $\{M_{\sigma\delta}\}$, which are linearly related to the $A^{(jk)}$ according to Eqs. (5) and (9). By choosing the $A^{(jk)}$ to be a set parameterized by $c = 8$ independent parameters, the set (10a) supplies $c = 8$ equations from which they can be determined. The α, β, and γ matrices then follow from Eq. (7b). Mathematical consistency puts some constraints on parameter choices, since the determinant of $M_{\sigma\delta}$ and its principal minors must be positive.

7. DISCUSSION

The scheme described in section 6 is particularly convenient, since the experimentally determined parameters are the populations P_σ^* of the various activity categories and these enable the matrix elements to be calculated from Eq. (10b). In principle any stable point of the system can be chosen as a representative point, and the corresponding P_σ^* used. If the system has more than one stable point, the model can be checked for consistency, since the matrix elements are assumed to be fixed for any activity pair, independent of initial conditions. To test the model, one can imagine perturbing the system and letting it settle down to equilibrium values, measuring the activity populations and calculating the matrix elements. If the system is again perturbed in a different way, it eventually again settles down to equilibrium values, which generally may be different from the previous ones. Again one can calculate the matrix elements from the observed populations and compare them with the matrix elements previously obtained. This procedure would give a rigid test of the model and indicate ways in which it might be refined.

In summary, the model developed by Gordon[7] et al., as modified by Torres and Trainor,[11] gives a view of colony behavior in terms of the emergent properties of a system of identical ants interacting in a highly parallel, distributed way. The model assumes, in accord with observations, that individual ants have the capacity to be in any one of eight activity states. Starting from any distribution of ants over these activity states, activity switching occurs and the populations change in time until an equilibrium state is achieved.

The model is capable of elaboration in several ways, particularly with respect to giving the matrix elements a time dependence of one or more types. A short-term time dependence can, in principle, be devised so as to account for the dynamics of the "daily round" described in Section 2. One could also conceive of extensions to the model which would allow a spatial as well as a time description of ant activities. Alternatively, one could approach spatial activity through a cellular automaton approach.[1]

ACKNOWLEDGMENTS

Financial support for the work in preparing this manuscript was provided by NSERC (the Natural Sciences and Engineering Research Council of Canada). Ongoing discussions with Drs. Deborah Gordon and Jose Torres are gratefully acknowledged. Thanks also to Dave Jourard for assistance with the computer programs and to Karl Schroeder for preparation of the manuscript.

REFERENCES

1. Goodwin and collaborators. Private communication.
2. Gordon, D. M. "The Dynamics of the Daily Round of the Harvester Ant Colony (*Pogonomyrmex barbatus*)." *Anim. Behav.* **34** (1986): 1402–1419.
3. Gordon, D. M. "Group Level Dynamics in Harvester Ants: Young Colonies and the Role of Patrolling." *Anim. Behav.* **35** (1987): 833–834.
4. Gordon, D. M. "Caste and Change in Social Insects." In *Oxford Surveys in Evolutionary Biology*, vol. 6, edited by P. Harvey and L. Partridge. Oxford: Oxford University Press, 1989.
5. Gordon, D. M. "Dynamics of Task Switching in Harvester Ants." *Anim. Behav.* **38** (1989): 194–204.
6. Gordon, D. M. "Notes and Comments on an Article by B. Cole." *The Amer. Naturalist* **137(2)** (1991): 260–261.
7. Gordon, D. M., B. C. Goodwin, and L. E. H. Trainor. "A Parallel Distributed Model of the Behavior of Ant Colonies." *J. Theor. Biol.*, in press.
8. Hopfield, J. J. "Neural Networks and Physical Systems with Emergent Collective Computational Abilities." *Proc. Nat. Acad. Sci. U.S.A.* **79** (1982): 2554–2556.
9. Kauffman, S. A. "Metabolic Stability and Epigenesis in Randomly Constructed Genetic Nets." *J. Theor. Biol.* **22** (1969): 437–467.
10. Oster, G., and E. O. Wilson. *Caste and Ecology in the Social Insects*. Cambridge: Princeton University Press, 1978.
11. Torres, T. L., and L. E. H. Trainor. "An Energy Function for a Model of Ant Colonies." In preparation (1992).
12. Trainor, L. E. H. "Remarks on Emergence in Physics and Biology." In *Mathematical Essays on Growth and Form*, Ch. 6, edited by P. Antonelli, 201–206. Edmonton: University of Alberta Press, 1987.
13. Trainor, L. E. H., B. C. Goodwin, and D. M. Gordon. "On Parallel, Distributed Models for Category Switching in Studies of Social Behavior." *J. Theor. Biol.* (1992): under revision.

Index